*Routledge Revivals*

# Alternative Agricultural and Food Policies and the 1985 Farm Bill

In 1985, the U.S. Congress confronted the difficult and complex task of developing a 5-year omnibus legislation allowing for lower commodity prices. But, policies predicated on the concept of agriculture as a unique sector of the economy became less and less appropriate to the highly interdependent, open agricultural economy throughout the 1980s. First published in 1985, this collection of 16 papers and related discussions contained in these proceedings is an important contribution toward understanding the issues, options, and dilemmas in U.S. agricultural policy. This is an ideal title for students interested in environmental studies, agriculture, and national policy.

I0093910

# Alternative Agricultural and Food Policies and the 1985 Farm Bill

Edited by
Gordon C. Rausser and Kenneth R. Farrell

RFF PRESS
RESOURCES FOR THE FUTURE

First published in 1985
by Resources for the Future, Inc.

This edition first published in 2016 by Routledge
2 Park Square, Milton Park, Abingdon, Oxon, OX14 4RN
and by Routledge
711 Third Avenue, New York, NY 10017

*Routledge is an imprint of the Taylor & Francis Group, an informa business*



**Publisher's Note**
The publisher has gone to great lengths to ensure the quality of this reprint but points out that some imperfections in the original copies may be apparent.

**Disclaimer**
The publisher has made every effort to trace copyright holders and welcomes correspondence from those they have been unable to contact.

A Library of Congress record exists under LC control number: 85229709

ISBN 13: 978-1-138-96287-3 (hbk)
ISBN 13: 978-1-315-65911-4 (ebk)
ISBN 13: 978-1-138-96290-3 (pbk)

# ALTERNATIVE AGRICULTURAL AND FOOD POLICIES AND THE 1985 FARM BILL


*Editors*

**Gordon C. Rausser**
Chairman and Professor
Department of Agricultural and Resource Economics
University of California, Berkeley, California

**Kenneth R. Farrell**
Senior Fellow and Director
National Center for Food and Agricultural Policy
Resources for the Future, Washington, D. C.

Printed by Blaco Printers, Inc., San Leandro, CA

Typeset by Vera Allen Composition, Castro Valley, CA

Library of Congress cataloging in publication data.

# ACKNOWLEDGMENT

Acknowledgment is made to Amor Nolan, Marjorie Teach, Angela Garcia, and other members of the Word Processing Unit for their untiring and often unheralded support; to Carole Nuckton for her unstinting editorial assistance; and to Gertrude Halpern and Ikuko Takeshita whose expertise in directing and coordinating the publication of this volume proved invaluable.

# TABLE OF CONTENTS

## ALTERNATIVE POLICY PROPOSALS

## INDIVIDUAL COMMODITY POLICY OPTIONS

## 9 Tobacco and Peanuts ........ 213

***Dale M. Hoover and Daniel A. Sumner***



## 10 The U. S. Sugar Program and Its Effects ........ 237

***Andrew Schmitz, Roy Allen, and Gwo-Jiun M. Leu***

## A GENERAL REACTION TO THE EVALUATION OF ALTERNATIVE PROPOSALS

# 1 SUMMARY AND CONCLUSIONS

**Kenneth R. Farrell**
**Gordon C. Rausser***

In 1985 the U. S. Congress will confront the formidable task of developing omnibus legislation to succeed the Food and Agriculture Act of 1981. It will be a difficult, complex task. Tied as they are to a long series of legislative precedents, to deeply embedded goals and objectives, and to the vested interests of many groups, current policies will not yield easily to change. American agriculture, however, continues to evolve in ways which undermine some of the very premises of past and current policies. Policies predicated on the concept of agriculture as a unique closed sector of the economy appear less and less appropriate to the highly interdependent, open agricultural economy of the 1980s.

There is a growing criticism and dissatisfaction with current policies within and outside agriculture. Some describe the current juncture as a watershed of policy choice and call for bold reorientation of current policies. Some, viewing agricultural policy as a continuum of evolutionary change, would continue with the prin-

*The authors are, respectively, Senior Fellow and Director, National Center for Food and Agricultural Policy, Resources for the Future, Washington, D. C., and Chairman and Professor of Agricultural and Resource Economics and Economist in the Agricultural Experiment Station and on the Giannini Foundation, University of California, Berkeley.

cipal elements of current policy while adjusting them to better accord with realities of the 1980s. On one point, however, most observers agree: the outcome of the 1985 legislative process could have important implications for domestic and global food and fiber markets during the remainder of the 1980s and beyond.

The collection of 16 papers and related discussions contained in these proceedings is an important contribution toward understanding the issues, options, and dilemmas in U. S. agricultural policy. No consensus nor collective policy recommendations were sought nor emerged spontaneously among the 150 economists and policy leaders who participated in the conference. By design, the conference focused primarily on agricultural commodity price policies and their relationships to other policies. That emphasis was chosen for two reasons. First, price policy has been the focal point of policy-related agricultural economic research for several decades. If there is any one area of policy studies in which economists could be expected to have reached some definitive conclusions, it is that of commodity price policy. Second, it seems clear that commodity price policy issues will again dominate the debate on agricultural legislation in 1985.

## *Past and Current Policies*

A common conclusion in several of the conference papers is that past and current commodity price policies have been characterized by failure—failure to accord with the principles of economic efficiency of resource use and, frequently, with failure to attain goals they were presumably designed to achieve. D. Gale Johnson argues persuasively that the commodity programs have failed to bring about long-term improvement in the aggregate level of farm income. The principal beneficiaries have been landowners due to the capitalization of program benefits in land values. Several authors and discussants stress the inconsistencies among price policies themselves and between price policies and other policy interventions and goals, particularly those related to natural resources and trade. Johnson, Gardner, Just and Rausser, Lesher, Schuh, and several authors of commodity papers point to serious distortions in resource use as a result of the continuous operation of complex programs which have failed to recognize the interdependencies within the agricultural sector and between the agricultural sector and other sectors of the domestic and world economy. Although grain reserves have been an important component of price and income stabilization policies for nearly 50 years, the authors generally conclude that, because of the nature of those programs and changes external to agriculture and domestic agricultural markets, there is little evidence to suggest that instability in agriculture has been reduced. Indeed, Lesher's detailed accounting of policy and program changes in the past seven years and the Just and Rausser conceptual framework lend credence to the likelihood that government policy itself has added additional uncertainty and instability to the sector.

There was considerable discussion about the nature and magnitude of changes in the economic environment of agriculture and perhaps tacit agreement that the capability of even "good" commodity policies to positively affect economic performance of the sector had become significantly constrained by economic interdependence in domestic and international markets. Schuh alleges that policymakers simply have not understood the economic world in which we now live. If pressed, he might extend his allegation to most agricultural economists!

Commodity price policies not only have failed in large measure to provide enduring improvement in the economic welfare of farmers as a group but also have impacted the welfare of other groups and yielded unintended or unforeseen side effects. Zilberman and Carter conclude that commodity policies have combined and interacted with monetary and fiscal policy, technological change, and off-farm earning possibilities to affect farm structure. The resource diversion or land control components of commodity policies benefit landowners relative to operators and make entry into farming more difficult; risk-reducing components of such policies appear to have benefited middle-size farms. Ray, Tweeten, and Trapp, while concluding that commodity programs on the whole have reduced variability of grain supplies and prices and thus benefited livestock producers, suggest that administration of those programs has become a source of instability in recent years. Heien points out that commodity policies directed to increasing farm prices are in conflict with consumer interests. He estimates that current policies result in a total cost per household of about $400. However, the net long-run effects of agricultural programs as a whole on consumer welfare are unclear. Benbrook, Crosson, and Ogg explore the linkages between commodity and resource conservation policies and conclude that inconsistencies and the sometimes negative effects of commodity policies on resource and environmental quality might be ameliorated by targeting conservation payments and closer linkages of commodity program benefits with conservation goals.

Many of the foregoing conclusions have been previously advanced. Leading agricultural economists having been saying as much for several decades, particularly those weaned on the paradigm of neoclassical economics. They undoubtedly will continue to draw similar conclusions as long as commodity price policies exist. Economists are frequently at their best in post hoc analysis and criticisms. They are useful social functions even if the advice or lessons learned are not always or even frequently heeded by policymakers. From the persuasive seven policy lessons of the past (articulated by Johnson) to Lesher's recounting of policy and program features of the past seven years, to the several carefully crafted commodity papers, and to the provocative papers by Just/Rausser and Schuh, there is much to be learned from past policies to apply in the design of future policies and programs. Why have commodity policies of the current type persisted given their uneven, sometimes capricious, and generally poor performance adjudged by economic criteria and the equity standards of some? The answer lies, in part, on the fact that public policies are not based on economic efficiency criteria alone, and, in part, in the structure of government itself and the policy process—circumstances which make it difficult to discard or achieve quantum modification at a discrete point in time even when policies are generally accepted as inadequate. Cochrane, in his conference address, put it more bluntly: "Agricultural policy in the 1980s is held in the vice-like grip of powerful farm commodity groups.... They have become the mechanism whereby large, already affluent farmers obtain monopoly privileges and gains for themselves—monopoly privileges and gains which become capitalized into land values contributing to the increased wealth of these large farmers."

## *Policy Options for the Future*

Authors of the major commodity-centered papers presented at the conference were asked to analyze the effects of several policy scenarios on selected performance measures. Those scenarios were to include (1) continuation of the Food and Agriculture Act of 1981; (2) adjustment of the Act of 1981 to make policy instruments sensitive to key variables in the domestic and international economic environment; (3) a radical alternative to the Act of 1981 with no specific commodity or land management policies but with some government intervention (e.g., direct support of farm income at the individual farm level); and (4) no government intervention, i.e., not direct price or income policies.

In a paper prepared following the conference (and included in these proceedings), Calvin, Foster, and Rausser offer six criteria by which the results of analysis of the foregoing scenarios might be compared: (1) "political or government failure," (2) farm income enhancement, (3) food security, (4) risk reduction, (5) flexibility of the agricultural sector, and (6) minimization of Treasury costs. Employing these criteria, they then draw generalized conclusions about the effects of seven policy alternatives: (1) free markets, (2) reinforced free markets, (3) revenue insurance, (4) flexible storage policies, (5) supply management, (6) demand expansion, and (7) direct income enhancement. These assessments provide useful insights into the effects of the specified policy alternatives. However, until the effects of the alternatives or combinations of alternatives on critical variables have been quantified (e.g., farm income and income distribution, trade volume and value, consumer prices and food expenditures among income groups, and Treasury costs), they are of limited value for policymaking purposes. Nevertheless, the framework is potentially useful as a means of ordering much needed policy analysis.

As McCalla notes in his review of papers presented at the conference, it is not possible to draw precise conclusions or comparisons of the effects of the four scenarios that the authors were requested to address because, in part, some scenarios are more apt for some commodities than for others and, in part, because authors interpreted the scenarios differently. However, some highlights from the several papers are noted.

Gardner, in his assessment of policy options for the grains sector, concludes that continuation of the 1981 Act would increase producers' annual incomes about $5.5 billion, increase Treasury costs approximately $5 billion, and reduce consumers' incomes approximately $3 billion annually relative to no programs. With modification of the 1981 Act, producers' incomes would be increased from $4.5 to $5.0 billion, Treasury costs might be from $3.0 to $8.5 billion higher, and consumers' incomes from $1.0 billion larger to $5.0 billion less than with no program. He then sketches the outline of a government-managed price insurance scheme and the use of put options as a means of privately insuring against price risk as an alternative to current policies. Just and Rausser focus on the grains sector and, with major attention to policy and market risk and uncertainty, argue for flexible market-conditioned storage policies as a means of reducing instability in farm prices and incomes and reducing political and government failures.

The papers on cotton policies (Firch), tobacco and peanuts (Hoover and Sumner), sugar (Schmitz), dairy (Babb), and livestock (Ray, Tweeten, and Trapp) document in some detail the effects of past policies on performance in the respective commodity subsectors. Most point to substantial policy failures, inadequacies, or unintended side effects. An exception is Firch who concludes that recent policies for cotton have worked reasonably well and with some modification could be made to operate well in at least the near future. However, Babb and Hoover/Sumner conclude that substantial incremental modification of dairy, tobacco, and peanuts is called for to better attune the programs to current and emerging markets, reduce distortions in resource uses, and constrain Treasury costs.

## *Concluding Remarks*

Summing up, conference papers and discussions ranged over a broad spectrum of issues surrounding agricultural price policies. If there was a single dominant point of agreement, it was that such policies have failed in many respects as measured against several policy goals and economic criteria. There is much to be learned from such failures which is applicable to designing of future policies. Such failures have been noted by economists for several decades. One can only speculate whether lessons from the past will be applied more effectively in 1985 than on previous occasions of enactment of quadrennial legislation.

With respect to policy options for the future, the conference yielded some interesting insights but no definite conclusions. There was a common theme that changes in policy were needed to adapt to the far-reaching changes occurring in the economic organization of agriculture and its economic environment in domestic and global markets. However, with the possible exception of Johnson, Lesher, Schuh, and Cochrane, none of the authors imparted a sense of urgency of the need for change. Most authors and discussants seemed to be in at least implicit agreement with Just and Rausser that future policies should contain greater flexibility to permit program instruments to be adjusted as changes in the economic environment occur. However, aside from the Just and Rausser specific proposal, there were few detailed suggestions of how such flexibility might best be designed and even less empirical evidence of the effects which such policy changes might engender. Little or no attention was focused on the manner and extent to which flexibility has been employed in recent years in the use of program provisions authorized in current policies.

The conference, by design, focused on commodity price policies and related effects. However, as noted by Cochrane and McCalla, there is much more to contemporary food and agriculural policy. Cochrane's 10 point policy agenda explicitly incorporated credit, food consumption, nutrition, regulatory, soil conservation, farm labor, and tax policies as parts of a comprehensive food and agricultural policy. To this, one might add research and education as well as environmental policies and, as McCalla suggests, marketing and water policies.

The conference was by and large a conference by economists for economists. That, too, was by design. It is useful on occasion to integrate and bring to bear on policy issues the collective judgments and research results of the profession. However, a greater air of reality might have been achieved by intermingling economists with political scientists, sociologists, and, perhaps, even psychologists as well as pragmatic policymakers. "Preaching to the converted" has its limits as a pedagological device for shaping public policy!

Finally, there are inferences which might be drawn from the conferences concerning the state of agricultural economic research and the preparedness of agricultural economists to provide input to the design of agricultural policy. Clearly, there has been a great deal of potentially relevant research on the effects of past policies. To be sure, many agricultural economists have familiarity with intricacies of past and current policies. It also is quite clear that many resources have been directed to empirical analysis of economic efficiency aspects of agricultural policies and programs.

However, there are apparent shortcomings in our research and preparedness. Aside from Zilberman and Carter, conference papers gave scant attention to the distributional effects of agricultural policies. Options to current policies were given much less attention than past policy failures. To the extent that alternatives to current policy were examined, it was with much less rigor and comprehensiveness. This suggests that much work remains to define and assess alternatives to current policies if our research is to be of maximum value to policymakers. Along the lines of the Calvin, Foster, and Rausser paper, there is need for more clearly specified policy analytic frameworks incorporating more explicitly policy goals and policy instruments to guide comparative policy studies. To this might be added the need for greater continuity and additivity in policy research. In the absence of clarity and comprehensiveness of frameworks to guide research—and continuity and additivity of that research—we should not be surprised when periodic conferences such as this yield inconclusive, disjointed results. The challenge is to move beyond the role of critics of past and present policies to more effective research and policy analysis to assist in the future design of improved public policies.

Despite these shortcomings, the conference yielded valuable papers and dialogue. In this volume, we grouped the papers into five major sections. The first provides the Historical Perspective (Johnson, Lesher, and Schnittker); the second, Alternative Policy Proposals (Gardner, Just and Rausser, Daft, Cochrane, and Calvin, Foster, and Rausser); the third, Individual Commodity Policy Options (Firch; Hoover and Sumner; Schmitz, Allen, and Leu; Babb; Ray, Tweeten, and Trapp; and McCalla); the fourth, Additional Dimensions of Food and Agricultural Policies (Zilberman and Carter; Penn; Schuh; Chambers; Benbrook, Crosson, and Ogg; Howitt; Heien; and Lane); and the fifth, A General Reaction to the Evaluation of Alternative Proposals (Johnson, Knutson, and Randall).

# HISTORICAL PERSPECTIVE

*2. The Performance of Past Policies: A Critique*

*3. Future Agricultural Policy: A Challenge for All*

# 2 THE PERFORMANCE OF PAST POLICIES: A CRITIQUE

D. Gale Johnson*

## *Introduction*

In the closing years of the 1940s, policy decisions were made that had a major impact on agricultural resource use and adjustment for the next two decades. During World War II and in the immediate postwar years, there was a lively debate about the future course of agricultural policy generally and of farm price policy particularly. There was considerable professional, academic, and political support for a policy of full production and moderate price supports. While the Agricultural Act of 1948 embodied this view, in the end Congress and the Administration opted for price supports at levels that quickly resulted in the accumulation of large stocks of the major farm crops.[1]

---

*The author is Chairman, Department of Economics, and the Eliakim Hastings Distinguished Service Professor of Economics, University of Chicago.

[1]Three excellent books provide discussions of the farm policy debate following World War II: Benedict (1953, Chapter 18); Cochrane and Ryan (1976, Chapter 5); and Tweeten (1979). Benedict also provides a detailed description of the development of agricultural policy through 1950. Cochrane and Ryan have produced an exceedingly valuable reference work covering the farm commodity programs since 1950.

The relatively moderate provisions of the Agricultural Act of 1948 were not scheduled to take effect until 1950. The outcome of the national election of 1948 led to new farm legislation that repealed the provisions of the 1948 Act before they became effective. The 1949 Act represented a victory for the proponents of high price supports. The election of a Republican as President in 1952 changed nothing because the Congress remained in Democratic hands. The inability of a divided government to agree on price policy meant that throughout most of the 1950s price supports were at high levels; and significant efforts to limit agricultural output were not made or, when made, were ineffective. The primary policy responses to the rapid accumulation of stocks held by the Commodity Credit Corporation (CCC) in the early 1950s were the Agricultural Trade and Development Act of 1954 and the payment of substantial export subsidies on several farm products.[2] While some efforts were made through acreage allotments and acreage diversions to limit the production of the major farm crops, the measures were ineffective.

There was fairly general agreement that the conflicts over agricultural programs that prevailed during the Eisenhower Administration between the Executive Branch and the Legislative Branch created the legacy of a set of high-cost programs judged ineffective in improving the income of farmers and responsible for the accumulation of enormous quantities of farm crops by the CCC. The Executive Branch pressed for lower price supports to discourage production and expand utilization while Congress voted to maintain price supports at relatively high levels. Programs that were adopted to restrain output were generally ineffective. When in the late 1950s some of the price- support levels were reduced, there was clear disappointment in Congress that output did not decline immediately and domestic use and exports did not respond in a year or two.

---

[2]On June 30, 1952, loans outstanding and commodities owned by CCC had a value of $1.5 billion; a year later the value was $3.4 billion; and on June 30, 1954, it was $6.0 billion (U. S. Department of Commerce, 1961). During the 1950s export subsidies paid on wheat averaged approximately one-third of the export price. Export subsidies were also paid on cotton and, to a lesser degree, on feed grains. In addition to explicit export subsidies, approximately 30 percent of agricultural exports in fiscal years 1956 and 1957 were exported through P. L. 480 under the Agricultural Trade and Development Act of 1954.

Thus, the 1950s ended with no apparent progress in finding a solution to the problem of excess resources engaged in agriculture.

The conflict and trial and error of the preceding eight years might have ended with the election of both a Democratic President and Congress in 1960; instead, dissension continued. President Kennedy believed the large governmental costs of farm programs were unsustainable; his proposed approach was to increase the government's role through strict mandatory supply management covering most farm commodities. But Congress refused to support the President in his call for mandatory controls. During the 1950s, Congress had persisted in its efforts to achieve a balance between supply and demand at unrealistically high prices through the use of voluntary measures. Farmers were induced to remove land from cultivation by payments. As surpluses in the hands of the CCC mounted during the last half of the 1950s, there was recognition that price supports had been maintained at unrealistic levels; and downward adjustments were made, especially for corn, as the decade ended (Cochrane and Ryan, 1976, Chapter 6). In spite of the presumed failures of the prior voluntary programs, Congress maintained its adherence to voluntarism.

President Kennedy desired mandatory controls because he believed that in this way government costs could be substantially reduced. The last thing Congress wanted to do was to give up credit for distributing substantial payments to farmers and, at the same time, accept the onus for approving mandatory controls that would displease the majority of their farmer constituents. In the end the conflicts between the Administration and Congress were resolved by a series of costly voluntary programs designed to limit crop production and a gradual reduction of almost all price supports for crop products to or below world market levels. With most crop prices at world market levels—a transition largely completed by 1964—farm incomes were maintained at politically acceptable levels by direct payments.

The major features of the compromise farm policies that evolved after 1955 were included in the Food and Agriculture Act of 1965. Although this legislation was modified in a number of ways by three succeeding farm acts (1970, 1973, and 1977), most of the basic concepts remained unchanged: (1) Price supports were maintained at levels permitting the market to allocate supplies among the various markets, (2) income support was achieved by direct payments, and (3) voluntary methods of supply management were used when deemed necessary (Cochrane and Ryan, 1976, Chapter 3).

## *Price-Support Levels*

Price-support levels established after 1965 were substantially lower than those prevailing during most of the 1950s both in real and in deflated values (Table 1). The support levels for wheat were reduced by more than half; the reductions for corn and cotton were nearly as large. Most of the time, price supports established for 1965 and through the late 1970s and early 1980s were low enough to permit U. S. farm products to compete in international markets—and U. S. farm product exports increased quantity and value.

After World War I, U. S. agriculture had apparently lost its comparative advantage. Starting in 1922, for two decades (except for one year), the United States imported more agricultural products than it exported. Exports exceeded imports during World War II and during the early postwar years. By 1950, the United States returned to a net import status which lasted until 1957, even with substantial export disposal under the partial guise of food aid. It was not until the early 1960s that commercial exports of agricultural products exceeded agricultural imports in value (Johnson, 1979).

The competitive pricing of farm crop products was associated with a significant growth in agricultural exports during the latter part of the 1960s and early 1970s. A variety of factors was responsible: the devaluation of the dollar, the expansion of grain imports by the centrally planned economies, increased food demand in the middle-income developing countries (including members of OPEC), and the rapid growth of farm output together with a slower growth in domestic demand for all farm output after 1973. The United States became the major supplier for grains in world markets after 1973, supplying more than 80 percent of the increase in world grain exports during the 1970s.

Dairy price supports were reduced between the early 1950s and the early 1960s by about one-third but were gradually increased thereafter and, by 1976, returned to the high levels of the early 1950s. Further increases in the late 1970s brought dairy price sup-

TABLE 1. Price Support or Loan Rates?[a,b]

| | Wheat | Corn[c] | Soybeans | Manufactured milk | Cotton[d] |
|---|---|---|---|---|---|
| | dollars | dollars per bushel | dollars | dollars per hundred weight | cents per pound |
| 1950 | 2.94 | 2.18 | 3.03 | 4.53 | 44.6 |
| 1951 | 3.02 | 2.17 | 3.39 | 4.99 | 44.8 |
| 1952 | 3.01 | 2.18 | 3.50 | 5.26 | 44.3 |
| 1953 | 2.97 | 2.15 | 3.44 | 5.03 | 45.0 |
| 1954 | 2.97 | 2.15 | 2.94 | 4.18 | 45.1 |
| 1955 | 2.70 | 2.05 | 2.65 | 4.10 | 44.9 |
| 1956 | 2.52 | 1.89 | 2.71 | 4.04 | 41.2 |
| 1957 | 2.43 | 1.71 | 2.54 | 3.96 | 39.4 |
| 1958 | 2.18 | 1.63 | 2.51 | 3.67 | 42.1 |
| 1959 | 2.12 | 1.31 | 2.16 | 3.58 | 39.9 |
| 1960 | 2.05 | 1.22 | 2.13 | 3.57 | 37.3 |
| 1961 | 2.04 | 1.37 | 2.62 | 3.88 | 37.7 |
| 1962 | 2.24 | 1.34 | 2.52 | 3.48 | 36.4 |
| 1963 | 2.01 | 1.18 | 2.48 | 3.46 | 35.8 |
| 1964 | 1.41 | 1.19 | 2.44 | 3.42 | 32.6 |
| 1965 | 1.33 | 1.12 | 2.39 | 3.44 | 30.8 |
| 1966 | 1.29 | 1.03 | 2.57 | 3.86 | 21.6 |
| 1967 | 1.25 | 1.05 | 2.50 | 4.00 | 20.2 |
| 1968 | 1.20 | 1.01 | 2.39 | 4.10 | 19.4 |
| 1969 | 1.14 | 0.96 | 2.05 | 3.90 | 18.4 |
| 1970 | 1.08 | 0.91 | 1.95 | 4.03 | 17.5 |
| 1971 | 1.03 | 0.86 | 1.85 | 4.06 | 16.1 |
| 1972 | 0.99 | 0.83 | 1.78 | 3.90 | 15.4 |
| 1973 | 0.93 | 0.78 | 1.68 | 4.10 | 14.6 |
| 1974 | 0.94 | 0.76 | 1.54 | 4.51 | 17.3 |
| 1975 | 0.86 | 0.69 | —[e] | 4.55 | 21.5 |
| 1976 | 1.34 | 0.90 | 1.49 | 4.87 | 22.2 |
| 1977 | 1.27 | 1.13 | 1.98 | 5.08 | 24.0 |
| 1978 | 1.24 | 1.05 | 2.37 | 5.19 | 25.2 |
| 1979 | 1.21 | 1.02 | 2.18 | 5.56 | 24.3 |
| 1980 | 1.33 | 1.00 | 2.22 | 5.67[f] | 21.3 |
| 1981 | 1.30 | 0.92 | 2.03 | 5.30 | 21.3 |
| 1982 | 1.36 | 0.97 | 1.92 | 5.00 | 21.8 |
| 1983 | 1.34 | 0.97 | 1.84 | 4.80 | 20.2 |

(Continued on next page.)

[a] 1967 dollars.

[b] Actual loan rates deflated by the GNP implicit price deflator. This deflator is considered more appropriate than others, such as prices paid by farmers for production items or the wholesale price index, since the deflator reflects the return to all factors of production and not just to purchased factors and products. The loan rates do not include any direct payments. With a base of 1967, the GNP deflator stood at 272.7 in 1983 compared with 306.0 for the index of prices paid by farmers for production items which includes farm-produced inputs as well as purchased inputs. The loan rates do not include the higher rates loaned on grain and cotton put in the farmer-owned reserve in 1978 and later. Thus, support prices for later years underestimate the support available to farmers. The loan rate is for the crop of the stated year; the deflator is for the calendar year, e.g., the 1982–83 loan rate for wheat of $3.55 is for the 1982 crop, deflated by the GNP implicit price deflator for calender year 1982.

[c] Other and lower loan rates prevailed in 1950 and 1956 through 1958 (Cochrane and Ryan, 1976).

[d] For 1950 through 1960, ⅞-inch middling basis; 1961 to date, 1-inch middling basis. Prior to 1971, loan rate was on gross weight basis; after 1970, on a net weight basis. Loan rates after 1970 should be reduced about 4 percent to be comparable to loan rates for 1970 and earlier.

[e] No price support.

[f] From April through September, 1980, as of October 1, 1980, the price support level increased in current dollars.

Sources: U. S. Department of Agriculture, *Agricultural Statistics*, *Wheat Situation*, *Feed Situation*, *Fats and Oils Situation*, *Cotton and Wool Situation*, and *Dairy Situation* (various issues).

ports to higher real levels than in the early 1950s. Consequently, it is hardly surprising that a substantial milk surplus has been created.

Since the early 1970s, price supports for wheat, corn, and cotton also increased significantly in real terms. Wheat price supports returned to the 1965 level—above the levels of the late 1960s and early 1970s when more wheat was produced than could be disposed of at prevailing prices. Corn price supports in recent years have also been equal to or higher than prices a decade earlier when there was an excess supply of corn at prevailing prices. Cotton price supports, too, are higher than a decade earlier. Thus, we should not be too surprised that wheat, corn, and cotton production is greater than today's markets will absorb at recent and current price-support levels. It is true that the rise in the foreign exchange value of the dollar has exacerbated the problem. However, it should be remembered that in 1977, 1978, and 1979 the dollar was at a very low level and yet farm crop output was greater than could be absorbed at the price-support levels.

As noted above, the competitive pricing of farm products was associated with direct payments to farmers for the major crops with

**TABLE 2. Market Prices, Supplementary Payments, and Export Subsidies for Wheat and Feed Grains, United States, 1968-69 to 1983-84**

| Year | Wheat: Prices received by farmers for corn | Wheat: Average payment on all wheat | Wheat: Export subsidy | Feed grains: Prices received by farmers for corn | Feed grains: Average payment on all feed grains | Feed grains: Export subsidy |
|---|---|---|---|---|---|---|
| | dollars per metric ton | | | | | |
| 1968–69 | 45.56 | 17.68 | [a] | 39.68 | 8.84 | |
| 1969–70 | 45.56 | 21.88 | 5.51 | 45.67 | 10.22 | 0.12 |
| 1970–71 | 48.86 | 23.65 | 8.45 | 52.36 | 10.39 | 0.61 |
| 1971–72 | 49.23 | 20.11 | 5.14 | 42.52 | 5.62 | 0.50 |
| 1972–73 | 64.66 | 21.04 | 9.38 | 61.81 | 10.27 | |
| 1973–74 | 145.12 | 10.21 | 1.41[b] | 100.39 | 6.28 | |
| 1974–75 | 150.27 | 2.10 | | 119.29 | 2.18 | |
| 1975–76 | 130.79 | 0.88 | | 100.00 | 0.62 | |
| 1976–77 | 100.30 | 2.49 | | 84.60 | 1.16 | |
| 1977–78 | 104.40 | 22.32 | | 79.50 | 2.81 | |
| 1978–79 | 120.50 | 16.77 | | 88.54 | 4.62 | |
| 1979–80 | 138.72 | 2.32 | | 99.16 | 1.04 | |
| 1980–81 | 146.80 | 5.28 | | 122.38 | 2.08 | |
| 1981–82 | 133.96 | 8.18 | | 98.38 | 1.70 | |
| 1982–83 | 129.55 | 8.52 | | 105.46 | 1.65 | |
| 1983–84 | | | | | | |

[a] Blanks indicate no subsidy.

[b] The wheat export subsidy program was suspended in August, 1972, and no new wheat export subsidy commitments were made; but actual subsidy payments were made when wheat on which there was a subsidy commitment was actually shipped.

Sources: U. S. Department of Agriculture, *Wheat Situation* and *Feed Situation* (various issues) and *Agricultural Statistics* (1973, 1977, and 1983).

such payments associated with acreage control and diversion programs. For wheat, direct payments were large relative to market prices from 1968–69 through 1972–73 and in 1977–78 and 1978–79. Direct payments for feed grains were more modest and applied to approximately half the output rather than to nearly all of it as in wheat and cotton (Table 2). (The data for 1983–84 would swamp any of the figures in Table 2 but are not included.)

The United States relied upon export subsidies for several commodities during the 1950s and 1960s. The reduction in price-support levels did not entirely eliminate the use of export subsidies. However, as shown in Table 2, the magnitude of export subsidies increased during the late 1960s and early 1970s. After 1973, explicit payment of export subsidies was halted and was not revived until the 1983 flour sales to Egypt. Implicit export subsidies—the export at a loss of products owned by CCC—continued for dairy products through the 1970s to the present.

From the vantage point of hindsight, it can be said that the 1976 election resulted in an interruption in a move toward market orientation. Before the 1976 election, the Ford Administration had announced substantial increases in the price-support levels for grains for both 1976 and 1977 (Table 1).

Not only were price-support levels increased for grains in both nominal and real terms, the relative loan rates for wheat and corn changed significantly—from a wheat/corn rate differential of about 20 percent to one of 50 percent in 1976 and 1977. The Agricultural Act of 1977 resulted in a significant increase in the price-support levels for corn and the other feed grains returning the relative wheat-to-corn price support relationships to a little less than 20 percent. However, the 1977 Act set the target price for wheat at substantially more than the loan rate and introduced the concept of cost of production into the determination of target prices. The wheat target price, relative to the loan rate established by the 1977 Act for 1978, resulted in approximately the same relationship that prevailed when target prices were first introduced in the 1973 Act. In 1974, the target price for wheat was 50 percent greater than the loan rate; in 1978, 45 percent. The 1977 Act also maintained the large discrepancy between the target prices for wheat and corn that was established in the 1973 Act. For 1974 and 1975, the target price for wheat was set 49 percent above that for corn, while farm prices for 1964 through 1972 differed by only 19 percent in favor of wheat. The 1978 target price for wheat, as established in the 1977 Act, was $3.00 per bushel, compared to $2.00 for corn; the wheat target price was 50 percent greater. Then, in the Emergency Agricultural Act of 1978, the target price for wheat for 1980 was increased to $3.40 per bushel while the corn target price was increased (by just 10 cents) to $2.10. Thus, the actual target price for wheat was 62 percent above that for corn.

The 1977 Act also provided for the continuation of target prices for barley and grain sorghum, with their target prices set to bear a reasonable relationship to the costs of production. But the cost of production estimates used by the Secretary of Agriculture to justify target prices for barley and grain sorghum were so high that the target price for 1977 for barley was set at 108 percent of the corn target price; the grain sorghum target price was set even higher—at 114 percent. For 1970–1979, the average farm prices per bushel for barley, grain sorghum, and corn were, respectively, $1.83, $1.83, and $2.02; thus, the average prices of barley and grain sorghum were only 91 percent of the corn price.

Why were these relative target prices set at significantly different levels than their relative market prices—not only for wheat compared to corn but also for barley and grain sorghum to corn? The three feed grains are such close substitutes that their relative prices vary rather little from year to year. Consequently, it is rather difficult to assume that their relative costs of production, in any meaningful sense, differ significantly from their relative market prices. Yet, Congress and the Secretary of Agriculture, in their collective responses to the political winds of the day, introduced substantial differences in the target prices with resulting distortions in resource allocations.

While the 1981 Act abandoned the direct use of cost of production measures to determine target prices, Congress still believed it had the capacity to set target prices at reasonable nominal levels for four more years (through 1985–86). So far, Congress has shown a quite remarkable reluctance to change the target prices in spite of overwhelming evidence that the target prices and loan rates are encouraging a level of grain output well beyond the level of demand at the market prices implied by the loan rates.

While the high ratio of the target price for wheat to that of corn was continued in the 1981 Act, the relative target prices for the feed grains were brought more nearly in line with their relative market values. In a paper that I prepared in 1980, I wrote:

> "As economists, we must recognize that the target prices and the associated deficiency payments represent the price the political system requires that we pay for support prices that interfere little, if at all, with the functioning of the market. It is evident that the political

price is not uniform—it is highest for wheat and modest or nil for corn and cotton. The grains that compete with wheat for resources, namely barley and grain sorghum, fall in an intermediate position.

"The growing disparity in relative target prices has reached a level that is cause for concern. The target prices can have an influence on resource allocation, both in terms of aggregate output and the allocation of resources among the various grains. We must hope that the current disparities do not grow but rather that differences are reduced.

"The most immediate effective limit on the level of target prices is the budgetary cost. Price support or loan levels are limited by the desire to export freely and by the rapid accumulation of stocks when levels are set too high. The output stimulating effects of high target prices become apparent rather slowly. And for a brief period, perhaps two or three years, the output effects can be contained by set-asides and acreage diversion programs. But recent experience indicates that expenditures required to elicit voluntary participation in acreage diversion programs can be substantial if significant output effects are desired. And it needs to be recognized that the magnitude of the required output adjustment is a function of the level of the target prices. Consequently high target prices impose two budget costs—the deficiency payments and the costs of achieving additional diversion" (Johnson, 1981).

## *Some Effects of the 1977 and 1981 Acts*

The rapidity with which grain stocks were rebuilt after their depletion in the early and mid-1970s should have been a warning that there was a real danger of creating excess production capacity in North America and other major western grain-producing countries. But this prospect was ignored. In fact, as Lesher (1984) put it earlier this year:

"At the time the 1981 Farm Bill was formulated, the main concern was that world food needs would outpace production. Many believed—inside and outside of government—that *the* agricultural issue of the 1980's was going to be how to produce enough for a starving world rather than surpluses. Many believed that the United States was the only country that possessed the potential to expand food production enough to meet world needs."

While some of us can argue that we saw a return to the long-run downward trend in real crop prices during the 1980s, there is little doubt that rigid price supports and unrealistically high target prices, especially for wheat and cotton, were simply not appropriate to the economic conditions that emerged in 1981–1983. True, if the exchange value of the dollar had not risen so sharply, the 1981 Act would have fared much better than it did. But after all the attention that had been given to the increased instability that emerged post-1972, it was wholly inappropriate that legislation with so much inflexibility had become the law of the land. Perhaps the new administration went too far in asking for full flexibility on the part of the Secretary of Agriculture in setting commodity loan rates or target prices. But the Congressional reaction in giving so little flexibility must be avoided in the 1985 legislation if U. S. agriculture is not to become a permanent large-scale ward of the government.

## *Changing Conditions: Unchanged Programs*

I argued above that beginning in 1961 there was a serious effort to permit the prices of most farm products to move to market-clearing levels, aided by direct payments to maintain farm incomes. However, the components of the farm commodity program remain more or less the same as those developed during the 1930s. Consequently, when price supports have been set at levels that induce the accumulation of large and unwanted stocks, the tools to reduce stocks and to increase prices have remained unchanged over the half century. Fundamentally, our commodity policies attempt to achieve their objectives of higher prices and incomes by either restricting output (supply) or expanding demand, domestically and/or internationally.

While our commodity programs have changed little if at all, American agriculture and the setting within which it functions have changed enormously over the past century and particularly in the past three decades. Consequently, even if supply limitation and demand expansion had been effective tools during the 1930s and 1940s, it does not follow that, given today's conditions, such efforts would be successful.

## *Lessons from the Past*

There are several lessons from our recent experience with commodity programs. These lessons that, hopefully, will influence legislation passed in 1985 and subsequent years can be summarized:

1. The level of price supports has a significant influence on the disposition of actual production—its allocation between domestic use, export, and stocks.
2. The level of price supports and/or target prices has a significant influence upon the quantity of U. S. farm output.
3. The level of price supports influences production not only in the United States but also in competing producing areas.
4. Supply management can have relatively little effect on the level of crop output unless there is a massive reduction in acreage harvested, such as with payment in kind (PIK) in 1983, involving a very high cost if compliance is voluntary.
5. The level of price supports or output prices and the effectiveness of supply management have almost no long-run influence upon the prosperity of agriculture.
6. The only farm resource whose price or annual use value is significantly affected by government programs, either through higher output prices or deficiency payments, is land. As a result, the primary effect of lower prices and the elimination of direct payments would be on the rental value and price of land.
7. The incomes of farm people, from both farming and nonfarming activities, depend on available alternative opportunities for the use of their labor and capital.

I would assume that there would be little argument about the first of the lessons, namely, that the level of our price supports has a significant influence on how the available supplies are distributed among end uses. We know that, when farm prices approach or go below the price-support levels, stocks increase. A comparison of carry-over stocks for soybeans and corn is instructive. Generally speaking, except for the last year or so of the Johnson Administration, the price support for soybeans has been significantly below farm prices. On several occasions during the past two decades, the price support for corn has been at or above the farm price. Except for the one period of time, soybean carry-overs have been a small percent of annual production and have followed a very different pattern than corn carry-overs. Table 3 presents data on carry-overs, price supports, and farm prices since 1961 for soybeans and corn.

During the early and mid-1960s, soybean stocks were from 5 percent to no more than 10 percent of production. But in 1967 and 1968, the farm prices of soybeans moved to or below the price support for soybeans which had been increased from $2.25 per bushel for 1962 through 1965 to $2.50 per bushel for 1966 through 1968. Year-end stocks increased rapidly, reaching 29 percent of 1968 production at the end of the crop year. The price support was lowered for 1969 and for subsequent years to $2.25 per bushel, and stocks fell rapidly declining to well under 10 percent of production by the beginning of the 1972 crop year as the margin between the farm price and the support level widened significantly.

The corn stocks followed the expected pattern when the farm price was at or below the price support. Only when farm price was above the price support and that margin widened, as it did following 1972, did corn stocks decline to less than 10 percent of production. Too little attention was given to the sharp increase in corn stocks in 1978 and 1979 (to 18 percent and 20 percent of production, respectively). The large corn stocks at the end of 1982 at 38 percent of production were reminiscent of the early 1960s.

Both the corn and soybean data show that stocks are very sensitive to the relationship between the farm price and the loan rate. This, of course, is what economists expect, but it apparently is not what politicians and some commodity groups are willing to accept as fact.

The second lesson, namely, that the level of price supports or target prices can have a significant effect on U. S. farm output, is also very well founded. If price supports and target prices are set significantly above market-clearing levels, output will respond in a positive manner; and the quantity produced will be greater than the quantity demanded at that price.

Our evidence on the elasticity of supply for major farm products is less robust than we would like it to be. However, the available estimates do not seem to be unreasonable. Tweeten (1979), for

**TABLE 3. Soybeans and Corn: Loan Rates, Farm Prices, and Carry-Over Stocks**

| Year[a] | Soybeans | | | Corn | | |
|---|---|---|---|---|---|---|
| | Loan rate[b] | Farm Price | Carry-over | Loan rate | Farm price | Carry-over |
| | dollars per bushel | | percent of production | dollars per bushel | | percent of production |
| 1961 | 2.30 | 2.28 | 8 | 1.20 | 1.10 | 46 |
| 1962 | 2.25 | 2.34 | 2 | 1.20 | 1.12 | 38 |
| 1963 | 2.25 | 2.51 | 4 | 1.07 | 1.11 | 38 |
| 1964 | 2.25 | 2.62 | 4 | 1.10 | 1.17 | 33 |
| 1965 | 2.25 | 2.54 | 4 | 1.05 | 1.16 | 20 |
| 1966 | 2.50 | 2.75 | 10 | 1.00 | 1.24 | 20 |
| 1967 | 2.50 | 2.49 | 19 | 1.05 | 1.03 | 24 |
| 1968 | 2.50 | 2.43 | 29 | 1.05 | 1.08 | 25 |
| 1969 | 2.25 | 2.35 | 20 | 1.05 | 1.16 | 21 |
| 1970 | 2.25 | 2.85 | 9 | 1.05 | 1.13 | 16 |
| 1971 | 2.25 | 3.03 | 6 | 1.05 | 1.08 | 20 |
| 1972 | 2.25 | 4.37 | 5 | 1.05 | 1.57 | 13 |
| 1973 | 2.25 | 5.68 | 11 | 1.05 | 2.55 | 9 |
| 1974 | 2.25 | 6.64 | 15 | 1.10 | 3.02 | 8 |
| 1975 | —[c] | 4.92 | 16 | 1.10 | 2.54 | 7 |
| 1976 | 2.50 | 7.32 | 8 | 1.50 | 2.15 | 14 |
| 1977 | 3.50 | 5.88 | 9 | 2.00 | 2.02 | 17 |
| 1978 | 4.50 | 6.66 | 9 | 2.00 | 2.25 | 18 |
| 1979 | 4.50 | 6.28 | 16 | 2.10 | 2.52 | 20 |
| 1980 | 4.50 | 7.57 | 18 | 2.25 | 3.11 | 16 |
| 1981 | 5.02 | 6.04 | 13 | 2.40 | 2.41 | 27 |
| 1982 | 5.02 | 5.65 | 17 | 2.55 | 2.38 | 38 |
| 1983 | 5.02 | 9.00[d] | 8 | 2.65 | 3.00[d] | 13[e] |

[a] October-September.

[b] Loan rate for No. 2 soybeans from 1961–1968; for No. 1, soybeans from 1969.

[c] No loan.

[d] Projected.

[e] If the corn crop had been a normal size, the carry-over would be about 7 percent.

Sources: Cochrane and Ryan (1976, pp. 179 and 249) and U. S. Department of Agriculture (1974 and subsequent years).

example, estimates that, as of about 1960, the short-run elasticity of supply of crops is 0.17 while the long-run elasticity may be as high as 1.56. For aggregate U. S. farm output, his estimates are a short-run elasticity of 0.25 and a long-run elasticity of 1.8. Tyers (1982) of the Australian National University estimates the long-run elasticity of U. S. wheat supply at 0.8 and the short-run elasticity at 0.45; for coarse grain, he estimates the long-run elasticity at 0.75.

What is clear, or should be, is that, even with very low price elasticities of supply, the increase in output induced by a price just 10 percent above the market-clearing level soon results in a substantial excess production that must be either stored, disposed of in some manner, or eliminated through output reduction. Assume that the price elasticity of supply for a price established for four years is 0.15 and that the elasticity of demand is 0.2. A price 10 percent above the market-clearing level would increase output by 1.5 percent for each of the four years and reduce consumption by 2.0 percent each year. Consequently, in four years the output increase would total 6 percent of annual output and the consumption reduction would be 8 percent of annual output; together, 14 percent of annual output could be in storage at the end of the four years. If the price increase were 20 percent, the excess in four years would be 28 percent of annual output. If the elasticities of supply are closer to those estimated by Tyers for grains and if we consider more reasonable elasticities of demand, it is easy to see how a 15 percent price increase above market-clearing levels could result in stocks equal to a third of annual output in two or three years.

The third lesson is little more than an extension of the second in that, when we increase our price supports, we encourage production in competing countries. The reason, of course, is that our price supports can have a significant effect on world market prices. As world market prices are increased due to higher price supports in the United States, the price signals to competitive producers in countries permitting international market prices to be reflected in domestic prices induce output expansion. Tyers estimates long-run elasticities of supply in Canada for wheat at 0.5 and for coarse grains at 0.7; in Australia, for wheat at 0.9 and for coarse grains at 0.6. Consequently, we should not be surprised that, when we increase international market prices through our price supports and output limitations, others respond by increasing their output, reducing the U. S. share of the market.

If further evidence is needed that by establishing prices above competitive market levels foreign output is encouraged, the U. S. tobacco program is a case in point. The tobacco program has been successful in increasing both domestic and world market prices of tobacco through effective control of U. S. output. In contrast to almost all other crops, output levels in recent years have been no more than they were two decades ago; some years, they were significantly less. While the United States held an umbrella over world tobacco prices, our share of world production of flue-cured tobacco (our principal type of export tobacco) fell from 64 percent in 1935–1939 to 18 percent in 1980–1982; our share of world exports of this tobacco declined from 83 percent in 1935–1939 to 21 percent in 1980–1982. Much of the loss of production dominance and of export share has occurred since 1960–1964 when the United States produced 40 percent of the world's flue-cured tobacco and 51 percent of world exports (Seagraves, 1983).

I doubt if anyone anticipated one consequence of the tobacco program, namely, that the United States would become the world's largest importer of tobacco. High U. S. prices of tobacco combined with poundage output allotments have meant that farmers have emphasized high-quality tobacco, foregoing the production of low-quality tobacco. Meanwhile, low-quality tobacco is imported; nearly one-third of all tobacco used in cigarettes is imported. In 1981, tobacco exports had declined to just 95,000 metric tons—a little more than half the net exports two decades earlier.

The fourth lesson is that the level of crop output is not affected significantly by voluntary acreage reductions unless there is a massive reduction in acreage harvested such as with PIK in 1983. The setting aside of up to one-fifth of crop acreage is required before output can be reduced by more than 3 percent to 5 percent. The slippage in voluntary programs is very great—farmers set aside their lowest quality land. Land on which there would be crop failure is counted in the set aside, and farmers use more yield-increasing inputs in response to the higher rewards associated with supply management programs. To induce a reduction in cultivated land of the required magnitude becomes a major budgetary drain. In fact, the drain may well be so large that voluntary supply management may not be a viable alternative.

The fifth lesson is, however, even more discouraging than the fourth. Even if supply management were effective in reducing farm production, there would be little long-run positive effect on the prosperity of agriculture. Assume that effective supply management made it feasible to increase price supports significantly, say, by 15 percent. As time passed, the restrictiveness of supply management would have to be increased to offset the loss of export markets and the yield-increasing measures undertaken by farmers. Thus, it would be necessary to idle more and more land; under these circumstances, there is not the slightest shred of evidence that the return to either farm labor or capital would increase for any period of time as a result of the higher output prices.

The sixth lesson emphasizes that increasing the demand for farm inputs, such as for labor or capital, has no measurable long-run influence on the return to that input unless its supply is limited. Limiting the supply of either labor or capital engaged in farming is not possible. Consequently, any increase in the return to labor through higher output prices will be quite temporary. This seems to be the most difficult of all the lessons to understand and accept.

The supply of labor to agriculture is quite elastic; in other words, any increase in the returns to labor, other things the same, results in an increase in the amount of labor offered (Johnson, 1973). This is true of both family and operator labor and hired labor. Consequently, raising output prices results primarily in increasing farm employment, not in any significant increase in the returns to farm families for their labor. A number of statistical studies document the responsiveness of the supply of labor in agriculture to differences in returns to farm *vs.* nonfarm labor (Trychiniewicz and Schuh, 1969, and Tweeten, 1979).

Hathaway, while Undersecretary of Agriculture, strongly supported the view that it was adjustments in the labor market and not in commodity programs that resulted in the unprecedented improvement in the relative income position of farm families after 1940. A lengthy quote seems justified (Hathaway, 1980):

> "During the war, the differential began to narrow. In the last half of the 1940's, per capita income of farm people averaged 60 percent of the per capita income of nonfarm people from all sources.
>
> "In the 1950's, however, farm income per person again fell behind—remaining mostly static while the per capita income of nonfarm peo-

ple rose by more than a third. In the last half of the 1950's, the per capita income of farm people was only one-half the per capita income received by people living off the farm.

"In the early 1960's, we would see the beginning of adjustment. By the end of the decade, per capita income on farms averaged above $2,000 compared with around $3,000 for nonfarm people. For the 5 years 1965 through 1969 people living on farms averaged 71 percent of the per capita income of people living off the farm.

"In more recent years, this percentage has risen to 85 or more—although this of course varies from year to year.

"So—the labor market did adjust. But the adjustments were difficult for many. Despite government efforts to deal with these difficulties, it appears in retrospect that no government policy or program was significant in aiding the adjustment or softening the pain of adjustment for farm people."

The seventh and last of my litany of lessons follows from the sixth. Incomes of farm people, from both farming and nonfarming activities, depend on the availability of alternative opportunities for the use of their human and physical capital. This point follows from the significant elasticity of supply of labor to agriculture. When labor and capital can flow reasonably freely—at least over a period of two to four years—between agriculture and the rest of the economy, returns increase at approximately the same rate as those in the rest of the economy. The real return to labor in the rest of the economy is the primary determinant of the income level of farm people—not the absolute level of that income but its changes over time.

How fully farming has been integrated into the general economy is well illustrated by the growing importance of nonfarm income to farm families. Some information on this point is presented in Table 4.

If further evidence is needed to support the view that increasing prices is an ineffective method of increasing returns to farm people and that it is alternative opportunities that count, a recent report of the European Commission is very relevant. As is well known, the European Community has held farm prices significantly above the level that would have prevailed in the absence of market intervention. Yet, the European Commission has reported numerous times that the commercial agricultural policies have been ineffective in reducing regional income disparities within the Community. One forthright statement comes from its 1980 report on the agricultural situation (Commission of the European Communities, 1981, p. 52):

> "During the period from 1962/65 to 1976/77, regional disparities in agricultural incomes (as measured by gross value-added per agricultural worker) increased in the Community. The ratio between the regions with the highest agricultural incomes and those with the lowest rose from 5:1 to 6:1.
>
> "Generally speaking, the regions with an above-average level of agricultural income are to be found in a favourable general economic context; the converse is true of regions with a low level of agricultural incomes."

It is hard to imagine a starker indictment of high output prices as a policy tool than what the Commission has written. As I argued more than a decade ago in *World Agriculture in Disarray* (Johnson, 1973): *High prices are not enough; if controlling output prices is to have any long-run effect on the return to labor and management in agriculture, output prices must increase continuously*. But this is not possible. Imagine what the cost to taxpayers and consumers would be to increase real farm prices by just 3 percent per year (approximately the long-run growth in per capita incomes in industrial economies); this would require doubling real farm prices in a little more than two decades. But if one wishes to reduce substantially the flow of labor out of agriculture, something like a 3 percent annual increase in real output prices would be required.

## Who Gets What?

*Farm Commodity Programs: Who Participates and Who Benefits?* by Lin, Johnson, and Calvin (1981) is a study of the distribution of direct payments and market price enhancements of the 1978 farm programs by size of farm. This detailed and competent study contains no surprises. The study estimates the short-run price effects resulting from the supply management program for 1978 and estimates the distribution of price benefits and direct payments among farms by size. In the aggregate it was estimated that direct payments were $1,754 million and price benefits, $912 million. Contrary to

**TABLE 4. Net Farm Income and Off-Farm Income of Farm Operator Families**

| Year | Net farm income | Off-farm income | Total income | Net farm income |
|---|---|---|---|---|
| | | million dollars | | percent of total income |
| 1960 | 11,518 | 8,482 | 20,000 | 57.6 |
| 1961 | 11,957 | 9,163 | 21,120 | 56.6 |
| 1962 | 12,064 | 9,904 | 21,968 | 54.9 |
| 1963 | 11,770 | 11,020 | 22,790 | 51.6 |
| 1964 | 10,492 | 11,637 | 22,129 | 47.4 |
| 1965 | 12,899 | 12,727 | 25,626 | 50.3 |
| 1966 | 13,960 | 13,882 | 27,842 | 50.1 |
| 1967 | 12,339 | 14,095 | 26,834 | 46.0 |
| 1968 | 12,322 | 15,466 | 27,788 | 44.3 |
| 1969 | 14,293 | 16,612 | 30,905 | 46.2 |
| 1970 | 14,381 | 17,617 | 31,998 | 44.9 |
| 1971 | 15,043 | 19,110 | 34,153 | 44.0 |
| 1972 | 19,507 | 21,265 | 40,772 | 47.8 |
| 1973 | 34,435 | 24,714 | 59,149 | 58.2 |
| 1974 | 27,309 | 28,135 | 55,444 | 49.3 |
| 1975 | 25,555 | 23,901 | 49,456 | 51.7 |
| 1976 | 20,132 | 26,681 | 46,813 | 43.0 |
| 1977 | 10,821 | 26,120 | 45,941 | 43.1 |
| 1978 | 27,651 | 29,704 | 57,355 | 48.2 |
| 1979 | 32,251 | 35,267 | 67,518 | 47.8 |
| 1980 | 21,505 | 37,660 | 59,165 | 36.3 |
| 1981 | 30,057 | 39,877 | 69,935 | 43.0 |
| 1982 | 22,051 | 39,431 | 61,482 | 35.9 |

(Continued on next page.)

**TABLE 4—Continued**

| | Net farm income per farm | Off-farm income per farm | Total income per farm | Net farm income |
|---|---|---|---|---|
| | value of sales, 1982 | | | percent of total income |
| $500,000 and over | 571,097 | 26,831 | 597,929 | 95.5 |
| $200,000–499,999 | 53,461 | 13,720 | 67,180 | 79.6 |
| $100,000–199-999 | 19,786 | 11,074 | 30,861 | 64.1 |
| $ 40,000–99,999 | 5,539 | 10,615 | 16,155 | 34.3 |
| $ 20,000–39,999 | 504 | 12,887 | 13,391 | 3.7 |
| $ 10,000–19,999 | -728 | 17,208 | 16,479 | -4.4 |
| $ 5,000–9,999 | -881 | 19,146 | 18,265 | -4.8 |
| $ 2,500–4,999 | -998 | 19,328 | 18,330 | -5.4 |
| Less than $ 2,500 | -465 | 20,758 | 20,292 | -2.3 |
| All farms | 9,959 | 16,430 | 26,386 | 37.7 |

Source: U. S. Economic Research Service (1983).

most studies of this type, the authors estimated the amount of income foregone in the set-aside acreage which amounted to $1,158 million or somewhat more than the price benefits. The net increase in cash farm income from the 1978 farm program was estimated to be $1,508 million. The cost to the government of the farm income and price-support program for fiscal year 1979 was $3.6 billion.

It was estimated that the largest 10 percent of the farms received 55.5 percent of the net benefits; the smallest 70 percent received just 8 percent. The largest 10 percent received average direct payments of about $12,000 plus $6,000 in price benefits, minus about $5,000 in foregone income. The smallest 50 percent of the farms received a net cash gain of approximately $460 per farm or about one-thirtieth of what the largest 10 percent of the farmers received. Note that benefits as measured did not accrue wholly to farm operators; a significant fraction was shared with landlords.

Another aspect of "Who Gets What?" relates to the distribution of benefits among those who supply production factors to agriculture. I have argued elsewhere that, if there are long-run income benefits from farm commodity programs, the primary beneficiaries are the owners of the land or of the rights to produce the product. Recent research provides definitive evidence that a large percentage of the increase in net farm income from output limitations and higher prices resulting therefrom go to those who own the right to produce, either through ownership of land or a quota. In the examples described below, output is limited by marketing quotas that are transferable, so it is possible to determine the cost of the right to produce. Where the rights to produce a commodity can be bought or rented, new producers must pay higher prices for land or buy or rent quotas to produce. They gain very little from higher market or deficiency payments already capitalized into the value of the land or quota. In acquiring the underlying asset—land, quota, or producing cows—required for engaging in agriculture, the new entrant pays the market price for the asset and will continue to earn a reasonable rate of return on the asset. This reasonable rate of return will be the same as what he would earn on an asset of equal risk in some other sector of the economy. Unfortunately, everyone who purchases capitalized assets after programs have been in effect can lose if the programs are discontinued, even though they gain little or nothing from the continuation of the programs.

The tobacco program, with its quota allotments, illustrates the point. For flue-cured tobacco, only a quarter of the total poundage quotas is grown by the owner of the land or a relative of the owner; for burley tobacco, just half of the tobacco is grown by the owner or a relative. The remainder of the quotas are rented out, either as a quota or as a quota rented together with the land. For flue-cured tobacco in North Carolina for 1977–1981, the average annual lease price was somewhat greater than 25 percent of the tobacco price. The same general relationship held for burley tobacco (Sumner and Alston, 1984, p. 12). Sumner and Alston estimate that the annual rental value of tobacco quotas (1981 dollars) was in the general range of $800 million to $1,000 million. This is what the owners of quotas would lose if the tobacco allotment program were abolished. The capital value of the quotas would be on the order of $4 billion if the discount rate used were as high as 25 percent. Consequently, there should be no surprise that there is strong support for continuation of the program.

Barichello (1984) presents estimates of the value of quotas to produce milk in British Columbia and Ontario. For British Columbia, in 1980 he estimates that the annual value of the quota to produce 100 liters of milk was approximately $21 (Canadian) or almost half the selling price of milk. For Ontario, where the price of milk was substantially below the British Columbia price, the annual value of the quota was about $11 (Canadian) or one-third of the milk price.

The estimated capital value of the quotas for producing broilers and eggs in Canada, where quotas are salable, has been put at $440 million for broilers and $280 million for eggs for 1980. Assuming a discount rate of 12 percent, these capital values imply that the annual quota income per producer was about $23,000 for each broiler producer and $15,000 for each egg producer (Arcus, 1981).[3]

While producers who were granted the quotas have benefited substantially, a farmer who wishes to enter dairy production or poultry production has to pay most of the anticipated future benefits by

---

[3]The discount rate of 12 percent may be too low and 25 percent may be more realistic according to estimates presented by Barichello (1984). But a doubling of the discount rate still means the annual rental value per farm was $11,500 for broiler producers and $7,500 for egg producers.

purchasing the quotas. Consequently, new producers gain little or nothing from the supply restrictions and higher prices.

## *Concluding Comments*

I have presented seven lessons from our experience with farm commodity programs over the past half century. The strong implication of these lessons, both individually and collectively, is that price supports, deficiency payments, output quotas, and supply management contribute little or nothing to the long-run rate of return to farm resources. This is directly true for labor; the return to labor is determined in the long run by the amount of human capital and nonfarm alternative earnings. Farm prices, whether high or low, have no effect on the return to farm labor; farm prices can affect the amount of farm employment but not the return to those who work.

But higher farm prices do increase the price of land transferring wealth to those who own the land. After a program of higher prices is in effect, the rate of return on assets is the same as before. The only difference is that more capital is required to purchase or rent the asset, say, to have an economic-sized unit. It has never been clear to me that there is any social gain created by increasing the price of land, or why we wish to have policies that increase the amount of capital required for a farm of sufficient size to provide farm people with real income comparable to that of nonfarm people.

Lessons from the past also indicate that holding an umbrella over world market prices, as we do with our price supports and supply management, benefits farmers in the rest of the world far more than it benefits our farmers. I find it hard to believe that our policymakers intend that the benefits of their policies be distributed in this way.

## *References*

Arcus, Peter L. *Broilers and Eggs.* Technical Report No. E/I 3. Ottawa: Economic Council of Canada, March, 1981.

Barichello, Richard R. "Analyzing an Agricultural Marketing Quota." Yale University, Economic Growth Center, Discussion Paper No. 454, March, 1984.

Benedict, Murray R. *Farm Policies of the United States, 1790–1950.* New York: The Twentieth Century Fund, 1953.

Cochrane, Willard W., and Mary E. Ryan. *American Farm Policy, 1948–1973.* Minneapolis: University of Minnesota Press, 1976.

Commission of the European Communities. *The Agricultural Situation in the Community: 1980 Report.* Brussels, 1981.

Gisser, Micha. "Needed Adjustments in the Supply of Labor." *Journal of Farm Economics* 49, No. 4 (November 1967):806–815.

Hathaway, Dale E. "Shifting Markets, Government Policies, and the Implications for Agricultural Financing." An address before the Farm Credit/ Universities Conference, Louisville, Kentucky, June 26, 1980.

Johnson, D. Gale. *World Agriculture in Disarray.* London: Macmillan Company, Ltd., 1973.

———. "World Agricultural and Trade Policies: Impact on U. S. Agriculture." In *Contemporary Economic Problems 1979,* edited by William Fellner. Washington, D. C.: American Enterprise Institute, 1979, pp. 293–324.

———. "Agricultural Policy Alternatives for the 1980s." In *Food and Agricultural Policies for the 1980s,* edited by D. Gale Johnson. Washington, D. C.: American Enterprise Institute, 1981, pp. 183–209.

Lesher, William G. "Farm Policy Update and Perspective." Speech given to American Farm Bureau Federation, Orlando, Florida, January 9, 1984.

Lin, William, James Johnson, and Linda Calvin. *Farm Commodity Programs: Who Participates and Who Benefits?* U. S. Economic Research Service, Report No. 474. Washington, D. C.: U. S. Government Printing Office, 1981.

Seagraves, James A. "The Life-Cycle of the Flue-Cured Tobacco Program." North Carolina State University, Department of Economics and Business, Working Paper No. 34, March, 1983.

Sumner, Daniel A., and Julian M. Alston. *Consequences of the Elimination of the Tobacco Program.* North Carolina State University, North Carolina Agricultural Research Service, Bulletin No. 469, March, 1984.

Trychiniewicz, E. W., and G. E. Schuh. "Econometric Analysis of the Agricultural Labor Market." *American Journal of Agricultural Economics* 51, No. 4 (November 1969):770–787.

Tweeten, Luther G. *Foundations of Farm Policy.* 2nd ed., rev. Lincoln: University of Nebraska Press, 1979.

Tyers, Rod. "Effects on ASEAN of Food Trade Liberalisation in Industrial Countries." Paper presented to Second Western Pacific Food Trade Workshop, Jakarta, August 22 and 23, 1982.

U. S. Department of Agriculture. *Agricultural Statistics*. Washington, D. C.: U. S. Government Printing Office, various issues.

———. *Wheat Situation*. Washington, D. C.: U. S. Government Printing Office, various issues.

———. *Feed Situation*. Washington, D. C.: U. S. Government Printing Office, various issues.

———. *Fats and Oil Situation*. Washington, D. C.: U. S. Government Printing Office, various issues.

———. *Cotton and Wool Situation*. Washington, D. C.: U. S. Government Printing Office, various issues.

———. *Dairy Situation*. Washington, D. C.: U. S. Government Printing Office, various issues.

U. S. Department of Commerce. *Statistical Abstract of the United States, 1961*. Washington, D. C.: U. S. Government Printing Office, 1961.

U. S. Economic Research Service. *Economic Indicators of the Farm Sector Income and Balance Sheet Statistics, 1982*. Washington, D. C.: U. S. Government Printing Office, October, 1983.

# 3 FUTURE AGRICULTURAL POLICY: A CHALLENGE FOR ALL

**William G. Lesher***

I appreciate the opportunity to begin this conference with a discussion concerning past, current, and future agricultural policies. This certainly is a distinguished group of economists and others interested in this topic. I am hopeful that the dialogue and debate that follow will add to developing more appropriate agricultural policies for the balance of the 1980s, something that is most important and needed.

As I looked over the list of participants and those planning to attend this conference, I noticed many who have spent most of their lives dealing with these issues. One or two of you may even remember some of the programs implemented not too many years after the enactment of the Agriculture Adjustment Act of 1933—our first comprehensive program for farm commodities. Thus, it is a humbling experience for me to speak before some who are much more experienced and wise.

My real policy training and experience only began in June, 1977, when I came to Washington, D. C., to work for Senator Richard G. Lugar of Indiana—then a newly appointed member of the Senate

*The author is Assistant Secretary for Economics, U. S. Department of Agriculture, Washington, D. C.

Committee on Agriculture, Nutrition, and Forestry. My feeling at that time was that, if I was going to work in a university on agricultural policy matters, I first should see where, how, and who made the policy decisions. I did not take the advice of one of my professors who suggested that one could only have peace of mind if he or she never witnesses sausage or public laws being made. During the past seven years, I have witnessed several farm policies being developed. Many of them were implemented under the broad banner of "helping" agriculture and "saving" the family farm. Here are some of the major events that have occurred since 1977:

- *Fall, 1977:* A farm bill was enacted that (1) mandated dairy price supports at the high levels which had been administratively established earlier that year and eventually set in motion production increases that led to a crisis in the dairy industry and (2) indexed target prices (entitlements) to the cost of production.
- *Winter, 1977, and Spring, 1978:* A new farm group called the American Agricultural Movement was formed whose members drove their tractors to Washington, D. C., in pursuit of 100 percent of parity, and Congress responded by (1) increasing the target price of wheat from $3.05 to $3.40 per bushel, (2) providing an additional $4 billion in FmHA credit for farmers, (3) raising the minimum cotton loan rate from 44 cents to 48 cents per pound, and (4) increasing Commodity Credit Corporation (CCC) borrowing authority from $14.5 billion to $25 billion to finance increased price-support activities.
- *1979:* A record amount of capital investment was made in the farm sector as high inflation rates provided the incentive for farmers to buy land and machinery that were escalating in price, while the U. S. dollar cheapened and exports rose.
- *January, 1980:* An embargo on U. S. agricultural exports to the Soviet Union was announced; later, measures to help ameliorate the effects of the embargo were initiated that included CCC grain purchases and changes in the farmer-owned reserve (FOR) to encourage additional entry of grain to the reserve and not the market. (Entry loan rates and release and call prices were increased as well, and a one-year waiver of interest was imposed.)
- *January, 1980:* An alcohol fuel program was announced which was intended to provide between $8.5 billion and $13 billion of

assistance to stimulate production of alcohol fuels over a 10-year period.

- *Fall, 1980:* Congress abolished normal crop acreage requirements designed to make acreage cutbacks more effective.
- *Winter, 1980:* Congress passed legislation to further soften the effects of the embargo by increasing commodity loan rates (corn, $2.25 to $2.40 per bushel; wheat, $3.00 to $3.30 per bushel; and soybeans, $4.50 to $5.02 per bushel), increasing the FOR entry prices for corn ($2.40 to $2.55 per bushel) and wheat ($3.30 to $3.50 per bushel), establishing a 4 million metric ton food security reserve.
- *Spring, 1981:* The Soviet embargo was lifted, and major concerns developed about having enough food to feed the world.
- *December, 1981:* A farm bill was enacted that carried forth the basic provisions of the 1977 bill but with some modifications.
- *Early, 1982:* The recession hit with full force, both in the United States and around the world, and demand for agricultural products dropped dramatically in the face of a record output.
- *1982:* Export enhancement measures were adopted including record volumes of export credit and credit guarantees.
- *1982:* Farmer-owned reserve entry levels for wheat and corn were raised to very high levels to attract more program participation; farmers responded by producing for the reserve.
- *1982:* As part of a budget-reduction package, a new dairy bill was developed by Congress in response to the dairy crisis which called for assessments on all milk marketed to pay for surplus removal without any reduction in the price-support level.
- *January, 1983:* With commodity prices depressed, record surpluses, and record farm program costs, the largest acreage reduction program in history was announced and implemented—the payment-in-kind (PIK) program.
- *January, 1983:* The Egyptian wheat flour sale was announced to counter other countries' subsidies.
- *August, 1983:* The new Union of Soviet Socialist Republics (USSR) long-term agreement was negotiated with higher minimums.
- *Fall, 1983:* Dairy farmers universally disliked the recently passed dairy assessment legislation; in response, new dairy legislation was enacted giving producers a paid diversion in exchange for scheduled reductions in price-support levels if surpluses continued.

- *Early, 1984:* New legislation was enacted to modify (primarily to freeze target price levels) in exchange for more attractive 1984 and 1985 crop program provisions, which included paid diversions, and abandoned tighter rules for summer fallow and haying and grazing.
- *Mid-year, 1984:* Major farm credit problems arose as drought and falling land values reduced producer equity, and lenders tightened up on farmers who had leveraged themselves in expectation of continued inflation. As farmers fell in arrears on payments, pressures built for billions of dollars in new credit or loan forgiveness.

After having gone through this time period at various stages of the policy-making process, I conclude that much of what has occurred has not necessarily helped agriculture that much, especially in the long run. Indeed, in my view some of these policy actions have been extremely harmful. Furthermore, I believe that the many persons involved in this process who have argued for these policies are, for the most part, well-intentioned even though they probably do not recognize the long-term consequences of their actions or are not willing to stand the pressures in the short term to see the long-term results.

I also feel that, in general, agricultural economists and policy analysts must accept some blame for the confusion over the consequences of various policy alternatives. While some have continued to make solid contributions to the policy debate, I do not believe that they are in the majority. I still read or hear about well-known agricultural economists that argue for higher loan rates, larger reserves, massive credit bailouts, or more stringent supply management programs—including mandatory controls—as a means of solving the problems facing agriculture. They argue for these points as if price—and, thus, economics—really does not matter in business decisions.

It seems that most present their case without really looking at the track record of these programs. We have a rich history filled with many types of programs since 1933. Some, such as Don Paarlberg, have analyzed these program impacts with clarity and economic facts. I assume D. Gale Johnson will do the same today at this conference as he has done for so many years. But in my view, Paarlberg and Johnson are in the minority. If some economists are confused about the impacts of farm programs, including the cost of PIK, I do not know why we should expect politicians, farmers, and farm and commodity groups to come together and agree on the appropriate future policies for agriculture.

In this chapter I am going to discuss the following points:

- Farm legislation that has been with us for the last 50 years has not been that effective except to help bid up the prices of fixed factors of production such as land. Such policies are especially inappropriate for today's agriculture that depends so heavily on world markets and in which about two-thirds of production is produced by one-tenth of our farmers.
- The unprecedented growth in demand for agricultural products that we experienced in the 1970s was a temporary move away from long-term growth trends and was produced primarily by spurts of growth in world economies and the adjustment of the world economy to higher energy prices through various fiscal and monetary policies.
- The 1970s set in motion several matters that will cause us problems for some time to come: investment in land and machinery that increased capacity to produce for a market that is not there; farm policy parameters that ratchet up with temporary upswings but which resist downward adjustments; credit problems for farmers as land values recede to values that can be sustained from farm production; and the movement of agriculture away from a domestic market orientation and toward export markets, thereby subjecting the sector to greater influence from outside sources such as macroeconomic and international factors.
- The 1981 farm bill, which extended the policies begun during the New Deal, is not serving agriculture well primarily because agriculture has changed so much in composition and orientation. Forcing an old policy on a new agriculture has not worked recently and will not work in the future.
- If the 1985 farm bill does not become better focused on the needs of modern agriculture, we risk losing one of the most efficient industries as it will go the way that other protected industries have gone.

In short, I believe that agriculture is too important economically and socially to allow another farm bill to come and go without a

Herculean attempt to better understand the consequences of alternative policies and adjust our policies to today's agriculture and economic climate. This will require that all involved reduce their posturing, examine and understand the facts, and make decisions based on what will be good for our future agricultural industry.

Agricultural economists and policy analysts have a very important role to play in providing the facts and analysis. As a matter of fact, all of us have much to lose if we do not. In Washington, D. C., economists have about as good a reputation as lawyers—so good, in fact, that in 1982, during budget deliberations, the Senate Committee on Agriculture, Nutrition, and Forestry almost passed an amendment to reduce the number of economists within the U. S. Department of Agriculture by one-half. Warning signals are being fired but will we respond? Future agricultural policy is going to be a challenge for all. We all have much at stake.

## *Farm Policy Review*

Public support for U. S. agriculture dates back to the creation of the Department of Agriculture—the people's department—in 1862. It was, indeed, the people's department as more than 55 percent of the American people were engaged in production agriculture. Agricultural policies at the time were, in fact, social policies—broad-based public responses to problems affecting a major segment of the U. S. population.

Early in the Department's history, it was recognized that it was in the public's interest to promote agricultural research and to disseminate useful research results to the nation's farmers. Improved seed varieties were developed, new techniques were tried, and improved machinery was tested under the auspices of publicly funded farm research and education programs. The American farmer responded by adopting this new technology at a breakneck pace and expanding productivity and output substantially.

It was not until the Great Depression that agricultural programs took on the added responsibility of price and income supports. Low commodity prices led to depressed farm incomes far below levels of the rest of society. Farm failures became widespread. There was a general fear that, unless something drastic was done, a large segment of the population would be turned out of their homes and off their land and that banks themselves could fail because their farm collateral would be worthless.

The sharply expanded farm programs put in place with the New Deal were designed to address these problems not only because they were farm problems but because they directly affected more than 25 percent of the U. S. population. The Agricultural Adjustment Act of 1933 gave the Secretary of Agriculture the tools to prop up sagging commodity prices and farm returns and, in the process, protect the sector from collapse. Credit programs were also used to help stem the tide of bankruptcies and enabled farm families to stay on the land and in their homes.

The New Deal's farm programs worked largely through the use of nonrecourse loans that continue even today as a major component of our commodity programs. When prices were low, farmers could use the commodities they produced as collateral for federal loans. If prices failed to rise above loan levels, which were set high enough to generate acceptable farm incomes during the year, farmers could forfeit their collateral to the CCC. In subsequent years, if prices rose due to crop failure or increased demand, the CCC could sell forfeited commodities back into the market to augment supplies. It became clear soon after implementing the price- and income-support programs of the 1930s that growth in farm productivity, encouraged by higher and more stable prices, worked to expand the sector's capacity to produce faster than the market could absorb. Setting loan rates high enough to support incomes (which essentially became the market price) exacerbated this excess capacity problem and led to (1) the accumulation of large government stocks which tended to further depress market prices, (2) a reduction in or at least slower growth in demand for farm products here and abroad, (3) reduced U. S. competitiveness in world markets, and (4) increased government costs.

Additional policy tools were developed over time to deal with the worst aspects of this excess capacity problem. Supply control programs were introduced in the 1930s to idle much of the increase in excess capacity that resulted from setting support prices above market-clearing levels. The programs have been voluntary; but cooperators were given the right to participate in other programs, such as the loan and disaster protection programs, and in some cases were actually paid to restrict acreage or marketings. During the 1950s and early 1960s, longer term acreage retirement programs of 5 to 10-year duration were used. The most commonly used supply controls in recent years have been land diversions including set-aside and the PIK programs.

With the passage of the 1973 farm bill, there was an attempt to separate income and price-support activities; and different programs were established to accomplish each of these goals. This legislation attempted to sever the link between loan rates (which historically had largely determined the market price) and farmers' incomes. Loan rates were to be set low enough to minimize at least some of the demand problems associated with artificially high commodity prices. With the market free of the loan rate linkage, federal payments were made to program participants to supplement incomes. These payments, called deficiency payments, equaled the difference between target prices—set by law or formula at high enough levels to protect farmers' incomes—and the loan rate or market price, whichever was higher. This notion of income support through deficiency payments has continued.

## *The 1981 Farm Bill*

Over the last three years, agriculture has experienced a series of unprecedented events that have brought current farm policy to the forefront of discussion. Since the passage of the 1981 farm bill, it has become clear that the present farm policy tools are not effectively dealing with the current situation. Many have asked: "How did this happen?" This is a good question, one for which I do not have all the answers even though I lived through the development and enactment of the bill. I do believe, however, that I can identify the major reasons why the 1981 farm bill, enacted in an era of concern about food shortages, was followed two years later by the largest acreage reduction program in U. S. history.

### The Setting

At the time the 1981 farm bill was formulated, the main concern was that world food needs would outpace production. Many believed—inside and outside of government—that the agricultural issue of the 1980s was going to be how to produce enough for a hungry world rather than how to deal with surpluses. Many also believed that the United States was the country that possessed the most potential for expanding food production in order to meet world needs. Growth in world demand during the 1970s was evidenced by an increase of nearly one-third in world grain consumption, a rise in oilseed consumption of over 50 percent, and similar growth in world consumption of other major farm commodities produced in the United States. In response, the volume of U. S. exports increased over 150 percent.

At the same time, there was great concern over inflation and rising production costs. Inflation was running at double-digit rates, the prime interest rate was around 20 percent, and inflationary psychology was influencing all business decisions.

Thus, the 1981 farm bill was developed in a climate of optimism for growth in exports and pessimism over controlling inflation as land prices were escalating beyond levels that most farmers could pay for out of production revenues. Even in this environment, the administration believed that conditions could change and in Spring, 1981, sent a bill before Congress that did not contain minimum commodity loan rates or target prices. It also had flexibility concerning the dairy price-support level. In other words, the bill the administration developed would have given the Secretary of Agriculture more discretionary authority for controlling price-support levels and farm program costs. At that time, it had become evident that indexed entitlement programs were causing a major part of the budget deficit problem.

While the 1981 farm bill that was eventually signed mandated higher loan rate minimums (though not escalated through the life of the bill) and annual target price increases of 3 percent to 5 percent, it appeared workable. At the same time, it seemed to represent a compromise between the administration's initial bill and the congressional concerns that support levels would be too low. It is important to note that the 1981 farm bill passed the House of Representatives by only two votes. It was perceived by many that it provided too *little* to farmers; thus, the closeness of the vote. Indeed, many in Congress and elsewhere called it a "do nothing" bill for farmers. At that time, the administration was criticized repeatedly for being too worried about budget deficits and not worried enough about farmers. Some economists suggested that farm incomes would drop because the price supports were too low.

## Unforeseen Events

In just over a year, the farm bill that was thought to be potentially workable at the time had become unworkable. The rigid price-support levels set in the 1981 farm bill have come back to haunt agriculture by working against it in a time of very competitive world markets. The target prices, thought to provide only moderate income support, turned out to be so high that they were inducing increased production both here and abroad.

How could all this occur in such a short time span? First of all, the world became gripped in the worst recession in modern history. With this came a downturn in world demand that saw U. S. agricultural export volume decline for the first time in 13 years. A sharp rise in debt worldwide, especially among middle- and lower income countries, restricted the ability of many countries to import since an increasingly larger portion of their income was required to service the debt. In addition, the rise in debt reduced available credit needed to finance imports. Thus, as the financial condition of many of our prominent foreign customers (such as Eastern Europe, Mexico, and Brazil) steadily worsened, their purchases from us plummeted. The U. S. export volume in fiscal 1984 is projected to be 15 percent below 1980—the fourth straight year of decline.

Meanwhile, the value of the dollar steadily increased as foreign investors sought the security and high rates of return from American investments. The increase in the value of the dollar meant that our products were becoming more expensive in terms of the local currencies of our foreign customers. Thus, American products became less competitive in world markets, and our volume of trade suffered. As an example, over the past four years, the dollar has increased in value by about 50 percent compared with the German mark. This means that, even if the price of a U. S. commodity had remained the same over the past four years, the price in German currency would now be 50 percent higher. Such effective price increases are bound to reduce demand.

To add to the problem, the downturn in world demand was accompanied by increased tariffs and subsidization practices by some of our major foreign competitors—most notably the European Community (EC). For example, the EC a few months ago increased its subsidy for wheat flour in order to take over a larger portion of the Egyptian wheat flour market—a market where the EC could not sell 1 bushel without massive subsidies. Other countries, such as Japan, have policies of limiting access of certain agricultural products through quotas and nontariff trade barriers.

Other factors beyond anyone's control, such as the weather, also compounded our problems during this period of stagnant demand. Unusually good weather worldwide during the 1981 and 1982 growing seasons resulted in record crops at home and good crops abroad. While bountiful harvests are welcomed, the record grain harvest of 1981 and 1982 came at a time of slack demand and resulted in the largest surpluses in history. The large surpluses continued to overhang the market until 1983, leading to reduced prices and incomes for many farmers.

## Payment in Kind: A Temporary Measure

By Fall, 1982, it became obvious that the traditional commodity programs were insufficient to deal with the huge surpluses on hand. In fact, certain aspects of these programs were encouraging more to be produced. With budget outlays soaring, paid cash diversion programs of the magnitude needed would have been irresponsible from a budget standpoint. Moreover, such programs would not have resulted in the necessary acreage reduction since the payment limitation would have restricted program participation. Voluntary acreage programs of the traditional size and type had proved ineffective in reducing supplies since yield increases largely offset reductions in acreage.

A special program was needed to specifically address the immediate needs of agriculture. At that time, the administration formulated the PIK plan and went before Congress to seek its approval. Congress, however, seemed more interested in adopting so-called "farm crisis" legislation from Fall, 1982, which would essentially have imposed mandatory controls. After failing to gain congressional approval for two changes in current law needed to operate the PIK program as it was envisioned, a revised version of the PIK program was implemented in January, 1983, under existing authority.

As a stopgap measure, PIK was the best alternative available to deal with the record surpluses overhanging the market. Simply stated, it did not short the market and was the least costly approach since it used government stocks and secured loans as payment for the acreage taken out of production. This contrasts greatly with a paid cash diversion approach which would have extended budget exposure and shorted the market.

The PIK program met its intended goals, namely, to reduce acreage and potential production while maintaining adequate supplies and putting farmers on a sounder financial footing. However, PIK, in combination with the worst drought in 50 years, reduced crop supplies more than initially expected.

Even at its inception, PIK was not billed as a long-term farm policy solution. It was a program designed to give a year or two of breathing space to decide what direction future farm policy should go. It avoided the need for precipitous actions from which the industry may never have recovered. It also proved to be a good drought assistance program and provided much needed relief to farmers without making additional outlays.

In short, for the past three years, the administration has been struggling with farm legislation that was designed for an economic environment that never materialized. Frustrated in attempts to obtain congressional approval for needed changes in legislative authority that was too rigid, the administration has attempted to hold the farm economy together with a series of actions designed to minimize the damage until this nation adopts new policies for the rest of the 1980s.

## *Farm Policy Issues for the Balance of the 1980s*

While farm policy tools have changed little in the last 50 years, agriculture and the forces that affect it have undergone dramatic change. The main question is: Will future farm policies be adopted with adequate consideration of these changes? In other words, how will future farm policies be modified or revamped to take into account:

- Midcourse corrections of rigid policies that lead the agricultural economy into trouble do not get implemented on time, if at all.
- Approximately 10 percent of our farmers produce about two-thirds of the output, earn most of their family incomes from the farm, and have average family incomes of about $85,000.
- Approximately two-thirds of our farmers produce only 10 percent of the output, receive about five-sixths of their family income from off-farm sources, and have average family incomes above the national average.
- As we pursue acreage reduction programs, the world expands production, neutralizing desired supply-demand or price adjustments and increasing the costs of our programs while reducing our market share.
- Farmers now purchase more than three-fourths of their production items from off-farm sources that are directly affected by inflation rates.
- Interest expenses alone account for almost 20 percent of total case outlays, up from less than 10 percent in the early 1970s, making interest rates a key variable in determining profits or losses.
- Although only about one-third of U. S. agriculture is covered by commodity programs, related sectors without direct support—e.g., the livestock industry—are indirectly affected.
- The top 13 percent of farms accounts for 45 percent of direct government payments, while 71 percent of the farms receives only 22 percent of the payments.
- Conventional acreage reduction programs seem to be ineffective because they cause both participating and nonparticipating farmers to increase their acreage bases and yields, partially negating the intended cutback in production and increasing the size and cost of government programs.
- Continuing gains in productivity projected for the 1980s, coupled with the expanded land base now available for use, imply a productive capacity that exceeds most projections for potential growth in demand.
- Exports are key outlets for our production, taking 60 percent of wheat production, 45 percent of soybeans, and 50 percent of cot-

ton; therefore, our agricultural sector is much more sensitive to outside forces such as the value of the dollar, world credit, and trade issues.

- Current commodity programs appear to be aggravating rather than reducing soil and water conservation problems by artificially stimulating the production of certain commodities regardless of market conditions or the suitability of the land and availability of water to produce that particular crop.

In assessing the future course of farm policy, we must be realistic about the performance of past policies. If our farm policies are working and we are merely experiencing an aberration, we do not need to adjust them. However, the facts strongly suggest otherwise. As we look at the basic components of today's farm programs (commodity loans, acreage adjustment programs, target prices, and reserves), it seems that the burden of proof is upon those who advocate a continuation of the policies of the past 50 years. Where is the evidence that they have been effective in their current form over any time period? A review of the evolution, rationale, and reality with respect to effectiveness of these programs raises some very fundamental questions.

## *Commodity Loans*

### Evolution

The authority for commodity loan programs, known as price supports, comes from the Agricultural Adjustment Acts of 1933 and 1938. Since that time, they have been the major farm policy tool. The use of loan programs has changed little since the origin of price supports even though the risks of using loan rates as an income-enhancement tool were apparent by the mid-1950s. Support prices still serve in varying degrees as price floors, income support mechanisms, price stabilizers, and marketing tools, depending on the commodity and the time period.

## Rationale

As a price floor, commodity loan rates have been used as a safety net to protect farmers from sharp drops in prices due to short-term weather or demand factors. It is argued that this helps to maintain the productive capacity of agriculture by providing some assurance that random and uncontrollable shocks to the market will be cushioned when market prices deviate substantially and temporarily below trend levels. Also, it is argued that, in years of short crops, stocks acquired under loan and purchase programs can be released as prices rise to help dampen upside price swings, lessening the short-term impact on dairy, livestock, and poultry feeders while buffering the rise in food prices. This, in turn, benefits consumers by assuring adequate wholesome food supplies at relatively stable prices. Thus, it is believed by some that effective use of price supports not only can help mute transitory price signals and benefit all (due to short-term weather or abnormal demand factors) but can also reinforce basic structural price signals from the market over the longer term.

As a marketing tool, it is argued that commodity loans allow farmers to obtain cash to satisfy immediate obligations to creditors while retaining control of the commodity. Rather than sell at depressed prices during the harvest season, the producer stores the commodity until later in the marketing year when prices are usually higher. This evens out marketings and tends to make supplies available at more stable prices throughout the year. Some argue that loan rates are valuable even when they are below market-clearing levels because they allow producers to use the cash flow and provide for more orderly marketing.

## Reality

While the use of price supports as a means to support market prices and incomes may have been partially legitimate in the past when agriculture was predominantly oriented to the domestic market, it does not appear to work now. When such rates are above the prices the market would otherwise dictate, the government becomes a substitute for the marketplace. Excessive production, reduced markets, and excessive stock accumulations are the results. Critics suggest that we could not have crafted a better policy for our competitors than the farm programs of the past 50 years. Let us look at some examples of each and try to separate fact from fiction.

**Excessive Production.**—When loan rates are set above variable costs, farmers are guaranteed a return. The reduction in risk is an incentive to expand production despite developments in the marketplace. Despite declining prices and growing surpluses, large acreage remains and production increases. Some argue that this occurs because farmers not only fail to respond to price signals but they produce more to "cover their costs." That, however, suggests a backward-bending supply curve (as price moves down, more is produced), defying rational human behavior. It seems that farmers do respond to price signals, but the signals are coming from the loan rates. It is estimated that loan rates now cover average variable costs for over 98 percent of all corn and wheat production. It is safe to assume that, if almost everyone's average variable costs are covered, then at least one-half of production is assured a return *well* over variable costs.

Some also argue that this has encouraged further acreage expansion including use of more fragile land—land that can be idled under acreage programs with little loss of total output. For example, in 1982 we witnessed record wheat and corn production which exceeded total use by over 2 billion bushels in spite of acreage reduction programs of 15 percent and 10 percent, respectively.

Moreover, it seems that the impacts are not confined to respective program crops. The higher loan rates on wheat are almost certainly the key incentive to expanded wheat and soybean double-cropping in the southeast. With most of the variable costs of wheat assured through the market price floor established by the loan program, producers in the southeast have tremendous incentives to plant wheat on acreage intended for soybeans later in the crop year. For example, soybeans double-cropped reached 11.5 million acres in 1982 compared with 3.2 million acres in 1978. The result is increased production of both commodities and at lower costs than in most other areas of the country. As new growth regulators (used successfully in field tests accelerating plant growth by as much as 10 days) are released, double-cropping soybeans could appear as far north as Kansas and northern Missouri. Since the economic feasibility will be influenced by the degree of price protection provided to wheat, soybean producers in the northern regions that cannot double-crop and only grow soybeans will be adversely affected by the wheat program. Therefore, some wheat producers, such as dryland wheat farmers, could be affected as others grow more soybeans and wheat.

The dairy program is another example of high price supports leading to excess production. In the Food and Agriculture Act of 1977, the minimum level of price support for milk increased from 75 percent to 80 percent of parity. In addition, the 1977 Act required a midyear adjustment to the level of price support. As a result, prices received by farmers increased substantially; during fiscal year 1981, milk prices averaged over one-third above fiscal year 1978. Consequently, dairy production increased continuously until this year when the diversion program and other factors brought about cutbacks. At the same time, CCC net removals of excessive production grew dramatically—from 1.1 billion pounds in fiscal year 1979 to 16.6 billion pounds in fiscal year 1983. Likewise, the cost of the dairy program rose from $244 million in fiscal year 1978 to $2.5 billion in fiscal year 1983.

This problem could be further aggravated by new production techniques. Tests at Cornell University reveal that injections of a growth hormone can increase milk production per cow by 15 percent to 40 percent. Inflexible price supports cannot adjust to these realities. It is estimated that each 5 percent increase in milk production, caused by the use of the hormone, would result in more than $1 billion in additional CCC outlays.

**Excessive Stock Accumulation.**—Rice serves as a good example of excessive stock accumulation caused by price supports above market-clearing levels. Ending stocks of rice increased from 16.5 million cwt. in 1980-81 to 71.5 million cwt. in 1982-83. In other terms, the ratio of stocks to total use rose from 11 percent to 58 percent over that two-year period. About 40 percent of the 1982 rice crop and 43 percent of the 1983 crop were placed under CCC loan, compared with only 17 percent of the 1980 crop. In California, over 56 percent of the 1983 rice crop was placed under loan. Moreover, nearly two-thirds of the rice placed under loan in 1983 was forfeited to the CCC, including nearly all of the California crop that had been placed under loan.

We have also witnessed similar situations for feed grains, wheat, cotton, dairy, and tobacco in recent years. For example, nearly one-fourth of the total 11 billion bushels of wheat and corn produced in 1982 were put under loan. Of the over 2-1/2 billion bushels put under loan, three-fourths was forfeited to CCC or entered in the grain reserve under extended contracts. By the end of 1982, the ending stocks of wheat and corn equaled nearly 50 percent of annual use compared to a 21 percent level at the end of 1980.

**Loss of World Markets.**—American farmers now depend on foreign customers for marketing about 60 percent of their wheat production, 45 percent of their soybeans and rice, and over 50 percent of their cotton. The United States accounts for nearly one-half the world grain trade and three-fourths the soybean trade. When our loan levels exceed world prices, our competitiveness erodes.

Moreover, our competitors expand their production based on the price floor we create with our support levels. Competitors *do* react to our price floors as is apparent from their production decisions for certain major commodities. As an example, our major wheat competitors expanded their acreage by 8 percent from 1981 to 1983 under the shelter of our price umbrella, while we unilaterally reduced our acreage by nearly one-fourth in order to enhance farm prices and reduce budget exposure. Furthermore, the Canadians, who have steadily expanded their production, openly expressed their displeasure when we lowered the wheat loan rate for the 1984 crop. Other countries have expressed similar concerns.

Rice, cotton, and tobacco are other examples of commodities where our support levels are reducing our share of world markets. For instance, in the past three years, the volume of U. S. rice exports has declined one-third, while our major competitor—Thailand—increased its exports by over one-third. Currently, U. S. milled rice is about \$150–\$175 per metric ton above the Thai price on the world market. Normally, U. S. rice commands only a \$70–\$80 premium due to quality differences. This means that our rice loans are now about \$2.00 per cwt. above the world market. Some suggest that it is not surprising that Thailand has now replaced us as the No. 1 exporter of rice.

In addition, for both cotton and tobacco, we have witnessed substantial growth in world production, while our exports have either declined or stagnated. In the last 20 years, world tobacco production has doubled and our share of world tobacco trade has dropped by about 50 percent. The U. S. tobacco exports have declined nearly 25 percent in just the last five years. In the case of cotton, the U. S. share last year fell to less than 12 percent of world production compared with more than one-half of global output in 1930.

**Loss of Domestic Markets.**—In recent times, honey support prices, established at the statutory minimum of 60 percent of parity, have risen well above world market levels. As a result, U. S. honey imports increased from 9 million pounds in 1970 to about 110 million pounds in 1983. Meanwhile, U. S. ending stocks climbed from around 35 million pounds in the 1970s to 166 million pounds in 1983—a stocks-to-use ratio of 60 percent compared with only 12 percent just five years earlier. Also, over the last five years, government acquisitions have gone from zero to an estimated 115 million pounds. Effectively, domestic honey is now being sold to the federal government and is being displaced in the domestic market by less-expensive imports.

There are other examples where high supports and the resulting impact on market prices led to market substitutes or alternate preferences. Synthetic fibers (substituting for cotton) and corn sweeteners (replacing sugar) are examples. Throughout the 1950s, 1960s, and early 1970s, cotton support prices were often above market prices. The artificially supported cotton prices during this time were a contributing factor in prompting the textile industry to shift to synthetic fibers that were, for the most part, less expensive. Cotton's share of total mill use went from 65 percent in 1960 to 25 percent in 1980. At the same time, total mill use of all fibers gradually trended upward.

## *Acreage Reduction Programs*

### Evolution

The Agricultural Marketing Act of 1929 established a Federal Farm Board with the authority to take action to prevent and control surpluses and to provide for more orderly marketing. The Board chartered a "stabilization corporation" which, in the face of the Great Depression, sought to maintain domestic wheat and cotton prices.

Having exhausted its funds in acquiring wheat and cotton stocks (which strained available storage space and depressed market prices) and facing a new wheat harvest that was twice the normal annual disappearance, the corporation retired from the market. In its last report, it provided a key recommendation:

> ". . . no measure for improving the price of farm products other than increasing the demand of consumers can be effective over a period of years unless it provides a more definite control of production than has been achieved so far."

## Rationale

Production controls became a major feature of agricultural policy as production increased due to support prices above market-clearing levels. The rationale was that farmers would receive price and income supports, consumers would be provided with a reliable supply of food, and production controls would be used to fine-tune supplies and help avoid surplus stocks and large government costs.

## Reality

For the first 30 years, acreage restriction programs, such as allotments, were used in an attempt to achieve production control. Such programs seldom achieved their intended results, however, primarily because acreage rather than output was restricted. Farmers rapidly increased yields on allotment acreage, and overproduction continued to be a chronic problem. By 1960, surplus stocks and large government outlays had generated considerable pressure for reform.

For most commodities, one of two paths was chosen: (1) continued high price supports with tighter controls on production that included both acreage allotments and marketing quotas or (2) separation of price and income supports with voluntary acreage reduction programs.

Peanuts and tobacco pursued the first path. The results have not been impressive. In 1930, the United States harvested 1.1 million acres of peanuts; by 1980, 1.4 million acres, up only slightly from 1930. Per capita production of major peanut products has remained virtually unchanged since the mid-1960s. Cheaper peanuts are available but only for the export market for crushing.

For tobacco, U. S. production has declined by about 10 percent over the past 20 years, while world production has doubled. Meanwhile, our share of world tobacco trade has dropped by about 50 percent due to the high and escalating support prices that have made U. S. tobacco less competitive in world markets. Pressures are now mounting to control cheaper imports. There is substantial evidence that the tight controls on these commodities did not benefit farmers; instead, the economic value of the quotas and allotments has been captured by landowners in increased land values and higher leases. Recent data indicate that there are approximately 550,000 tobacco allotments. While there are no precise figures available, estimates suggest that approximately 150,000 farmers actually produce tobacco. Most of these farmers are paying the allotment holders for the right to produce tobacco under the program. The value of the allotment is a production expense for the farmers that are not allotment holders since it is a cost not related to the actual cost of producing tobacco.

The second path was chosen for most other major commodities. However, the facts suggest that voluntary acreage reduction programs have been largely ineffective for a number of reasons.

First, farmers who choose not to participate in the voluntary acreage reduction programs tend to expand acreage for two reasons: (1) they hope to capitalize on the higher market prices expected to result from the programs and (2) they wish to increase the size of their base for the purpose of future programs. Experience with wheat provides a good example. In 1981, when no acreage reduction program was in place, 81 million acres of wheat were harvested. In 1983, 28.2 million acres of wheat base were idled; yet, harvested acreage was only 20 million acres less than in 1981. While this still represented a 25 percent reduction in harvested area, the 1983 production was only 15 percent smaller than the 1981 crop due to higher yields.

Some observers have suggested that the tightening of the rules and regulations for the voluntary acreage-reduction programs would make them more effective. Cross-compliance (authorized in the Food and Agriculture Act of 1981 for use with set-aside programs) and offsetting compliance are usually suggested. Cross-compliance requires a producer to participate in all acreage-reduction programs for which he or she is eligible in order to obtain program benefits in any one of them. Offsetting compliance requires that farmers with multiple farm units participate in the acreage-reduction programs on all units to be eligible for program benefits on any one unit.

It seems that cross-compliance, offsetting compliance, or any other measure that would tighten up the acreage reduction programs, however, are almost always met with strong opposition by some in Congress. In 1980, producers were expected to retain plantings of all program crops within a historically determined normal crop acreage (NCA) in order to receive program benefits. When this measure began to take effect, Congress mandated that producers be allowed to receive deficiency payments even when plantings exceeded NCA. More recently, Congress has required that this administration count summer fallow as conservation-use acres (CUA) and that the CUA rules for 1984 crops be no more restrictive than in 1983 when, some would say they were already fairly liberal. While it may be possible to tighten up certain programs, it seems more appropriate to focus on whether it is realistic to expect such rule changes to stay in place very long.

Second, about 12 percent of U. S. farms now market about two-thirds of all farm products. Thus, the participation of large units in acreage reduction programs seems essential. However, the law limits to $50,000 per year the amount of deficiency and diversion payments that an individual may receive. The payment limitation, therefore, makes participation less attractive to large units as the benefit to be gained seldom outweighs the potential value of foregone production. If they do choose to participate, the required acreage devoted to conservation is adjusted for the payment limitation; and the producer may plant other crops on this residual acreage. This occasionally leads to surpluses for other crops that do not even have support programs.

Third, voluntary acreage reduction programs are ineffective because of increased yields on acres remaining in production. When a reduction program is in effect, farmers leave their best acreage in production and intensify cultivation on that acreage. Research shows that, for every 10 percent of acreage base idled, average yield on the remaining acres increases 3.5 percent for cotton, 3 percent for corn, 2 percent for wheat, and 1 percent for grain sorghum.

Rapid gains in technology continue to improve overall yields—current corn yields are 17 percent higher than 10 years ago—and new cropping rotations, such as wheat and soybean double-cropping, continue to be adopted.

The inability to make the voluntary acreage programs effective is leading to a number of proposals that suggest our programs should begin moving in the direction of tighter controls. Just recently, hearings were held in Congress on proposed legislation that would liberally define the acreage bases in the current years and then make it very difficult for anyone to change or establish new acreage bases at a later date. However, some are suggesting that such an approach is the first step to establishing acreage allotments for the major commodities, thus permitting program benefits to be capitalized into the bases. Nevertheless, the legislation seemed to have fairly good support.

# *Target Prices and Income Transfers*

## Evolution

Income transfers to farmers date back to the Agricultural Adjustment Act of 1938 which directed the Secretary of Agriculture to make parity payments to producers of wheat and other "base" commodities (corn, cotton, rice, and tobacco) if and when funds were provided through appropriations. Then, during the 1960s, complying wheat producers received wheat certificates and direct payments determined by the difference between the market price and the announced per bushel value of the wheat certificate on a designated portion of production. These programs were forerunners of the current target price-deficiency payment programs.

The Agricultural and Consumer Protection Act of 1973 introduced the present concept of target prices and deficiency payments. Under that legislation, target prices were established for wheat, cotton, corn, barley, and sorghum. Since 1973, the target price-deficiency payment program has been extended to include rice and oats. Deficiency payments are paid to producers whenever the announced target price is above the average farm price for a designated portion of the marketing year. The deficiency payment rate equals the minimum difference between the target price and the average farm price or the loan rate and, thus, generally varies inversely with the level of market prices. Only those producers participating in announced acreage reduction programs are eligible to receive payments, with total cash payments to any single producer currently limited at $50,000 per year.

## Rationale

The current target price-deficiency payment program grew out of policies set forth in the early 1960s. By 1960, accumulation of commodity stocks had reached record levels—over 2 billion bushels of corn and 1.5 billion bushels of wheat—as a direct result of setting price supports above market-clearing levels and the inability to control production through voluntary programs. It was envisioned that separating price and income support objectives of commodity programs would permit reductions in loan rates to increase exports, strengthen demand for grain domestically, eliminate surpluses, and reduce the need for often ineffective and complicated production control programs. The nonrecourse loan rate was to provide a price floor and would become operative only with excessively large crops or extremely slack demand. Farmers' incomes were to be supported at minimum levels whenever prices fell below target levels.

## Reality

There were some good reasons why the target price-deficiency payment concept was initiated; perhaps, there still is some reason to provide some type of income transfer to some producers. But past experience indicates that the target price-deficiency payment program has had several unintended consequences such as raising land prices and production costs, reducing our competitiveness in world markets, and encouraging farmers to expand acreage even under surplus conditions. Some complain bitterly about program benefits going to large producers as well as smaller ones without regard to need.

Let us look at the results. The inflationary period of the 1970s taught us that basing target prices on costs of production can add fuel to an inflationary spiral. Including land costs in the determination of target prices can lead to an increase in land prices affecting production costs which, in turn, leads to a further increase in target prices. Higher target prices then lead to higher land prices as farmers attempt to profit from higher target prices by expanding acreage—either by purchasing or renting more. Many argue that this sequence of events occurred in the early 1970s as the 1973 Act established target prices that would be increased, based on changes in the Index of Prices Paid by Farmers adjusted for changes in productivity. This is probably why the 1977 farm bill indexed target prices to cost of production increases excluding land costs.

Even when target prices are not directly tied to land price or cost, some say they may still contribute to an increase in land prices. For example, some argue that the target price may cover total costs for some producers but not for others since production costs vary among producers, especially those in different regions. If target prices more than cover production costs for some geographically concentrated group of farmers, land prices and production costs in that location will be bid up, with the benefits going to existing landowners and not to new entrants or renters. As a result, production costs could go up making these farm products less competitive on world markets.

I believe that there recently has been a new twist in the use of target prices that has significantly added to our inability to compete in international markets. It is indirect but, nevertheless, important to discuss during any debate on future farm bills. Let me explain the reasoning.

In the beginning, target prices, which provided deficiency payments to farmers when prices fell, seemed to work fairly well. We were competing in world markets, and direct payments were made to supplement incomes. However, most farmers prefer to obtain their prices through the marketplace and not the U. S. Treasury. Congress sensed this and began raising target prices. The resulting budget exposure was a lever over the Executive Branch to force production cutbacks in order to get market prices higher and eliminate direct payments from the U. S. Treasury. Target prices set above market-clearing levels, such as those mandated in the 1981 farm bill, are an effective guarantee that acreage reduction programs will be required. Simple mathematics reveal that the target price-deficiency payment program, by itself, could cost U. S. taxpayers in excess of $10 billion annually given the current differential between target prices and loan rates. Faced with the budget exposure and political pressure resulting from spending in excess of $10 billion annually on deficiency payments, not to mention the costs of CCC loans and purchases, storage payments, and export programs, there is little choice but to implement production control programs to ensure that market prices are nearer announced target levels. Whenever market prices are not near target prices, some get on their soapboxes and call for even greater acreage cutbacks, claiming that the administration has "mismanaged" the programs.

This situation can lead to a real problem. For example, as target price escalations force greater cutbacks and prices move up, other countries produce more and take over our markets. As our markets dwindle, even larger cutbacks are needed. For example, even after the 1983 rice crop was dramatically reduced because of PIK, analysis suggests that rice stocks near the end of 1985 will build back to excessive levels even if a 40 percent acreage cutback is put in place. Even then, the price of rice will not move above the loan rate with such a stock buildup.

Unfortunately, the rest of the world can see the same forces at work. They see price floors to protect downside risk, and they see budget pressures that ensure production controls in the United States designed to push market prices higher. Reduced risk, smaller competing U. S. supplies, and potentially higher market prices are an open invitation to competitors to increase their production. Moreover, the facts seem to suggest that high target prices encourage our farmers to increase acreage bases and yields on acreage remaining in production. If loan rates are covering 98 percent of average variable costs, then target prices well above loan rates are providing a large incentive to build bases simply for the purpose of receiving benefits from future government programs. Some farmers may even expand acreage by plowing up highly erodible land. For example, cropland in the Delta and southeast has been increasing at the rate of 5 percent per year since the beginning of the 1980s. This growth in utilization of more erodible cropland during a period of excess supplies has been a topic of considerable congressional discussion over the past several months. The obvious concern is the incentive for producers to expand acreage in order to obtain larger program benefits.

The commodity programs may also be stimulating an unwise depletion of our water resources. Irrigation has nearly doubled in the last three decades and is now practiced on over 50 million acres. The largest increase recently has occurred in the northern Great Plains through the use of groundwater. Since federal irrigation investments have steadily declined since the mid-1960s, much of this expansion is coming through private investments. The primary factor encouraging this irrigation development appears to be the price-support programs since the major use of the newly irrigated land is for corn, cotton, wheat, and other program crops. From 1929 to 1978, 60 percent of the increase in irrigated acreage occurred on the principal program crops. We are mining and depleting our water resources to produce more of the agricultural products for which we already have burdensome supplies.

If there is a need for an income-transfer mechanism, it seems that it must not interfere with the signals from the marketplace which can lead to increased production costs, reduced competitiveness, increased production abroad, and larger surpluses in the long term. I believe we need a clearer understanding of how much income transfer is desired and who are the recipients we are trying to assist. At this time, about 45 percent of direct government payments are going to the 13 percent of the farmers who have farm sales of over $100,000. The balance goes to the remaining 87 percent of the farmers who earn most of their income from off-farm sources. Who are we trying to help, and what is our objective? If we can agree on what the objectives of the income transfer are, perhaps we can transfer income more effectively with less disruption to the marketplace.

## *Commodity Reserves*

### Evolution

Prior to the creation of the FOR, the CCC was essentially the nation's "reserve" program. When supplies became excessive, farmers forfeited their grain under loan and the CCC became the owner.

When markets tightened and prices rose above the loan rate by a certain amount, the Secretary of Agriculture had authority to sell the stocks back onto the market. It was the disposition of stocks back into the marketplace that almost always made farmers angry.

To address this problem, the FOR concept was developed in the 1977 farm bill. Under this approach, a farmer could enter into a contract with the CCC to hold grain off the market for up to three years. Incentives were paid to farmers for storing the grain. The reserves were to remain in place until prices rose to a specified level at which time both parties would be relieved of their obligations. At that point, the incentives stopped and farmers could do whatever they chose with the grain; then the Secretary of Agriculture could not be accused of taking an action that would depress the market. In fact, some argue that farmers are likely to be more careful sellers than the CCC.

## Rationale

The rationale behind a public reserve or stock program is to remove some of the inherent instability associated with commodity supplies and prices. However, as in most government programs, reserve policies have generated controversy.

Some farmers and farm organizations tend to favor building reserves when supplies are excessive, while most processors, marketers, and consumers would rather see the additional supplies come onto the market at lower prices. On the other hand, during periods of tight supply, producers are reluctant to see commodities drawn out of the reserve and onto the market, while users and consumers applaud the additional supply. Proponents of a reserve policy argue that it is this process of adding to government stocks during time of excess and drawing them down when markets are tight that (1) reduces some of the instability and uncertainty in commodity markets and (2) enhances our nation's reliability as a supplier. They argue that these benefits flow to society; and, without a reserve policy, the private market would not provide them.

## Reality

The realities of a reserve policy—at least the types of reserves we have had since 1977—are most difficult to assess. Of the major domestic farm program elements (loans, target prices, production controls, and reserves), the merits and demerits of a reserve are, in my judgment, the weakest. I have not seen research results and analyses that are persuasive in either direction. It is quite difficult to single out effects of the reserve from all other factors.

It really has not been demonstrated that our reserve policies have or have not led to a significant amount of stability. Critics argue that instability has increased due to the overly complicated rules, "triggers," and other parameters that are out of line with market realities. Proponents say that the degree of stability possible has been lessened because of these same factors. There seems to be some validity in both of these arguments.

The question of whether the private market would provide any or all of the necessary reserves for a satisfactory degree of stability has not been answered. Critics argue that, if publicly held reserves were eliminated, the private trade—processors, exporters, and others—would increase their holdings of some of the nation's stocks against future uncertainty. Unfortunately, it is difficult to test this argument. In the past, the government has always intervened and added stocks when excessive supplies have developed. The private sector has never been given the opportunity to demonstrate whether they would hold more stocks if the government did not. Logic would suggest that they would.

Moreover, the larger question of how much stability is needed has not been answered. If a reserve is needed for a certain level of stability, then another unanswered question arises concerning the appropriate size of the reserve to achieve the level of stability desired.

While it seems difficult to answer some of the questions concerning the pros and cons of a reserve, it does seem that the reserve has not performed that well as a policy tool. As with many government programs, the reserve seems to have been pushed beyond what it was initially intended to do: specifically (1) remove some instability and (2) improve the U. S. reputation as a reliable supplier. For example, in early 1980 the reserve was used in an attempt to offset some of the negative effects of the Soviet grain embargo. The incentives to place grain in reserve were increased, at first administratively and later by Congress. Again, in 1982, under the burden of huge supplies and depressed prices, the incentives to farmers to participate in the reserve program were increased. The results in both cases were as expected. Grain flowed into the reserve rather than onto the market. Reserve carry-over stocks of corn went from 185 million bushels in 1980–81 to 1.6 billion bushels in 1982–83. Reserve stocks of wheat increased from 360 million bushels in 1981–82 to 1.1 billion in 1982–83.

Hence, it seems that the reserve graduated from the role of removing some market instability to that of supporting prices. But, as reserve stocks mounted in the face of excess supplies and CCC costs escalated, it soon became obvious that drastic actions were needed to reduce the size of the reserve. A full circle was made with PIK. While we do not necessarily know what the benefits of a reserve policy are, it appears that, unless managed wisely, it is a policy tool that can certainly get us in trouble.

While there are many questions concerning the usefulness of a reserve policy and the difficulties of managing it, many still contend that there may be some role for a public reserve policy in agriculture—primarily due to the inherent nature of commodity markets which are affected by the vagaries of weather as well as many other factors. They argue that a reserve is needed if we want to rebuild our reputation as a reliable supplier and guard against too many resources entering or leaving agriculture during abnormal and sporadic price movements. However, I am not sure economists and policy analysts have enough information to lead policymakers toward a more appropriate reserve policy. Should we adopt a comprehensive reserve policy or adopt simple loan extensions for all commodities as we do currently for cotton? This is the type of question that still goes begging for an answer. Some suggest that, if some of these questions cannot be answered, it may be better to have no public reserve until its impacts can be further analyzed for the risks are too great.

## *Beyond Farm Programs: A Growing Interdependency*

Much of this chapter discusses farm policy and its impact on agriculture. But as agriculture has changed, other policies have also become important for the sector's well-being. Let us briefly examine a few of these.

## Macroeconomic Policy

One of agriculture's most important features, from a policy perspective, is its integration into the broader economy here and abroad. In fact, the linkages between the farm and overall economy have become so intertwined that conditions and policies beyond the farmgate probably affect agriculture's well-being as much if not more than traditional farm commodity programs.

Compared with 1930, fertilizer consumption has increased fifteenfold, feed concentrate usage has doubled, and average tractor horsepower has become 10 times greater. Farmers now purchase more than three-fourths of their production items from outside the sector compared with about one-fifth in 1930. The result is that a farmer's well-being is now much more sensitive to the general level of prices beyond the farmgate. Total annual farm expenditures are now about $150 billion compared with a total of $5 billion to $10 billion in 1930. Today, a 1 percent change in the general inflation rate results in about a $1.5 billion change in farm production expenses.

Not only does inflation push up production costs, but it also pushes up land values. Land values, caught up in the inflationary spiral of the 1970s, increased at an average annual rate of nearly 15 percent. Since the *real* cost of borrowing was relatively low during this period, purchasing land was one way farmers and others could experience asset appreciation each year. Some farmers felt that the way to be successful was to borrow and buy. Many bankers shared this view, and land prices increased far beyond levels expected, given the levels of commodity prices. But land values have fallen steadily since 1981 as a result of reduced returns and general disinflation within the economy. While this had to occur, it was particularly painful for highly leveraged operators who had used increased equity as collateral for securing loans.

While American farms have greatly expanded in size, farmers have increasingly used debt to finance capital purchases. Interest expenses now account for nearly 20 percent of farmers' total cash outlays, or roughly $20 billion, compared to less than 10 percent in the early 1970s. A 1 percent change in interest rates on all farm debt outstanding currently translates into a $2 billion to $3 billion change in net farm income.

High relative interest rates have an obvious negative impact on U. S. agricultural exports. Recent events provide ample evidence. As interest rates rose at home starting in 1979, the value of the dollar climbed over 30 percent, raising the prices of our commodities relative to our major export competitors. Our share of the world market fell—especially for wheat and rice—as some of our customers began purchasing from our competitors. Over the last two years, the U. S. Economic Research Service estimates that up to $6 billion in foreign sales of agricultural products and possibly $2 billion–$3 billion in farm income were lost as a result of the stronger dollar. Furthermore, for every 10 percent change in the value of the dollar internationally, agricultural exports are estimated to change by at least 5 percent.

In addition to a strong dollar, the recent worldwide economic recession has also contributed to more than a 10 percent drop in volume of agricultural exports during the last three years. The tremendous buildup in world debt under double-digit inflation was a significant contributor to the suddenness of the downturn in the world economy in 1980. Debt in developing countries accumulated at the rate of 20 percent per year, more than quadrupling during the 1970s. High interest rates reduced the ability of these countries to service their debt and forced many to reallocate foreign exchange earnings once spent on imported agricultural products to interest payments on foreign debts. Some have suggested that every 1 percent change in U. S. interest rates changes the debt burden of the developing countries by $4 billion. These developing countries account for over one-third of U. S. agricultural product exports.

Thus, the economic well-being of agriculture is becoming more and more determined by forces outside the farmer's control. Farmers have always lived with unforeseen circumstances such as weather, pests, and disease. Today, their well-being is also determined by monetary and fiscal policy and the interest and inflation rates they influence. Other sectors of the economy also are affected by monetary and fiscal policy. However, agriculture's dependence on exports to absorb up to one-third of its production and its reliance on borrowed capital make agriculture's well-being particularly sensitive.

## Tax Policy

Over the years, general tax policy has shaped the development of key industries in the U. S. economy. Indeed, it is the basic industrial sector which is the focus of much of the current investment tax policy adjustment. But agriculture is also affected because it is one of the most capital-intensive sectors of the economy.

For example, many argue that the accelerated depreciation and more liberal investment tax credit provisions of recent years have contributed to capital investments in farming and the tremendous buildup in farm debt. Perhaps, it is argued, these provisions are even partially to blame for some operators becoming highly leveraged financially by overextending themselves in relatively "good" years. In addition, some say these provisions, coupled with labor laws making hired workers more expensive, make investment in machinery attractive and generally lead to higher productivity.

Other tax laws, with respect to capital gains, seem to encourage investment in agriculture. For instance, expenditures incurred in the development of certain farm assets, such as dairy, livestock breeding herds, orchards, vineyards, and the like, are considered capital expenditures. But farmers are not required to capitalize these costs and deduct them in years in which they generate income. Instead, farmers may deduct the full amount of expenditures in the years in which they are incurred. Income from the eventual sale of such assets, however, is treated as capital gains which is taxed at a lower rate than ordinary income. Further, through prudent management, taxes can be reduced by developing the assets, using the deductions to offset other income. Such practices may have led to the large buildup of cattle herds in the early 1970s and, more recently, increased dairy herds and grape vineyards.

Finally, some say that tax policy probably has exerted upward pressure on the value of farmland and equipment because of the attractiveness of farmland as a tax shelter. Since production costs are partially offset by tax benefits offered through cash accounting and other measures, the argument is that the investment is less costly in real terms than it appears and inefficient resource allocation is encouraged.

Tax policies, in general, have made significant contributions to encouraging investment capital to enter areas of key importance to the U. S. economy. While it has enhanced the growth in the economy, in general, it is important to insure that these impacts are not in conflict with specific problems within any particular sector. If the problems of agriculture are an excess of resources, then some say any tax policies that might aggravate the situation should be analyzed. However, no tax policy is neutral and should be kept in mind as any current or alternative tax policies are considered or analyzed in the future.

## Credit Policy

Beyond the general concern with the impacts of monetary policy on interest rates and inflation, the agriculture community is directly influenced by farm credit programs that are not always consistent with the objectives of commodity programs. Federal farm credit programs were established in the early part of this century when commercial sources of credit were relatively unavailable for farmers. At that time, there was a shortage of banks in rural areas, and commercial banks were less familiar with the needs of the farming sector compared to those of their industrial clients. Now, however, commercial credit is more easily available to farmers.

The state of the agricultural economy has changed significantly since the various federal farm credit programs were started. Farm incomes no longer lag as far behind those of urban people. Rural economies have changed greatly since the Depression era. Therefore, farm credit policies may need review along with other farm policies.

For example, when acreage reduction programs are attempting to restrict production to decrease surpluses, federal credit programs may encourage expansion in capacity. During the 1960s, an average of 55 million acres per year were removed from agricultural production while, at the same time, the FmHA loans for real estate were increasing at a 20 percent annual rate. In the late 1950s, the corresponding figures were 26 million acres annually and 10 percent for FmHA real estate loans.

More recent examples of conflicting policy periods are 1971–72, when idled acreage averaged 50 million acres and growth in FmHA real estate loans was about 8 percent, and in 1978–79 when idled acreage averaged 17 million acres and growth in FmHA real estate loans was about 38 percent. In 1983, 79 million acres were idled, but FmHA real estate loans grew only about 4 percent.

One, therefore, has to question the consistency of commodity programs and credit programs since the early 1960s. With new farm legislation required in 1985, perhaps more consistent credit policies also are needed. Unfortunately, not enough is known about the long-term effects of government credit. Some argue that it is necessary, while others argue that excessive agricultural credit adds to surpluses and worsens the farmer's economic position. This issue is another challenge for economists to tackle.

## Trade Policy

Given our dependence on international markets (if we want to remain that way), it is important for the agricultural sector to have a consistent, broad-based trade policy. Part of such a policy must he future domestic farm policies that recognize our growing reliance on international markets and do not interfere with the operation of the market and encourage additional foreign production. In addition, we must recognize that policies permanently aimed at protecting commodities from any type of unsubsidized import competition are shortsighted. We must realize that many countries, particularly those developing countries suffering from large foreign debts, have only one way to gain much-needed foreign exchange—by expanding exports. With over 40 percent of our agricultural exports going to developing countries, it is in our interest to promote policies which allow them market access and, thereby, generate additional income with which to buy our agricultural products.

American agriculture must be concerned also with policies adopted for other sectors of the economy which may have a detrimental effect on our ability to expand exports. In recent years, a number of industries have sought import relief through a combination of duties and quotas. In some instances these actions have had direct impacts on our ability to move agricultural products into foreign markets. If we believe exports are truly the lifeblood of American agriculture, we must be prepared to promote freer trade in all sectors of the economy.

## *Summary*

I believe the most important thing that the agricultural industry can do at this time is to decide, in broad terms, where it wants to go with regard to future farm policies. The fundamental questions are: Do we want to adopt tightly controlled programs with above-market clearing levels of price support, such as at 90 percent of parity, and ultimately reduce the sector to producing only for the domestic market? Or do we want to move toward more market orientation and compete internationally by producing higher volumes at market prices? Once this decision is made, then programs can be devised and implemented to move in the appropriate direction.

Some may think that these questions are already answered, but they are not. A consensus has not yet been reached in the farm community. Furthermore, such a consensus will not come quickly. In fact, I am not sure a consensus can be reached by October, 1985, when the new farm bill should be put in place. We even could be looking at reverting to permanent legislation for at least the 1986 crop year. I believe it is a possibility as the industry moves back and forth trying to find its center of gravity.

If the industry finally decides to adopt a more market-oriented approach, some say it will take time and that a transition program will need to be developed and put in place. The one-third of agriculture that depends on farm programs is probably more dependent on them now than in the last 10 years. Asset values and equity positions and, thus, the ability to borrow money are tied to the programs. That is why some believe that a type of interim program may be needed if this approach is taken. On the other hand, if the industry decides to move toward more strict production controls and higher price supports, this, too, may take some time to put in place. Some say that in this case consideration should be given to mandatory controls and some type of program to help the industry shrink in size.

I also believe that, whatever approach is taken in farm policy for the rest of the century, other policies need to be consistent with the approach adopted. Presently, there seems to be more interest in developing trade, credit, and natural resource policies for inclusion in new farm legislation than at any time in the past.

In conclusion, there is need by all concerned parties to become involved in order to find the answers to the problems confronting agriculture—to help develop farm policy that is responsive to future changes in agriculture and the domestic and world economies. Agriculture is the largest and most important industry in our nation, and it is imperative that we use whatever means possible to ensure it a healthy future.

In this day when everyone seems to be looking for new growth industries in which the United States can become competitive, such as in high technology, it is ironic that we could lose the most efficient industry we have—agriculture—through our own doing. I have outlined many important issues and raised many questions that need answers. This is the challenge before all of us who are involved in agriculture, especially agricultural economists, since many of the most important questions still go begging for answers.

# 2 • 3 DISCUSSION

## John A. Schnittker*

I hardly know where to begin—faced with 40 pages of selective history, some good analysis of the depths to which agricultural policy has fallen and not a little self-justification in Lesher's paper, and a seven-part "litany of farm policy lessons" we should have learned by now from Johnson. Lesher seems to approach the subject of food and agricultural policy as if he were a spectator rather than a major participant.

It is easy to dismiss Lesher's implicit argument that U. S. farm programs have never been effective in dealing with the problem they were designed to handle. Clearly, they coped very well in the 1930s. It is unarguable, however, that price- and income-support policies and programs for grains, cotton, rice, tobacco, and dairy products are now obsolete, badly in need of amendment, and candidates for early replacement by relevant new policies.

Johnson has aptly identified a series of historical turning points when past policies took shape out of the crucible of human events and economic circumstances and made a difference. I call attention

*The author is President, Schnittker Associates, Washington, D. C., since 1972.

to this point because we may now be at such a juncture in history when change is possible if only the right catalysts are used.

Past turning points for agricultural policy have included:

- The aborted Agricultural Act of 1948 which would have begun to reduce the high World War II price support levels while continuing the mandatory acreage control programs for principal crops based on the 1938 Act.
- Replacement of the 1948 Act by the rigid Agricultural Act of 1949 which prevented or slowed reductions in support levels. This was made possible partly by farmers who voted for Truman and partly by the politically inept Thomas E. Dewey.
- The legislative marathon of 1961–1964 when a series of emergency programs finally culminated in the Agricultural Act of 1965 with reduced support levels, payments made only on a part of total production on the farm, and voluntary (paid) acreage reduction. This was the first "market oriented" legislation reducing the role of government in farm markets compared with the 1950s.
- Finally, the unfortunate escalation of target prices based on cost of production calculations beginning with the 1977 Act. This was followed by arbitrary target price and loan rate escalation in the 1981 Act which has helped cause many of our present problems.

I agree with Lesher and Johnson that the 1981 Act has a lot of flaws, but it is not the only culprit. I am sorry that Johnson and Lesher passed up the opportunity to call attention to the inept design and reluctant administration of farm programs since 1980; the persistent failure to use authority that was available to the extent possible to limit surpluses, expand exports, and reduce public costs; and the erosion of cost effectiveness in acreage adjustment programs which could barely be justified in the 1960s and 1970s when payments to farmers were low. All economic justification has been lost in the 1980s when we are paying farmers $2.00 or $3.00 for every dollar's worth of production prevented by farm programs.

We are again at a turning point in farm policy history with indicators on all sides that business as usual is not an option. This conference was organized to help us find a now sense of direction for the 1985 legislative struggle. I wish Johnson and Lesher had told us more about where to go and how to get there based on the compound policy mistakes of recent years, the results of those errors, and the broader economic and political factors that are now so crucial to agricultural policy. There is still time for that since it is at least a year before legislation will be seriously considered. My experience with economists and as an economic consultant tells me, however, that we spend too much time on repetitive analysis of the past and too little time on rebuilding for the future. I want to quote myself from a 1960 conference when I sensed that it was a critical year. I said at a conference in Ames, Iowa, that "for persons interested in agricultural policy, the 1960s would be a good time to make your work count—harvest years when the opportunities to be influential were at a peak." Next year will also be that kind of year.

From Johnson's litany of lessons from the past, we begin to draw the elements of prescription for a transition to a more effective and less wasteful agricultural policy. It reads essentially like a set of principles rejecting the policies, attitudes, and administrative decisions of recent years. These principles include the following:

1. Reduce the level of hypocrisy in farm policy, admit that it has been designed and operated in recent years mainly for big farmers, and redesign policies to meet future needs.
2. Maintain and use the authority to reduce loan and other support levels in order to encourage resource adjustment and to be competitive in world markets.
3. Have the patience to wait for world demand to increase and for the rate of expansion of production in other countries to decline avoiding policies that would lead to greater trade distortions.
4. Reconcile ourselves to transitional programs that can lead us step-by-step to a new long-run policy position in, perhaps, five years geared to the kind of agriculture we now have—not to the farm economy of our fathers.

This transition may be a very difficult time analogous to the 1961–1964 period in farm programs. We must "inch" our way to the future closing the costly loopholes, ending the contradictory policy of encouraging farm production with target price payments—holding it back with acreage reduction at the same time, and stopping the abuse of farmer-owned reserve. We must run the program if we are to be finished with them eventually.

We will also have to deal sympathetically with a degree of decapitalization of commercial agriculture that will accompany a more realistic farm policy in the next five years. Johnson and Lesher have called attention to the fact that past and present programs have built and now support an artificial asset value and farm cost structure that may be sustainable only at very high annual cost to the federal government.

How can we get to the Promised Land? And who will be our Messiah?

We will get there only by a series of crises including, possibly, a number of one- and two-year programs or even a chaotic year or two of operations under old permanent laws. We should not exclude consideration of a temporary return to mandatory acreage reduction in some commodities to limit public costs while controlling surpluses.

We will get to the promised land only if we:

- Are helped by the silent leadership of an urban and consumer opposition rejecting much of present farm policy because of its fiscal and distributional effects.
- Are aided by the pressure of a federal budget crisis in 1985 which no longer permits us to retain the open-ended (entitlement) financing of farm programs.
- Renew our commitment to public administration so that the U. S. Department of Agriculture, once again, administers farm programs strongly and competitively instead of passing the buck to Congress.
- Get strong leadership from the Executive Branch, including a Secretary of Agriculture who is willing to fight and a presidential veto strategy, if necessay, to deal with Lesher's frustration in having to administer programs for which the administration is responsible but which it rejects on principle.

As to a Messiah, he or she is still to be found and designated. This person, in the position of Secretary of Agriculture, must seize the initiative from Congress and the Agriculture Committee while maintaining their confidence. He or she must limit the reach of the specialized interest groups which have dominated farm policy and which will demand even more in 1985 without unduly threatening the leadership of these groups. It will be a thankless job but one well worth doing in 1985 in the public interest.

# ALTERNATIVE POLICY PROPOSALS

# 4 POLICY OPTIONS FOR GRAINS

**Bruce L. Gardner***

The Democratic National Committee (Democratic National Committee, 1984) summarized the results of a recent series of farm policy hearings as follows: "When all was said and done it came down to one word: price." The dominance and persistence of this sentiment are apparent in the fact that U. S. agricultural policy has for 50 years consisted of commodity programs intended to support prices. A case can be made that this focus on price is misguided. Some of the policy options that will be considered here are ways of pursuing objectives of farm policy without the price fixation of past programs.

The options to be considered are: (1) continuation of commodity programs in basically the form established by the Food and Agriculture Act of 1981, (2) elimination of all commodity programs, and (3) substitution of price insurance programs for current price-support programs. The discussion will focus on commodity policy for grains.

---

*The author is Acting Chairman and Professor, Department of Agricultural and Resource Economics, University of Maryland.

# *Continuation of 1981 Act Programs*[1]

The probability is high, say, .95, that the existing program structure for grains will continue. The probability is almost as high, however, that there will be a reduced level of real support for farmers within that structure. The best to be hoped for from the grain producers' point of view is a continuation of levels of support in 1981 through 1984 programs. This is one suboption to be considered here. The second suboption is reduced levels of support, and under that heading there are several further options in program structure that should be considered: primary reliance on production control, market-oriented loan rates with income support through target price/deficiency payments, elimination of target prices with higher loan rates, and primary reliance on export expansion. These will be discussed in turn.

## Continuation of Current Levels of Support

There is ambiguity in what "level of support" means and great difficulty in measuring it. As the quotation from the Democratic National Committee suggests, the idea is usually expressed in terms of commodity price, but this is not the real bottom line for farmers. Their real gains are economic rents generated for resources they own. For the present purposes, this is identifiable with net farm income[2]—not with price. For example, a mandatory production control program could conceivably raise the price of wheat yet reduce the net income of wheat producers. It is important for farmers' assessment of their own interests in farm legislation that they consider policy effects on net income—not just on price. Schuh

---

[1]The phrase, "1981 Act program," is used loosely to refer to the grain policy regime that appeared under the 1977 Act not as originally enacted but as modified in the spring of 1978 in response to farmers' unrest. It is the boosts in support prices that occurred then, along with the reinstitution of set-asides and the conversion of the farmer-owned reserve to a price-support scheme, that set the parameters of policy on their current course.

[2]Net farm income includes opportunity costs of owned resources and so is conceptually quite different from economic rent. However, given that no supply and demand functions in the grain sector are either perfectly elastic or perfectly inelastic, any policy that increases (decreases) net income will increase (decrease) economic rents of farmers.

1984) and Tweeten (1982) are among those who have recently warned of pitfalls of this type.

It is necessary to spell out the main features of current programs. The legislative language is rather opaque, and the program provisions are complex enough that disagreement may arise over what is fundamental and what is a mere detail. The essential features of the programs for the grains (particularly wheat and corn), as revealed by events since these programs were placed in the present form by the Food and Agriculture Act of 1977, are as follows. The farmer's incentive to produce is established principally by the target price. This incentive price has been high enough that, with normal yields, production exceeds consumption plus exports, so a tendency toward excess supply exists. Market prices (i.e., world trading prices) are thus reduced by the program in classic Brannan Plan fashion (Wallace, 1962). At the same time, the Commodity Credit Corporation (CCC) loan program places a floor under the market price for U. S. producers by the government's removing stocks from the market sufficient to maintain the market price at the loan level. Because this divergence between the target price and market price increases budget outlays and the threat of large CCC stock buildup is chronic, there is continuing pressure for acreage-reduction programs which we have had in every year, except 1980 and 1981, since the 1977 Act became law. The typical result is depicted in Figure 1 where PT is the target price ( $4.45 for wheat in 1984) and PL is the loan rate ( $3.30 for wheat in 1984).

A new provision in the 1977 and 1981 acts is the farmer-owned reserve (FOR), under which farmers can be paid a subsidy for holding grain off the market during low-price periods for release during high-price periods. The FOR could be, and initially was, expected to function as a pure stabilization program. However, the urge to use the FOR entry price as a market price support above the CCC loan rate proved irresistible. Consequently, the FOR served principally to place the support price ($P_L$ in Figure 1) higher than the CCC loan program would have provided and, hence, intensified the stock buildup and pressures for production control.

The 1981 Act programs for grains have made producers better off. The gains have not been predominantly through FOR subsidies or payments for voluntary diversion. Apart from the payment-in-kind program (PIK), these payments seem to have been not much more than sufficient to compensate farmers for their costs of compliance. The real gains are the deficiency payments and higher market prices that the programs have generated. I know of no econometric estimate of these effects. The principal difficulty is judging what grain prices would have been in 1981 through 1983 in the

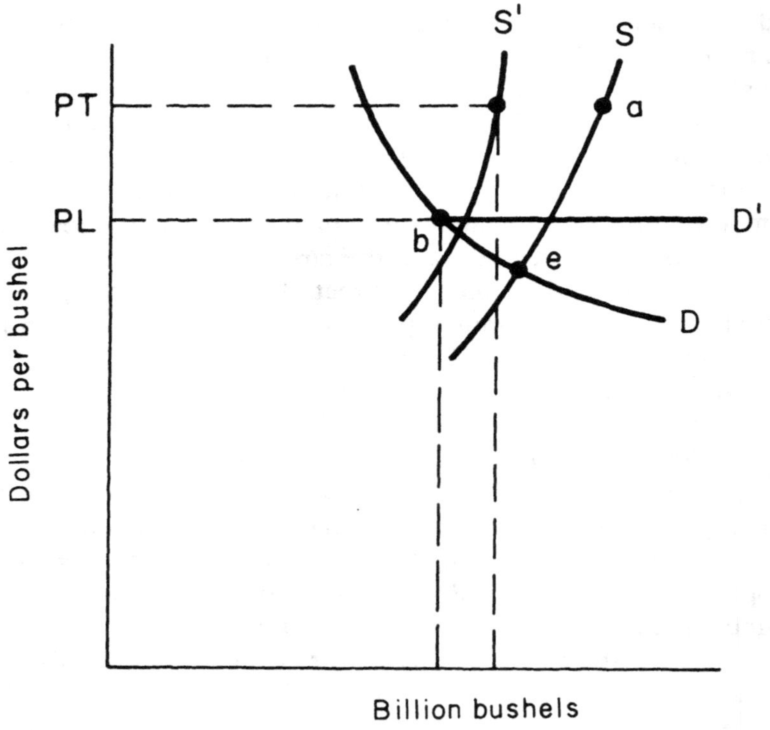


FIGURE 1. Schematic Representation of Typical U.S. Corn Program, 1978–1984

absence of the programs. The supplies actually on hand would have sold for lower prices if the government had not acquired stocks; yet, if there had been no target price guarantee, the desired production would have been somewhat lower.

For corn in 1981 through 1984, the target price was probably not a strong production incentive because the market price has been higher except in 1982 (although the target price still had expected value as price insurance). So the net effect on prices was mainly the effect of acreage diversion. For the 1983 PIK program, the U. S. Department of Agriculture predrought estimate that PIK would reduce production by 1.8 billion bushels seems appropriate on an ex ante basis. The 1982 diversion program was perhaps one-tenth as effective, suggesting for the four crop years (1981 through 1984) an average output reduction of .5 billion bushels. The four-year average is used rather than taking PIK as a one-year aberration on the view that the seeds of PIK were established by the policies of the preceding years and by market conditions; so, if 1981 through 1983 policy and market parameters were continued indefinitely in a stochastic stationary state, a PIK would be reimposed every few years. (An implication is that, unless market conditions strengthen, a continuation of 1981 Act policy parameters would result in endless recycling of production controls—PIK karma.)

The price effects depend on the elasticity of demand which existing econometric work and Figure 2 suggest is about –.5. Therefore, the average 1981 through 1984 production cutback, which amounts to about 7 percent, increased the price of corn by about 14 percent.

For wheat in 1981 through 1984, the situation is quite different. Target prices have uniformly been above market prices and have, in turn, been supported by CCC and FOR loan programs. At the same time, there have been significant production control programs. Therefore, policy has encouraged a significant increase in supply (movement along the supply curve) as well as a leftward shift in supply as in Figure 1. This means that observed price-quality combinations are not on either the no-program supply or no-program demand curve, making it even more difficult to estimate the no-program price for wheat than it was for corn.

Because the export market is so important for wheat, estimates, such as Tweeten's (1982), of an elasticity of demand for wheat of about −1 are plausible. Supposing that the elasticity of supply is .3, we have the following. If 1981 through 1984 production controls reduced output an average of 15 percent from what it would otherwise have been at a target price that averaged $4.00, we add the 15 percent back on to the observed 2.7 billion bushel average production to get 3.3 billion bushels at a point on the no-program supply curve at the $4.00 price (point a in Figure 1). The implied constant-elasticity supply curve is $Q = 2.18P^{0.3}$. On the demand side, we have an average disappearance of about 2.55 billion bushels at an average farm price of $3.60 (point b in Figure 1). The implied constant-elasticity demand curve is $Q = 9.18P^{-1.0}$. Solving for equi-

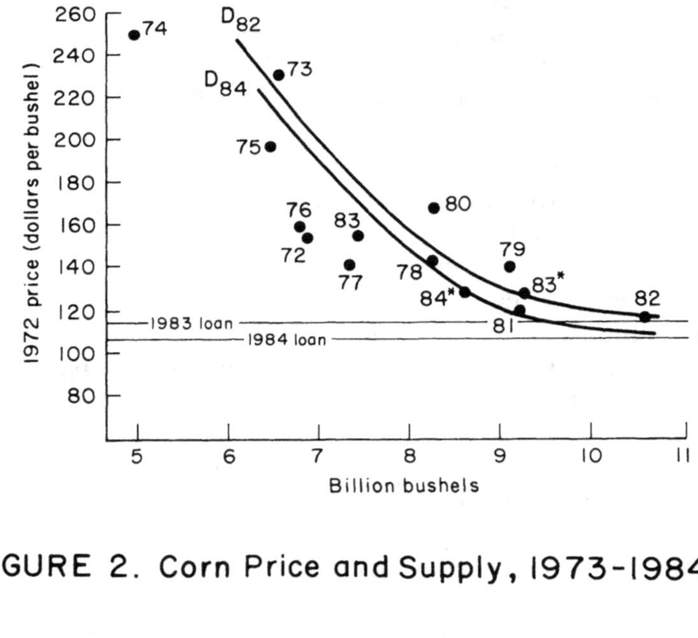


FIGURE 2. Corn Price and Supply, 1973-1984

NOTE: **Prices are average prices received by farmers for each crop year, deflated by the gross national product deflater.**

***The U. S. Department of Agriculture estimate as of March.**

librium P, the no-program price estimate (point e in Figure 1) is \$3.02. This would be a relatively short-run decline in price. The long-run demand elasticity for U. S. wheat is probably well above −1 in absolute value. If it is −2, the estimated no-program price is \$3.26 per bushel. This suggests a magnitude of producer price effect of 20 to 28 percent comparing 1981 Act programs to no program. The net farm income effects are somewhat less, however, because some of the producer price increases are offset by increased prices of agricultural inputs not owned by farmers and because the incentive price is not quite as high as the target price due to costs of qualifying for participation in some programs.

For grains as a whole, the preceding discussion suggests, as an order of magnitude, that returns for owned farm resources (including land owned by absentee landlords) amounted to about 15 percent of crop revenue. This would be about \$5.5 billion for a typical year in 1981 through 1984. It is doubtful that any greater level of support can be obtained politically—so the continuation of the 1981 Act programs, as administered in 1981 through 1984, is the best that can be hoped for from farmers' point of view.

It is likely that, even if real target and loan rates could be maintained at 1981 through 1984 levels in the 1985 Act, the benefits to farmers would not be as great in 1986 through 1989 as they were in 1981 through 1984. This is because long-run demand and supply curves are more elastic than are short-run curves. Therefore, more intensive production control efforts would be required to attain the same price results which could erode net producer gains and/or increase budgetary costs quite substantially.

It is possible that a resurgence of export demand—perhaps because of a weaker dollar, a crop failure abroad, or a stronger world economy—could strengthen markets. In this case, the 1981 Act programs would not be so costly and, consequently, so politically vulnerable. At the same time, however, the programs would not offer much to farmers but would be "slack variables" as in 1980.

Overall, it seems likely that the 1985 legislation, assuming it continues the current program structure, will generate slightly reduced real support levels. The arguments will be over nickels and dimes on these support prices, but they will be far from trivial. The order of magnitude suggested by the discussions of Figures 1 and 2 is that every 2.5 percent across-the-board cut in grain target and loan prices costs grain farmers about $1 billion and saves consumers and taxpayers about $1.5 to $2 billion in a typical year. If real support prices (particularly target prices) were cut 15 percent, the net benefits to producers would largely be eliminated.

The preceding discussion suggests that real support-price cuts would accomplish the same results as would scraping the programs completely. This is an exaggeration for, even if there were no payments in a typical year, the programs would place a floor under producers' prices. This has an insurance value that is likely to be substantial. Before considering the risk elements of grain programs, let us look at a few particular options for moving away from 1981 Act programs while maintaining the levels of producer price supports in that act.

## Deficiency Payments with Low CCC Loan Rates and Without Production Controls

This is the Brannan Plan idea of the late 1940s. Using Figure 1 as a proximate guide, in a typical year output at price PT would be given by the unrestricted supply curve, S, and would clear the market at the correspondingly lower price on D (assuming PL is lowered this far). For corn, the order of magnitude of these effects is as follows. Assume that PT is $3.00 (20 percent above a no-program price of $2.50 at an equilibrium output of 8 billion bushels). Then, if the elasticity of supply of corn is .3, this program causes output to rise by 6 percent to 8.48 billion bushels. Assuming the elasticity of demand for corn is –.5, the market price falls by 12 percent (30 cents) to $2.20. Deficiency payments are $(3.00 - 2.20) \times 8.5 =$ $6.9 billion. The net gain to farmers compared to having no program is approximately $8.24 \times .5 =$ $4.1 billion.

These figures are, of course, conjectural although based on the (quite mixed) available econometric evidence. It could be argued that the elasticity of demand for corn, given a four-year period of adjustment, is much more elastic—primarily because of the export market. Suppose that it is –2.0. Then price falls only to $2.42; but the budgetary costs are still fairly high at $4.9 billion.

A similar approach for the other feed grains and wheat would raise the probable budgetary costs for all grains perhaps 50 percent to $7 billion to 10 billion. Although it is a big drawback of the Brannan Plan option, it has the advantage of generating net benefits to farmers which are not much less than the budgetary costs (especially with the elastic-demand scenario); it is, therefore, a relatively efficient mechanism for accomplishing the sine quo non of any farm program. Moreover, domestic users of corn pay lower prices than with no program, and the U. S. Department of Agriculture is not faced with the task of establishing and enforcing acreage reduction agreements.

It can be claimed, as an advantage of this approach, that it encourages exports. However, it does this only as a consequence of an implicit subsidy in which the U. S. taxpayer pays part of the foreign consumers' bill to U. S. farmers. Balance-of-payments considerations notwithstanding, this should be counted as a net loss. Moreover, there is a risk to the overall trade environment as our competitors see this subsidy for what it is and threaten retaliation.

## Primary Reliance on Production Controls

Among some farm groups, this approach is being seen as the best choice for 1985. It is implicitly the choice of those who suggest simply setting a "reasonable" legislated price which all buyers must pay. The amount sold at this price is the same quantity that, under production controls, would yield the reasonable price as a market-clearing price. Using Figure 1 as a guide, if PT = $3.00 is the reasonable price, the production cutback is to the quantity at which PT cuts D. For corn, using the same parameters as above with the –.5 demand elasticity, a 10 percent output reduction (20 percent acreage reduction with 50 percent slippage) would be necessary to raise the market-clearing price to $3.00. This would mean reducing corn production to 7.2 billion bushels. If output were reduced by means of a voluntary paid diversion program, producers could be made just as well off as under the Brannan Plan approach. The payments would have to be large enough to pay for renting idled land and associated fixed factors. Suppose that these amount to $1.25 per bushel plus administrative costs. Gains to farmers would be about 7.2 × .5 = $3.6 billion with corn users paying this much more for their corn.

This program is less efficient than the Brannan Plan in that the \$1 billion in idled land and associated resources is a deadweight loss. This loss could, in large part, be avoided by a mandatory supply control program in which producers are not paid to reduce acreage but are free to shift land to other (nongrain) crops. This approach affects nongrain producers, e.g., soybeans or cotton, because it tends to drive down their prices. By contrast, production control with idled acreage tends to increase the prices, at least of soybeans, as a substitute in demand for feed grains.

A more damaging criticism of any long-term acreage control scheme is possible if, as seems to be the case, the long-run demand for U. S. grains is elastic. For example, using the parameter value of −2.0 from the earlier corn example, the production cutback necessary to achieve a 20 percent rise in the corn price is 40 percent. This size cutback implies a program for a typical year even larger than the 1983 PIK program. If done on a voluntary basis, the budgetary costs would be 3.2 billion bushels × \$1.25 per bushel (using the cost-of-idled-resources figure from above), i.e., \$4 billion for corn. This would be almost entirely deadweight loss. Moreover, farmers' net gains would be reduced to about 5.8 × .50 = \$2.7 billion.

If mandatory unpaid supply control were used, it is possible that farmers could be worse off than under no program. This would occur if farmers did not have good alternative uses for resources taken out of corn production. If, for example, land and associated inputs valued at \$1.25 per bushel yielded an average of only 60 percent of this value (75 cents) in the best alternative use, farmers would lose 3.2 × .50 = \$1.6 billion on the transferred acreage; and their net gains would be reduced to 2.7 − 1.6 = \$1.1 billion.

Schuh (1984) suggests that production controls have become counterproductive owing to elastic demand for grains. This means, as I interpret the idea, that the markup of support price over the no-program price is higher than a grain-producing monopolist would impose. Indeed, it is so much higher that monopoly rents are less than the loss of producers' surplus from the necessary production cutback; that is, in Figure 3, we are at a position where area b exceeds area a. It should be noted, however, that this result requires demand elasticities substantially higher in absolute value than −2. Also, even if long-run elasticities of demand for grain are substantially higher, as is probable for wheat at least, this does not imply that *no* production control approach can work. It only means that a smaller price target and facilitating production cutback are necessary to keep redistribution to producers reasonably efficient in terms of holding down deadweight losses. *Some* degree of production cutback pays for any monopolist facing a demand curve that is less than perfectly elastic; similarly, some production cutback makes farmers better off as long as the United States is not a pure pricetaker in world markets.

Generally, contrary to my sense of the prevailing sentiment of agricultural economists, such as Schuh (1984) and Tweeten (1982), I see trade considerations as favoring the production control over the Brannan Plan approach. It is true that elastic demand, for whatever reason, favors the Brannan Plan on efficiency grounds. But, given a high elasticity, the fact that it results from an export market favors production controls because it permits the United States to finance its redistribution to farmers, in part, from foreign consumers. The drawback of the Brannan Plan approach in this regard was discussed earlier.

## No Deficiency Payments with Higher Loan Rates

Apparently there is some sentiment for a program that would avoid direct government payments to farmers by using the loan rate to support market prices while abandoning deficiency payments. If the loan rate were raised to the level of PT in Figure 1, the program might be attractive to producers. However, this program is analytically closer to the production control approach (with paid diversion) than to the Brannan Plan. Indeed, the price-quantity equilibrium for high loan rates and production controls would be identical except for the CCC stocks that are built up under the former. If stocks lead to production controls, as they have in the past, the high loan rate option would be only a more costly roundabout method of introducing the option previously discussed.

## Export Promotion

It is a well-established result in trade policy that an export subsidy is less efficient than are alternative means of aiding domestic pro-

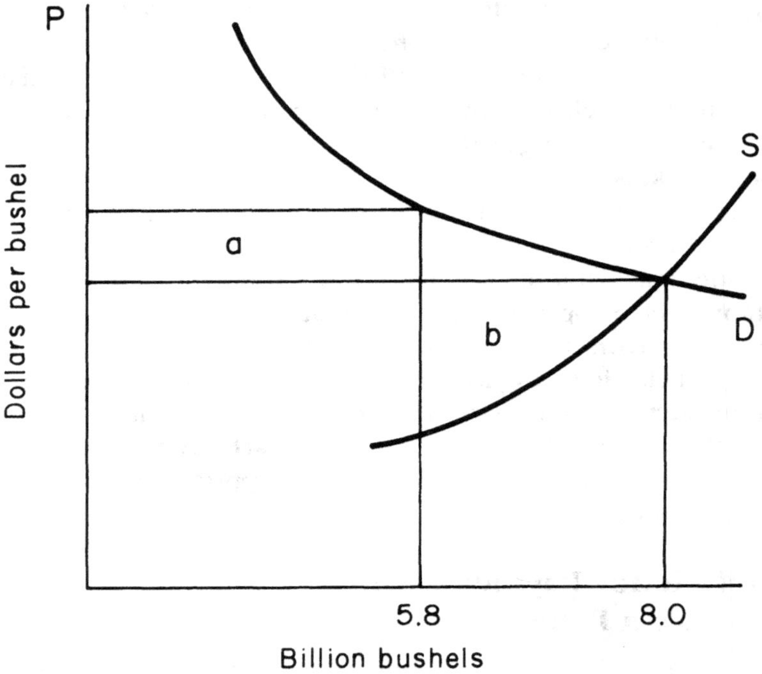


FIGURE 3. Hypothetical Producer Gains from Production Control in Corn

ducers of exportables. So, would not the expenditure of public funds in export sales promotion be a better alternative? Consider the best of all worlds in which spending a few thousand dollars to inform the Chinese, say, about our soybeans resulted in doubling U. S. soybean exports. This would add, perhaps, 25 percent to the total demand for U. S. soybeans, bidding up the price of soybeans by about \$1.80 per bushel (assuming an elasticity of supply of .8 and a base price of \$5.00). This would mean about \$2.5 billion in increased soybean revenue, but it would also mean that soybean users would pay about $1.5 billion more for soybeans. Still, producers would gain more than consumers lose so that, by the criterion of efficiency, this policy is attractive.

We must consider the promotional expenditures actually necessary to double soybean exports. Could it be done for less than a billion dollars? Senator John Glenn proposed this amount of additional spending on farm export promotion in his agricultural program so, if he had won and could have been persuaded to spend it all on soybeans, we would have had a nice experiment.

My overall view of suboptions is that, *if we are going to continue the 1981 Act programs at their current level of real support* (10 to 20 percent above the no-program prices), the most efficient redistribution coupled with general stabilization benefits would be to maintain the present mix of simultaneous target price incentives and voluntary diversion programs but to remove all trigger prices from the FOR (leaving a simple storage subsidy). Perhaps, we should forget about export promotion; on this, also, there is little evidence to support a strong opinion.

## *Elimination of Commodity Programs*

Even the most efficient administration of the 1981 Act grain programs has the problem that, when administrative expenses are included, it costs taxpayers/consumers $2.00 or more for every dollar gained by producers.[3] The efficiency of redistribution could be increased by a smaller scale support system, e.g., one to increase producers' incomes only 5 percent above the no-program situation.

A more fundamental issue is why taxpayers/consumers should redistribute income to farmers or, more particularly, to owners of rent-earning agricultural assets (mainly land). This is especially relevant given the evidence that the average commercial farmer is better off economically than is the average nonfarm citizen. This is, however, a political choice—not one we can resolve as economists. Although it is tempting to believe that there must be a more efficient

[3]Rosine and Helmberger (1975) estimated that, from 1949 to 1972 (although the situation varied greatly from year to year), taxpayers/consumers lost $3.00 for every dollar of net gains to producers excluding administration costs at the midpoint of the range.

way to redistribute income, such as cash transfers, it is not clear that there is a more efficient way to target transfers to the group that benefits from grain programs.

An argument against a *laissez-faire* approach to the grain markets is that prices would be too unstable. Theoretical and empirical work by agricultural economists suggests that stability can be increased by better price and crop information and that this is likely to be a beneficial area of governmental action. However, this is not a matter of commodity programs. The key stabilization aspect of commodity programs is the storage of grain in low-price years for use in high-price years. The large body of literature on this topic is quite mixed on whether governmental activity in this area can, in general, be expected to improve the joint well-being of producers, consumers, and taxpayers, i.e., to have a positive benefit/cost ratio. The empirical evidence is also mixed on whether governmentally directed storage under the CCC and FOR programs has, in fact, contributed to stability under the 1977 and 1981 acts. If there has been net stabilization, it seems insufficient to justify the cost and scope of our current grain programs. Moreover, social losses owing to risk aversion can be reduced by insurance programs which do not directly manipulate commodity markets. Thus, instability problems do not necessarily call for commodity programs even if they do call for public action.

The overall reasons for and against eschewing intervention in commodity markets are as follows. The pros are (1) maintaining prices different from unregulated market prices results in either surpluses or shortages, both of which are socially costly; (2) even well-intentioned efforts to improve the functioning of markets tend to be dominated by narrow interest-group politics; (3) historical experience with interventionist policies is chastening, especially in recent years, in that hardly any groups seem to like the results; (4) the federal budget would be reduced by nonintervention reducing either the deficit or taxes or freeing scarce federal funds for more productive purposes; and (5) the government is tending to get too deeply involved in farmers' decisions on production practices for the farmers' own good, and the administrative costs are themselves a significant resource drain.

The cons are (1) it is not well established that unregulated markets allocate the national agricultural resources well; (2) there is no free market internationally because of the policies of other countries and, therefore, our government must intervene in self-defense; (3) there is no competitive market domestically because of middle-persons' market power—therefore, government should intervene to redress the balance; and (4) intervention is a response to democratic political forces that should be respected as the will of the majority.

While points (2) and (3) of the reasons against eschewing intervention may call for some governmental activity, it should be in the area of trade negotiations and antitrust policy—not commodity policy. Point (1) calls for further research by agricultural economists. Point (4) may be, but probably is not, well taken as a judgment in politics; in any event, it is irrelevant to the *economic* case for *laissez-faire* policies—thus, the negative reasons having crumbled to dust, the positive reasons, however weak, prevail.

## *An Alternative Approach: Put Options on Grains*

A less radical alternative to traditional commodity programs is a price insurance scheme. The risky variable of ultimate concern to farmers is net income, but there are too much measurement difficulty and moral hazard to make low net returns an insurable event. Also, since all-risk crop insurance is already available for grains, only the addition of price insurance is needed to enable a farmer to buy protection against the disaster of low revenue for reasons outside his control. The practical meaning of moral hazard is that the insured risk should be outside the control of the insured. That is, the insurance would pay an indemnity if the insured's state-averaged farm prices or a U. S. indicator price, such as that of December corn futures, fell below a level specified in the insurance contract.

The congressionally mandated Task Force on Farm Income Insurance suggested that such price insurance could be sold as an "endorsement" or add-on to existing all-risk crop insurance contracts. The price insurance could then be marketed through the existing Federal Crop Insurance Corporation institutions. The governmental tie would probably be important because of the need for reinsurance by the writer of price insurance policies. Risk pooling within the set of grain producers would be less for price than for yield insurance because everyone would experience low prices in the same year. The indemnity payments would be similar to deficiency payments under current law except that farmers would have to pay for the right to receive them.

The best prospect for price insurance independent of governmental insurance and, hence, further from political pressures to subsidize premiums or place irrelevant conditions on the policy-holding farmers is put options. Put options already exist for sugar futures and could be marketed by the grain exchanges for futures. A $3.00 put on December corn, which gives the purchaser the right but not the responsibility to sell corn at $3.00 per bushel in December, pays off in the way a $3.00 price insurance policy would; but, instead of an insurance premium, we have a market-determined premium or price paid for puts on the exchange which trades them. The reinsurance in this case is decentralized among those who want to bet that price will not decline or those (millers) who have commitments to buy grain in the future and can hedge by writing puts. (If the price falls, they will gain on their grain purchases but lose on their puts.)

It is possible that farmers would not be willing to pay actuarially justified premiums for put options. If so, it would be a telling blow to the argument that the farmers' need for assurance against low prices constitutes a major failure of *laissez-faire* policies in the grain markets.

It could be argued that what farmers need is multiyear assurance against low prices to justify the large capital investments they undertake. However, this will never be achieved even by government-guaranteed prices which, after all, have been in a downward trend in real terms for the most of the post-World War II period. No policy will be able to insure a favorable rate of return to a long-term investment in land or fixed capital.

It is true, also, that price insurance or put options will do nothing to help a farmer who sells at low prices in a year when average farm prices are not low. This, too, is a feature shared with current programs, e.g., deficiency payments. Generally, policies concerned with market events will not aid those who have unusual difficulty coping with these events. This is particularly germane in current circumstances when perhaps 5 percent of the commercial grain producers are in serious financial trouble with debts near or exceeding their assets. Here, where it really counts, U. S. policy has always been essentially *laissez-faire.* Even the measures taken in the Economic Emergency Loan Programs since 1978 have involved more postponement of realization of losses rather than real income transfers to the hard-pressed farmers.

The preceding discussion is summarized in tabular form in the following table.

**TABLE 1. Summary**

| Effects compared to no programs | Options for grains | | | | |
|---|---|---|---|---|---|
| | 1981 Act | Modification[a] (1) | Modification[a] (2) | No program | Price insurance |
| Producers' price (percent) | +15 | +15 | +15 | 0 | 0 |
| Market price (percent) | + 5 | 0 | +15 | 0 | 0 |
| Producers' incomes (billion dollars)[b] | + 5.5 | + 5 | + 1.5 | 0 | + ⅔ premium subsidy[c] |
| Consumers' incomes (billion dollars)[b] | − 3 | + 1 | − 5 | 0 | + ⅓ premium subsidy |
| Budgetary cost (billion dollars)[b] | − 5 | − 8.5 | | 0 | minus premium subsidy |

[a] Modification (1) is the target price with a low loan rate and no production controls; modification (2) is voluntary (paid) acreage controls. With mandatory controls, producers' incomes and budgetary costs would both be about $3 billion higher than shown.

[b] Dollar values are rough conjectures based on midpoints of elasticities discussed earlier in the text.

[c] The incidence of two-thirds of the premium subsidy going to producers' incomes and one-third going to the consumers' incomes assumes that the elasticity of demand for grains is twice the supply elasticity. If no premium subsidy is paid, there is no redistribution.

## *Conclusion*

This paper has omitted systematic discussion of the goals of farm programs. A choice among the policy options discussed can be made only with reference to the goals they are supposed to serve. I would choose the *laissez-faire* approach in the commodity markets option because the goal that makes sense to me is maximizing the national income per person (maximizing the sum of the consumers' and producers' surpluses); and I think it likely that past commodity programs, even "stabilization" programs, have created distortions that have reduced the national income.

If the objective is to increase returns or economic rents to farm labor, land, and investment, the 1981 Act programs are preferable to no programs. The estimate used in this paper was a gain of $5 billion to $6 billion annually to grain producers and landowners in 1981 through 1984. However, it is not apparent that all farmers have benefited from these programs or that beneficiaries have gained by nearly as much as federal budgetary outlays and the scale of intervention might suggest.

If the objective is the type of general goal stated in the preamble to some legislation—viz., to assure consumers of an adequate supply of wholesome food at reasonable prices and farmers of reasonable returns—then all the options discussed are about equally irrelevant. The probability of an inadequate or unwholesome food supply seems about the same under all of them. With respect to reasonable prices and returns, if one accepts that the best criteria for them are prices and returns which equate supply and demand, we have another reason for *laissez-faire* policies.

Price insurance has promise for the grains as a vehicle for compromise between 1981 Act programs and no programs and for a transition between the two. In transition, such insurance could move from full subsidy of premiums (or of put options) to half subsidy and to no subsidy over a period of several years.

## *References*

Democratic National Committee, Democratic Agriculture Council, Press Release, Washington, D. C., April, 1984.

Rosine, J., and P. Helmberger. "An Aggregative Analysis of the U. S. Farm Sector." Department of Agricultural Economics, University of Wisconsin, Research Bulletin No. R2733, June 1975.

Schuh, G. E. "U. S. Agriculture in the World Economy." *Farm Policy Perspectives*, U. S. Senate, Committee on Agriculture. Washington, D. C.: U.S. Government Printing Office, 1984, pp. 47–61.

Tweeten, Luther G. "Policy Alternatives for Wheat." Department of Agricultural Economics, Oklahoma State University, November 1982.

Wallace, T. D. "Measures of Social Costs of Agricultural Programs." *Journal of Farm Economics* 44, No. 2 (May 1962):33–65.

# 5 UNCERTAIN ECONOMIC ENVIRONMENTS AND CONDITIONAL POLICIES

**Richard E. Just**
**Gordon C. Rausser***

## *Introduction*

Since the introduction of commodity-specific policies in the United States, a number of agricultural policy crises or disequilibria have arisen. Agricultural policy disequilibria emerge when significant changes occur in the agricultural economy which the government is attempting to influence. These policies are generally structured on the basis of perceived conditions at the time of policy formulation. If and when these perceived conditions prove false, a policy disequilibrium arises; and pressure mounts for changes in policy instruments or even changes in the mix of policy instruments (Rausser, 1982).

Numerous cases of policy disequilibrium have occurred in individual agricultural commodity systems. Two dramatic illustrations, however, have cut across all major agricultural commodities for

*The authors are, respectively, Professor; and Chairman and Professor of Agricultural and Resource Economics and Economist in the Agricultural Experiment Station and on the Giannini Foundation, University of California, Berkeley.

which policies have been defined. The first occurred in 1972–73 when the magnitude of increases in farm product and food prices surprised almost everyone in the public and private sectors. The move to flexible exchange rates, the rapid expansion of international markets, and the decreasing barriers between the agricultural economy and other domestic economic sectors all resulted in significant changes in conditions in the agricultural sector. These dramatic changes came as a surprise to many farmers who had learned to operate under a set of policies that, in effect, isolated the agricultural sector from the world economy as well as from the general domestic economy. These policies were changed, in part, because huge government stocks had accumulated during the many years that the U. S. government held price supports above market-equilibrium prices. The U. S. Treasury exposure of carrying public stocks became unbearable in the early 1970s. As a result, the "Soviet grain deal" appeared as a savior for the policy disequilibrium that existed. Owing to the desire to reduce Treasury costs and the unexpected increase in Soviet demand, the U. S. government liquidated public stocks which then exposed the economy to the risk of large agricultural price increases. From the standpoint of officials who were struggling to contain inflation, governmental stocks were liquidated prematurely and thus failed to provide the stabilizing influence for which taxpayers supposedly had been paying for so long.

Another agricultural policy disequilibrium of similar significance arose 10 years later. The 1977 Food and Agriculture Act established the farmer-owned reserve to again create a buffer stock to protect against rapid price escalation and to strengthen agricultural markets that were in decline from the 1972–73 boom. Because these objectives were accomplished during the term of the 1977 Act, the basic policy was perceived as working well; and the 1981 Food and Agriculture Act, which modified the provisions of the 1977 Act only slightly, was passed with little controversy. However, with substantial quantities of stocks in the farmer-owned reserve, the addition in stocks from the 1981 and 1982 record crops was considered excessive for its stabilizing and food-security objectives (Spitze, 1984). In addition, the 1982–83 economic conditions of the world economy and the U. S. general economy were almost the exact opposite of conditions 10 years earlier—the dollar was appreciating rapidly rather than depreciating, real interest rates were rising significantly rather than falling, central banks throughout the industrialized world were maintaining a tight rein on money supply rather than a loose rein, and inflation was virtually nonexistent. As in most policy disequilibria, the policy-setting process did not view the conditions that arose in 1982–83 as possibilities. With the accumulation in public stocks of more than 1 billion bushels of wheat and over 2.5 billion bushels of feed grains and the associated escalation in Treasury outlays, stronger voices of criticism surfaced; and some stopgap, crisis-driven policy provisions had to be enacted. This was first evidenced by the introduction of the Farm Crisis Act of 1982 which, if it had been passed, would have mandated controls on production. This Act, of course, was placed on the back burner and was ultimately replaced by the payment-in-kind (PIK) program. That program was precipitated primarily by the unacceptably low income for U. S. farmers, unacceptably large downward price movements, and large costs of carrying governmental stocks (including those subsidized in the farmer-owned reserves). Apparently, however, the huge Treasury costs incurred by the implementation of the PIK program were unanticipated and have led to yet a further policy disequilibrium as increasing public interest has been focused on the associated subsidization of the farm sector. Partially as a result of this crisis, Congress passed a bill in the spring of 1984 lowering the 1984 target price for wheat and freezing the 1985 target prices for wheat, feed grains, cotton, and rice at the 1984 levels thereby departing from the prescribed adjustments set out in the 1981 Food and Agriculture Act.

One proponent of this piecemeal "fire fighting" approach to public policy stated recently: "The goal of public policy is to find a responsive, workable solution to urgent, difficult, hurting problems" (Spitze, 1984). However, in examining the evolution of agricultural policy during the past several decades, one can easily draw the conclusion that the fire-fighting approach to policy formulation produces policies appropriate for conditions that have already materialized rather than for those which may exist after new policies are enacted. Cochrane and Ryan (1976) have warned that setting agricultural price and income supports four years in advance (as the recent major agricultural acts have attempted to do) can only lead to trouble in a highly uncertain world. If market conditions deteriorate, target prices or loan rates may turn out to be set too far above market-clearing levels and thus result in unpredictably large Treasury outlays and, perhaps, a large buildup in government-controlled stocks. On the other hand, if market conditions improve unexpectedly, a policy may fail to prescribe an orderly liquidation program from accumulated stocks.

The agricultural sector faces continual and often unanticipated changes in weather conditions, the national and international economic situation, and in political forces. Such unexpected changes are likely to continue. In such an environment, one must consider whether agricultural policy should incorporate prescribed changes to each of the wide variety of conditions that can possibly occur. With built-in prescribed changes, agricultural policy can react quickly to changing market conditions and, thus, policy can become more appropriate for current market circumstances than for conditions that existed prior to passage of the most recent agricultural act. The purpose of this paper is to examine the notion of conditional policies which incorporate such prescribed changes as a means of addressing policy disequilibria and policy crises. In this paper it is argued that conditional policies, if appropriately formulated, reduce producer uncertainties and Treasury cost uncertainties. Moreover, such policies can reduce the potential for governmental failure, i.e., enactment of policies that ultimately turn out to be inappropriate. Conditional policies cannot, however, totally eliminate government failure because not all potential economic changes can be identified; but the wider the variety of circumstances for which prescribed changes are incorporated into mandated conditional policy adjustments, the lower should be the probability of a policy crisis that manifests governmental failure.

This paper begins in the following section with a consideration of the rationale and objective of governmental involvement in the agricultural sector. The following sections consider policy uncertainties faced by farmers and their effects on economic efficiency of the agricultural sector, Treasury cost uncertainty faced by government, the need for correct market signals to farmers for expansion or contraction of production activity, the structure and formulation of conditional policies, the concept of policy-equilibrium rules, operational guidelines, and a specific policy proposal. The final section presents some concluding remarks.

## *Policy Rationale*

As stated in the enabling legislation, the general purpose of U. S. agricultural policy is "to provide price and income protection for farmers, assure consumers an abundance of food and fiber at reasonable prices, continue food assistance to low-income households, and for other purposes." Given this general purpose, the rationale for government involvement in the domestic agricultural sector includes a number of perspectives, *inter alia*: (1) farmers are, in some sense, an economically hard pressed—if not deprived—group; (2) the principal reason for farmers' economic status is their relatively disadvantaged position in the marketplace; and (3) in the absence of governmental intervention, there would be an intolerable degree of instability in commodity markets adversely affecting not only farmers but also consumers of food and fiber (Langley, 1983).

The most persuasive rationale for an active U. S. agricultural policy, given recent experience, is the market failure associated with risk and uncertainty. The other rationale—namely, that farmers are in some sense an economically hard-pressed if not deprived group and that farmers do not have sufficient economic power and thus are in a relatively disadvantaged position in the marketplace—can be dismissed. Without doubt, what poverty exists in agriculture cannot be remedied effectively through price supports or other commodity-oriented farm programs. The same observation holds with regard to the potential marketing power of large corporations and food processors in agriculture vis-a-vis farmers.

For the purpose of this paper, the market-failure justification for governmental intervention is assumed to be the unacceptable level of riskiness in the production and consumption of U. S. agricultural products. This rationale for governmental intervention is based on the stochastic character of both commodity prices and production and arises from the inability of farmers to trade their risk adequately to other agents of the economy. Inherent instability results from the significant dependence of production on weather patterns; the inelastic nature of aggregate demand; rapid technological change; asset fixity and atomistic behavior; and the significant integration of U. S. agriculture into international markets influenced by supply and demand fluctuations in other countries, changes in trade policies, and variations in exchange rates.

Furthermore, the inherent riskiness and uncertainty of the U. S. agricultural sector can be increased by unstable fiscal and monetary policies. For example, a flex-price specification of agricultural commodity markets and a fixed-price specification of labor, manufacturing prices, and the like cause highly volatile real rates of interest and exchange rates (resulting from unstable fiscal and monetary policies) to lead to overshooting in agricultural sector markets (Freebairn, Rausser, and de Gorter, 1982). These "macroeconomic externalities" introduce further instabilities into a sector that already is unstable. The tendency of the private sector (unencumbered by governmental intervention) to yield different results than are regarded as socially optimal has been recognized at least tacitly by policymakers.[1]

## *The Private Cost of Policy Uncertainty*

Many instances in U. S. agricultural policy can be cited where the political system has failed, i.e., policies that have been established to accomplish some task have led to contrary results. Such outcomes can be avoided by appropriately designed conditional policies. Before defining such policies and outlining their principal characteristics, however, it is useful to distinguish two often overlooked important concepts that should guide policy formulation: the private cost of policy uncertainty and the public cost of Treasury cost uncertainty. The first is discussed in this section; the second is discussed in the next.

For U. S. agricultural policy, conventional wisdom has long held the view that unstable markets can and should be stabilized by conscious economic policies of the federal government. The previous section argues that the principal incentive for governmental intervention in commodity markets is the basic instability of the private sector. However, the conventional view that government can positively influence the future of an economic system that is perceived to be well understood neglects the fact that past government poli-

[1] Farm policies, such as price stabilization schemes and crop insurance, are designed to affect directly the ability of the agricultural sector to cope with and respond to the capricious nature of its physical and economic environment.

cies have failed because such perceptions proved ultimately to be inappropriate. Policy analyses and formulation in such a conventional framework "sweep under the rug" the instability and imperfections in the political-administrative system (Rausser and Stonehouse, 1978). Political or governmental failure may, in fact, introduce more instability and uncertainty through the form and shape of governmental intervention than the instability or uncertainty that existed in the private sector prior to such intervention (Just, 1981). In essence, policymakers must recognize and treat explicitly instabilities and imperfections in both governmental and political-administrative systems.

Formally, unanticipated changes in policy are an additional source of risk and uncertainty that may affect farmers adversely. Uncertainty about which policy alternatives will be enacted, whether in terms of policy mixes or settings on the levels of policy instruments, is a policy-induced risk (or simply, a policy risk) faced by the private agricultural sector. To be sure, different policies lead to different expectations of commodity prices, availability of credit, cost of inputs, terms of trade, etc. As a result, the risk effects of changes in policies must be considered in terms of their implications for economic efficiency of the agricultural sector.

Agricultural policy uncertainty has typically resulted from two sources: (1) legislative changes and (2) adjustment of policy instrument levels promulgated by administrative officers under given legislative authority. As an example of the latter, the 1977 "act" enabled the U. S. Secretary of Agriculture to base adjustments in deficiency payments on a discretionary allocation factor. This factor was unknown to farmers when they enrolled in the program. Legislative change, however, is the more important type of uncertainty because it often involves major and abrupt changes and thus can have significant implications for physical or human capital investments or disinvestments.

Policy uncertainty is a crucial issue simply because of the adjustment costs private agents incur in adapting to new policy eras. This is also normally reflected in farmers' incurring costs to maintain sufficient flexibility to respond to new, unanticipated changes in policy. Moreover, once a particular set of policy controls is authorized by law, the levels of the controls change from one production period to the next. Under the 1981 legislation, price-support levels, target prices, and set-aside requirements are subject to change annually with varying degrees of lead time. The Secretary has explicit power to impose varying levels of set-aside requirements in each new production period depending on the outcome during the previous crop year. In the face of this short-run uncertainty, farmers must make a number of decisions, the outcomes of which have long horizons. This includes the purchase of land, machinery, equipment, and livestock. When future levels of policy controls cannot be anticipated, the farmer experiences increased risk and must provide for such risk by maintaining a more flexible business organization.

To be more concrete, the effects of policy risk can be illustrated by two classical expectation-formation patterns, namely, adaptive and rational expectations. Suppose that a farmer forms perceptions of price distributions adaptively in response to his experience and the flows of market information. As a new policy is instituted, the farmer will begin to observe prices from a new distribution. Thus, his perception of risk will increase initially if the new distribution is different from the old one. As time passes, the perception of risk will diminish gradually as information from the previous distribution decays in his adaptive process. Thus, a policy change could increase risk for the farmer until effects of the new policy are observed over a sufficient period of time. Also, decisions immediately following policy changes may be subject to substantial allocative errors because perceptions have not yet adapted to the new policy regime. With frequent changes in policy, this may never be possible.

Now, suppose a farmer forms perceptions of price distributions rationally. In this case, the period of policy formulation imposes additional risk on the farmer although risk may not be excessive once a new policy is finally instituted. As Congress considers alternative policy controls for a new agricultural act or as the Secretary of Agriculture or Congress is deciding how to revise existing controls, a farmer with rational expectations will determine a price distribution under each policy alternative and then attach subjective probabilities to each policy choice. Thus, his subjective risk will be greater the larger the divergence of policy alternatives under congressional consideration. By contrast, the adaptive-expectations decision-maker does not perceive increased risk in the policy formulation period although he may turn out to be poorly adapted to a new policy once it is instituted. Reality, most likely, lies between these two specifications of adaptive and rational expectations; thus, policy risk has adverse implications for farmers both before and after policy changes.

These notions have been examined empirically by very few analysts. One of the few empirical studies is presented by Just (1981), who examined the cost of policy risk for the adaptive-expectations paradigm. He compared the effects of the farmer-owned grain reserve program on prices, quantities, and real income for grain and livestock markets relative to the case of no farmer-owned reserve. Using a 34-equation, nonlinear, simultaneous equation model of the U. S. wheat, feed grain, and livestock economy, his results suggest that a serious problem with the farmer-owned reserve in its first few years was its sharp departure from the previous policy and the long time period required to adapt to the new policy. The adjustment problem occurred not only for grain farmers but also for livestock producers who faced grain prices that were significantly different from those that would have existed in the absence of the farmer-owned reserve in its early build-up phase; that is, initial price adjustments differed from long-run equilibrium levels and thus caused false price signals to producers. These false price signals caused substantial maladjustment in the livestock industry because of long lags in livestock production that fed back into the grain markets causing maladjustment to persist for some time. These results led to the conclusion that the practice of changing agricultural policies substantially every four years imposes unnecessary costs on the agricultural sector. The livestock industry can be continually in a state of trying to adjust to new policies because of its inability to adjust instantaneously. Because of the close link between livestock and feed grain markets, these problems "feed back" and cause sustained maladjustment of grain producers as well.

The above costs pertain only to the postpolicy change period and, thus, ignore any risk imposed on the agricultural sector in the prepolicy change period because of uncertainties about what future policies were going to be. Economic inefficiencies resulting from unrealized anticipations about new programs can be just as important as these results indicate for inefficiencies owing to incorrect anticipations about the effects of a program with given provisions. These considerations point to the importance of designing policies that incorporate clear, conditional adjustment to changing economic conditions. Thus, farmers can anticipate adjustments in policy controls through their own subjective assessment of future economic conditions. Furthermore, if the prespecified adjustment rules respond smoothly rather than abruptly, large changes in investment are not induced which cause years of similarly large oscillatory adjustments in related markets.

Another major source of instability in the political administration of U.S. agricultural policy is international in dimension. Policy must answer to many masters including *inter alia*, the U. S. Department of Agriculture, the U.S. Treasury Department, and the U. S. State Department. Policies are altered periodically as a result of changes in policies of other countries and/or of U. S. policies toward specific countries. To be sure, one of the largest shocks on record for U. S. grain markets occurred in the early 1970s when the Soviet Union purchased substantial quantities of U. S. grain rather than lowering their internal standards for meat consumption. In an attempt to reduce the risk for U. S. farmers and consumers of future shocks from Soviet grain policy, the United States negotiated a long-term trade agreement effective October 1, 1976. Nevertheless, another substantial shock occurred on January 4, 1980, when President Carter suspended delivery to the Soviet Union of any U. S. grain exceeding the minimum of 8 million metric tons that was specified in the agreement. This embargo occurred at a time when the Soviet Union already held contracts for delivery of 21.8 million tons of grain. An embargo of soybeans in 1973 caused similar problems, and some would argue that it led to the rapid growth of soybean production in Brazil and Argentina.

The effects of these changes in international policies could have been dampened by appropriate design of domestic policies. If existing policy is revised in a piecemeal fashion, farmers may suffer both from international policy risk and from policy risk associated with unanticipated corrective measures. When the Soviet embargo was announced, for example, the effects on U. S. agriculture and possible compensatory policies for U. S. farmers were highly uncertain. Later, the Secretary of Agriculture increased loan rates, release levels, and call levels for wheat and corn in order to mitigate the effects of the embargo on U. S. farmers. While this change closely followed the embargo, it illustrates the tendency for U. S. agricultural policy to respond to immediate needs in a piecemeal fashion.

Thus, farmers must bear policy risk about the response of the government to various situations. In additon, new risks for farmers may arise from a lack of familiarity with the new policies and their instruments.

## *U. S. Treasury Cost Uncertainty*

Evidence of a policy disequilibrium often appears in the form of huge and unanticipated increases in the Treasury cost of maintaining a particular commodity program. As argued in the introduction, the Soviet grain agreement in 1972 was viewed initially as a "savior" for policies of the 1950s and 1960s that had resulted in huge carrying costs of large public stocks. Similarly, a policy disequilibrium for the dairy industry was precipitated by huge costs of maintaining public stocks.

Formally, Treasury cost uncertainty may be defined as uncertainty about unanticipated government costs of implementing policies such as commodity-support programs. For many programs, Treasury exposure is not well bounded. The open-ended nature of many commodity-support programs, in terms of Commodity Credit Corporation (CCC) acquisitions or deficiency payments, has led to policy disequilibrium owing to unexpected increases in Treasury costs. Time and again, unanticipated Treasury cost exposure has resulted in changes in the levels of policy instruments and in the introduction of new policy instruments. For example, the rapid accumulation of stocks held by the CCC in the early 1950s resulted in the Agricultural Trade and Development Act of 1954 and in the payment of substantial export subsidies on several farm products. The rapid accumulation of stocks in the late 1960s and the associated carrying costs resulted, in part, in the Soviet grain deal in 1972. The significant fall in world sugar prices in 1982–83 resulted in the reimposition of import quotas. Large public stocks of wheat, corn, and other feed grains held in various forms and depressed prices and incomes of U.S. farmers led to the PIK program in 1983. Similarly, the huge governmental stocks of cheese and other manufactured dairy products resulted, in part, in the dairy PIK program. We could go on and on with examples.

The level and risk of Treasury costs emanating from alternative agricultural policies is, indeed, an important dimension of governmental behavior. It explains the potential for administrative instability of food and agricultural policy. A key normative issue is whether or not government should operate as a risk-neutral decision-maker with respect to Treasury cost uncertainty. Several schools of thought exist regarding this issue (Just, Hueth, and Schmitz, 1982). One maintains that, since the government is very large and undertakes many projects, it can spread or absorb risk easily and, hence, should act as a risk-neutral decision-maker in absorbing risks from the private sector. A second school of thought argues that risk should be discounted by the public sector although not at market rates. Rather, a national policy should be established on appropriate rates of discount for both expected effects and risk. A third school argues that public-sector risks should be discounted at private rates because private individuals generally bear the risk if not directly, then indirectly, through taxes. Which of these three schools of thought is appropriate depends on the conditions surrounding specific policy decisions. Namely, before the government can act as a risk-neutral decision-maker, one of the following conditions must hold: (1) the benefits (or costs) of a policy must be spread over a large number of individuals, the project must not affect private risk from market activities, and the project must not involve public goods; or (2) the project must be small and have benefits that are independent of economic benefits (Just, Heuth, and Schmitz, 1982).

Obviously, government intervention in U. S. agriculture satisfies neither of these conditions; thus, the government, with respect to Treasury cost uncertainty, should act as a risk-averse decision-maker. This means that government should be sensitive to Treasury cost uncertainty in policy formulation and, in fact, use private risk-discounting factors associated with individuals who actually bear the risk, namely, taxpayers.

## *Market Signals and Policy Disequilibria*

Another important consideration in agricultural policy formulation is that, if policymakers choose to isolate farmers from market signals, they must ultimately "face the music" in the form of large increases in Treasury costs of program ineffectiveness as the economy diverges from perceptions of market conditions at the time of policy formulation. When such events occur, enormous incentives for policy change accumulate and thus induce policy risk. The Nelson and Cochrane (1976) simulation study, for example, shows that government policy programs that kept farm prices and incomes higher than they otherwise would have been from 1953 to 1965 provided economic incentives to expand output sufficiently to keep farm prices lower than they otherwise would have been from 1968 to 1972. Prices were lower, in large part, because of huge stocks carried by the U. S. government which at some point had to find their way back to private markets. As Johnson (1984) noted in his presentation:

> "What is clear or should be is that, even with very low price elasticities of supply, the increase in output induced by a price just 10 percent above the market-clearing levels soon results in a substantial excess production that must either be stored, disposed of in some manner, or eliminated later by output reduction."

If the long-run price elasticity of the U. S. wheat supply is near 0.8 (with the short-run elasticity near 0.45) and if the long-run elasticity for coarse grains is near 0.75, then for reasonable estimates of the elasticity of demand "it is easy to see how a 15 percent price increase above market-clearing levels could result in stocks equal to a third of the annual output in two or three years" (Johnson, 1984). In an earlier paper, Johnson (1981) noted:

> "The most immediate effective limit on the level of target prices is the budgetary cost. Price-support or loan levels are limited by the desire to export freely and by the rapid accumulation of stocks when the levels are set too high. The output-simulating effects of high target prices become apparent rather slowly. And, for a brief period, perhaps two or three years, output effects can be contained by set-asides and acreage-diversion programs. But recent experience indicates that expenditures required to elicit voluntary participation in acreage-diversion programs can be substantial if significant output effects are desired. And, it needs to be recognized that the magnitude of the required output adjustment is a function of the level of the target prices. Consequently, high target prices impose two budget costs—

the deficiency payments and the cost of achieving additional diversion."

In the early 1980s, the rigid price supports and unrealistically high target prices (particularly for wheat and cotton) were totally inappropriate for the economic conditions that emerged from 1981 to 1983 even though these policies would have been quite appropriate for the economic conditions of 1972 and 1973. However, from 1981 to 1983, the exchange value of the dollar rose sharply and, as a result, destroyed the hope for effective policy performance under the 1981 Act. As Johnson (1984) has noted, "After all the attention that had been given to the increased instability that emerged post-1972, it was wholly inappropriate that legislation with so much inflexibility had become the law of the land."

What this means is that neither private agents nor policymakers can be allowed to be insulated from market signals. Such insulation creates the necessary conditions for policy disequilibrium which, in turn, leads to large Treasury cost uncertainties and policy uncertainties. Market signals can, of course, be modified; but they cannot be grossly distorted without adverse longer term implications. This general observation holds not only with respect to agricultural and food policy but with respect to all forms of government intervention.

In summary, because market conditions change, inflexible policies cannot be expected to filter market signals appropriately. What is needed are self-adjusting policies (conditional policies) that permit smoother, more orderly adjustments in producer prices that transmit at least some appropriate adjustment signals between farmers and consumers. As argued in the remainder of this paper, policy formulation along these lines could reduce farmers' policy risk, reduce governmental Treasury risk, and provide for more orderly investment and growth.

## *Conditional Policies*

Conditional policies are simply policies that follow a formal specification for change in policy instruments or instrument levels as a result of changes in economic conditions. A number of conditional policies have been enacted at various times. For example, under the

Agricultural Adjustment Act of 1938, the Secretary of Agriculture had discretion in establishing the nonrecourse loan rate between 50 and 72 percent of parity of wheat and cotton. The specific formula-regulated loan rate for corn was to be 75 percent of parity if the supply was not expected to exceed domestic consumption plus exports for the year and 52 percent of parity if the supply was expected to exceed domestic consumption plus exports for the year by 25 percent. From the standpoint of an efficiency norm, however, these conditional policies are nonoptimal. They did not prove to be sufficiently adaptive for an economy with rapidly changing technology, input markets, farm scale, etc.

Conditional policies can be of many types, forms, and shapes. Policies can be discretely conditional or continuously conditional. In either case, some policies can be allowed to change intraseasonally while others can change only interseasonally. Numerous studies show that optimal policies must be conditioned on the economic environment (Rausser and Hochman, 1979). Wallace (1962) was able to demonstrate this result by evaluating three discrete policy alternatives: a Cochrane-type plan for production quotas, a Brannan-type policy of price subsidy, and an input-restriction program for reducing agricultural output. His results show that, as supply and demand elasticities change, the policy set that minimizes social cost or inefficiency losses changes.[2]

Without formal specifications of conditional policies, agricultural programs contain potential flaws such as Firch (1984, p. 29) has pointed out regarding the cotton program, namely:

> "There is no mechanism in the current legislation that will insure that the real loan and target prices will move down over time at the

[2]The same point is made in the framework developed by Becker (1980) and applied to agricultural commodity markets by Gardner (1983). The most efficient redistribution policy depends upon the elasticities of supply and demand. For example, inelastic demand favors production controls, while inelastic supply favors a deficiency-payment policy. For some elasticities, the Cochrane plan dominates the other two plans; whereas, for other elasticities, the Cochrane plan is dominated. What this analysis shows is that, if export-demand elasticity is not constant from one year to another but depends critically on production shortfalls in major importing countries (such as the Soviet Union or China), in one year some policy instrument should be set to zero whereas, in another year, that same policy instrument should be at an active level (nonzero).

> natural rate of decline of commodity prices. Without this device for moving the real program prices down over time, it is quite possible that the program prices will rise relative to free-market prices and the program could be a major drain on the U. S. Treasury and a major distorter of prices and resource allocations."

Similarly, Babb, in the context of the dairy program, argues that programs "featuring price supports above longer term, market-clearing levels in the absence of some form of supply management are not feasible without continuing purchases and out-of-Treasury costs. Higher support levels might result in less variation in price and production and reduce the number of dairy farm exits, but they would also introduce economic distortions."

In the last few years, the favorable production years of 1981 and 1982 in cotton, wheat, corn, and other agricultural commodities combined with an unusual set of related market conditions (financial markets, exchange-rate markets, and general economic trends) to create significant policy disequilibria that motivated the PIK program of 1983. Alternatively, a set of conditional policies for target prices, loan rates, set-asides, diversion payments, and the various dimensions of the farmer owned reserve program (e.g., interest-rate subsidies, storage payments, call prices, and release prices) could have been used to specify adjustments to such conditions and thus could have avoided this policy disequilibrium or crisis. In theory this can be demonstrated formally. Past history, however, suggests that, whenever conditional policies were specified, they were defined in a peculiar manner from the standpoint of economic efficiency. For example, the parity concept used in 1938 is unresponsive to most major market signals. Thus, substantial Treasury cost uncertainty was incurred which ultimately led to policy uncertainty. From the standpoint of rent-seeking behavior and the demand to transfer income or wealth, such conditional policies may be reasonable; but if such political behavior defines conditions of policy disequilibrium, it, too, must be considered in formulating appropriate conditional policies.

In addition to the parity-conditional specification, the 1973 Act specified a procedure for adjusting initial target prices in 1974 and 1975 for wheat, feed grain, and upland cotton in accordance with measures of the cost of production. Similar specifications were provided in the 1977 Act. These conditional policies were attacked on numerous grounds. However, political intervention never actually allowed the conditioning formulae to operate. For example, target prices determined by the 1977 Act were changed occasionally by legislative action throughout the life of the bill. Also, target prices were never allowed to fall from the level of the previous year even though estimated production costs declined.

The failure of the 1977 Act in formulating target prices conditionally on cost of production (as evidenced by alteration of the provisions) can again be explained by inappropriate formulation of the conditioning factors. Fundamentally, the cost of a productive resource that is not perfectly elastic in supply cannot be determined independently of product demand because of pecuniary diseconomies of scale. Thus, an increase in the final demand for the commodity will increase the expected value of the particular inputs utilized. As Gardener (1981, p.125) has facetiously pointed out: "... if the price of wheat were set at $15 per bushel, the prices of scarce wheat-growing resources would be bid up enough to make the cost of production $15."

The 1981 Act abandoned the direct use of the cost of production measures to determine target prices. Nevertheless, Congress revealed in the specification of that Act that it believed it had the capacity to set target prices at reasonable nominal levels four years ahead through the crop year 1985–86. The resulting fixed annual escalation of target prices on crops created significant policy inflexibility. As market prices have declined and inflation has slowed, the drawbacks of this inflexible escalation in the target price have emerged.

Another example of conditional policies for the major commodities was introduced for the 1966 upland cotton crop. The procedure was delineated in the 1965 Act in which the loan rate for upland cotton was tied to a moving average of world prices and/or U. S. spot-market quotations compounded by a minimum level. This conditional policy proved to be effective. As noted in Cochrane and Ryan (1976), this procedure worked to reduce cotton loan rates significantly compared to those in previous years and was successful in regaining important export sales. This conditional policy provided the basis for eliminating the need for domestic mill and export subsidies.

In addition to the conditional target-price specification, the Food and Agriculture Act of 1977 stipulated a procedure whereby legislated loan rates could be reduced by the Secretary of Agriculture whenever necessary to maintain domestic and export markets. Specifically, loan rates for wheat and feed grains were allowed to be reduced for the next marketing year by a maximum of 10 percent if the national average price received by producers for the commodity in the current marketing year were not more than 105 percent of the loan level of the current year. A lower limit was placed on the amount that loan rates could be reduced over the life of the bill. This provision was also contained in the 1981 Act. The 1981 Act also introduced a conditional policy for determining soybean loan rates based on a five-year moving average of past market prices (excluding the high and low years) with a lower bound of $5.02.[3]

The 1981 Act required rice loan rates to be adjusted by the same percentage change as are rice target prices. For example, when the rice target price is increased by 5 percent, the rice loan rate is also increased by 5 percent. The rice loan rate can be adjusted downward if the Secretary determines that the loan rate established by the formula would discourage exports and result in excessive domestic rice stocks.

For each of these conditional policies, however, much policy risk remains. In point of fact, the potential adjustment cost faced by farmers resulting from unanticipated changes could well lead to substantial inefficiencies under each one of these conditional policies. That is, most of these policies give the Secretary of Agriculture short-run discretion in setting policy instruments. Thus, farmers can only anticipate the policy instruments within certain bounds. Furthermore, farmers cannot be sure that government will not decide to revise these policies in major ways with relatively short notice as on other occasions.

The above observations hold, also, for the discretely conditional policies that have recently been passed by both the House and the Senate for the 1984 and 1985 crop years. For example, in the 1985 feed grain program, corn is specified to have a target price of $3.00

---

[3]An exception to this rule is when the loan rate does not exceed the market price by more than 5 percent in which case the downward adjustment discussed above is applicable with an absolute minimum of $4.50.

per bushel. However, if the U. S. Department of Agriculture estimates that the corn carryout as of September 30, 1985, will exceed 1.5 billion bushels, the 1985 program would include a total acreage cutback of 5 to 20 percent and a combination of acreage reduction and paid diversion of which not less than 5 percent is in advance paid diversion. The payment rate would be a least $1.50 per bushel. In additon, any acreage reduction over 15 percent is to be proportioned equally between acreage reduction and paid diversion—meaning that paid diversion would be a maximum of 7.5 percent. Similar discretely conditional policies are specified in this legislation for each of the other major crops. In each of these instances, significant Treasury cost uncertainty and policy risk remain. The discrete intervals and specified bounds do not preclude the possibility of large and unexpected increases in Treasury costs. Moreover, given the discretionary features of these discretely conditional policies, governmental failure can well emerge again and lead to unanticipated increase in Treasury costs. The threat of these outcomes suggests further policy risk.

An appropriate specification of conditional policies should base the policies on factors that signal crises or major policy disequilibria. After all, these are the factors that historically have led to the need for major and unanticipated policy reformulation. For example, the actual triggering factors should include Treasury costs or excessive governmental stocks in addition to farm price or income levels. These manifestations of policy disequilibria are influenced by changes in external conditions such as exchange rates, inflation rates, real rates of interest, and the like. Therefore, conditional policies should be specified to change either directly or indirectly in accordance with exchange rates and the real rate of interest. Given the present state of knowledge, however, conditional specifications in terms of these indirect casual variables would likely involve too much "noise" from imprecisely estimated parameters. Hence, specification of policies in terms of the resultant variables—Treasury costs or public stocks, farm price, and farm income levels—is likely to be more successful. Furthermore, conditional policies must respond to these conditions over a sufficiently broad range of values of these variables. Policy rules with one or two discrete steps (such as the 1938 parity rule) may be able to react appropriately in the short run or with little variation in economic conditions but are likely to prove increasingly inflexible.

## *Policy Equilibrium Rules*

At an operational level, conditional policy mechanisms must be kept as simple as possible so their effects can be assessed more easily by the policymakers and farmers. On the other hand, changes in the policy must also be orderly and frequent enough to keep policy instruments closely related to current economic conditions. This is in sharp contrast to recent policies that involve setting a particular level for, say, the loan rate and then, when it appears to be too far off-line, making a substantial revision. Experience suggests that this piecemeal approach will always be necessary when specific levels, say, loan rates are determined only after existing levels appear to be in disequilibrium. For example, loan rates were relatively high in 1977 and 1978; but if sufficient inflation had occurred, the release levels would have become too low. The commodity policy actually acted more like a simple price support in early years in which case economic welfare analysis clearly implies a net loss for society as a whole. On the other hand, if sufficient inflation had occurred, the release level would have acted as a price ceiling in the absence of set-asides—at least until the farmer-owned reserves were depleted. Economic welfare analysis also clearly implies a net loss for society as a whole in this case.

Which of these cases occurs depends on which direction the general economy turns, but in neither case does the policy serve very well to meet the general policy rationale of providing (real economic) stability. In either case, the agricultural economy would be driven farther and farther from equilibrium thus eventually precipitating a policy crisis. As a result of this type of piecemeal policy adjustment, the program can become a destabilizing influence or, at best, can promote economic inefficiency by artificially holding prices up immediately after loan-rate revisions and then artificially holding prices down after inflation and just before new revisions are made. A superior approach would be to change loan rates more frequently and in smaller amounts in accordance with observed and anticipated changes in equilibrium price levels. Then, prices could be stabilized near equilibrium or efficient price levels rather than at price levels distorted first one way and then another.

These considerations suggest that policy instruments must be revised frequently to keep policy in equilibrium—in other words, to avoid policy crisis or disequilibrium. As argued earlier, however, frequent revision of policy instruments can impose excessive policy risk on farmers if the revisions cannot be well anticipated. To avoid imposing undue policy risk on farmers, agricultural policies must specify in advance how the specific controls of the program will be changed in response to market conditions. In this fashion, farmers can more easily anticipate such changes through their own assessments of future market conditions. Thus, better investment decisions should be possible than when farmers are left to speculate about future policy control levels.

In order for agricultural policies to specify such adjustment in advance, policymakers must work out policy equilibrium rules, i.e., define the responses to be taken in various "would be" policy disequilibria. Observable conditions which signal major policy disequilibria should influence revisions of the controls and, thus, define the policy equilibrium rules. These conditions should include government-related stocks and Treasury costs. If the rent-seeking activity of interest groups is a factor affecting policy equilibria, the major variables affecting their activity must also be included. Thus, inflation of food prices and farmer income levels is also a likely candidate for inclusion. Target prices and deficiency payments combined with loan rates supposedly avoid low farm incomes while release and call levels avoid rapid food-price inflation. However, acceptable levels of farm income and consumer prices must be allowed to change with the rate of inflation.

Similarly, acreage set-aside programs are introduced, in large part, to avoid excessive governmental stocks, in whatever form they may be held, and the accompanying high Treasury costs. Hence, the set-aside requirements could be keyed to the level of accumulated reserves.

Alternatively, loan rates could be tied directly to reserve stock levels. As noted above, the 1977 Act admits this possibility by authorizing the Secretary of Agriculture to reduce loan rates 5 percent if he finds that stocks are excessive; but, unfortunately, the discretionary nature of this conditional policy does not have favorable implications for policy risk. Revisions in the loan rate should not be made in a piecemeal manner that is difficult, if not impossible, to anticipate without substantial unnecessary policy risk for farmers. Conditioning the loan rate on the level of governmental reserves could be done in such a way as to reduce policy risk by using a continuous policy-equilibrium rule. For example, the loan rate could be increased (decreased) 1 cent per bushel for every 3 million bushels the government reserve is below (above) some specified target level. If farmers could anticipate this adjustment process when making their investment decisions, agricultural production should attain greater economic efficiency with less risk. In fact, with more efficient investment in the agricultural sector, lower prices may lead to the same levels of income.

Smooth and orderly changes in the release or call levels is also appropriate. If farmers believe that the loan rate and release levels establish a price corridor, the program itself alters their probability distributions by preventing "low" and "high" prices. The revised expectations may, in turn, generate ex ante production responses. In this context, government policy may offer reduced benefits when prices are nearer normal levels; and costs of providing some stabilizing influence are, indeed, very reasonable. On the other hand, in extreme price situations, the costs of achieving stability swamp any benefits that may be derived (Just and Schmitz, 1979).

Under inflexible policies, unexpected market developments can and do trigger large increases in reserve levels. As a result, Treasury costs can increase substantially. Conditioning price-control levels explicitly on stock levels would ease this burden. This conditional specification could be allowed to vary interseasonally and operate in accordance with a prespecified scale.

To the extent that conditional policies are allowed to change in a smooth and predictable manner, grain farmers will be able to make more informed long-run investment decisions. By contrast to the current formulation of inflexible policies, the farmer will be better able to determine how much grain he will want to plant over the planning horizon that might be covered by a long-term investment in machinery. Similarly, the investment inefficiencies currently imposed on the livestock sector will be reduced since these agents will be allowed to form more accurate long-term expectations of grain prices and, thus, position their long-term investments in their herds more efficiently. In this manner, governmental intervention would not be imposing additional uncertainty on farmers. In fact, the policy could then act to achieve the underlying policy rationale of reducing the market effects of existing, inherent uncertainties.

Governmental ownership of grain reserves is viewed with considerable skepticism because it concentrates power in a few individuals who make governmental buy/sell support/set-aside decisions. Similar concerns hold, perhaps to a lesser degree, regarding other policies such as meat import quotas. The conditional policies advanced here, however, avoid some of these problems because the role of the government is controlled mechanically by the conditioning variables. In this fashion, short-run randomness introduced through the political process is minimized. On the other hand, the success of mechanical rules for adjustment of policy instruments over long periods of time (sufficient to obtain the benefits of reduced policy risk) depends on appropriate conditioning of the rules. If the equilibrium rules do not prescribe an appropriate change for some type of "would be" policy disequilibrium, the policy rule may sooner or later require change. However, the wider the set of adverse conditions to which the rules respond, the less likely and less frequent should be those required changes and, thus, the less should be the policy risk faced by farmers.

The theoretical rationale for these kinds of conditional policy equilibrium rules has been developed elsewhere. Just and Schmitz (1979) have shown that the optimal governmental adjustments in stocks can be specified as a smooth function of the difference between target prices and observable prices. More recently, Meyer (1984) has shown that the only policy rule that is preferred by all agents, regardless of their utility functions, is the linear variation of price. This stochastic-dominance result implies that a self-adjusting policy specifying a simple linear function for the difference between target prices and observable prices is preferable to the usual approach of inflexible setting on loan rates, target prices, etc. Empirical justifications have also been offered for smooth and continuous policy equilibrium rules. They include the work of Cochrane and Danin (1976), Danin (1975), and Zwart and Mielke (1976). These studies demonstrate the potential benefits for society of governmental intervention with built-in responses to market conditions. It should be noted, however, that these empirical studies consider only the short-run effects. They do not take into account the longer term investment efficiencies associated with reduction in policy risk. Hence, empirical studies that take into account the longer term implications should result in even greater dominance of conditional policies geared to economic conditions.

## *Operational Simplicity*

To minimize policy risk, equilibrium policy rules must be understood easily and anticipated by all actors in the U. S. food an agricultural system. This suggests, for example, that optimal stochastic or adaptive control formulations with their complete decision rules cannot be utilized to determine the conditional policies that should be imposed. Such formulations are simply too complicated and would put weights on too many state variables in determining optimal "feedback rules" (Rausser and Hochman, 1979).[4] Optimal stochastic control feedback rules may also be dominated by simple conditioning rules merely because of an inability to properly specify and estimate the system.

As noted previously, the recent 1984 legislation passed by both the House and Senate offers some discrete conditional specifications for a number of policy instruments. Some of these conditions give stepwise adjustments with a number of steps that almost approximate a smooth rule. However, these conditional specifications not only allow unanticipated discretion to the Secretary of Agriculture in application but the discrete jumps lead to policy risk measures which exceed those that would result from self-adjusting policies admitting changes in a smooth and orderly fashion. One must also question whether rules with many steps are more easily understood than are linear rules stated in the simple examples above, e.g., a 1-million-bushel public stock transaction for every 1-cent change in price.

---

[4]Of course, if the stochastic control formulation were specified properly and recognized the cost of adjustment and the cost of information collection and monitoring by private agents, a derivation of the optimal feedback rules would, presumably, be simple, easily understood, and anticipated by private agents. Because of the inability to recognize or measure properly these costs of adjustments, the optimal stochastic control approach is not feasible in this context.

## A Policy Proposal

The task remains to specify a policy that balances Treasury cost uncertainty with private market risk in a manner that minimizes policy risk and transmits sufficient market signals for efficient long-term, private-sector adjustments while maintaining an operational simplicity understandable by farmers and policymakers. The policy proposed here to satisfy these concerns is a significant departure from that of the 1981 Act. It involves a difference in the set of policy instruments as well as a difference in how the levels of the policy instruments are set. The policy proposal is motivated by the understanding that the rationale for commodity-specific agricultural policy is the inherent instability in agricultural production and marketing and that concerns for equity or the need to transfer wealth can be accomplished more efficiently by means other than commodity-specific policies.

Given this perspective, a number of the policy instruments included under the 1981 program are inappropriate. First, the release and call levels associated with the farmer-owned reserve are inappropriate because they cannot be set conditionally in a smooth and graduated manner. They establish a fixed price band intraseasonally and, thus, lead to excessive Treasury cost uncertainty and market distortion in abnormal conditions while providing no benefits in normal conditions. The loan rate suffers from the same problems. Even when varied intraseasonally, a loan rate cannot induce smooth behavior by individuals in accordance with market prices, i.e., once a farmer has his crop under loan, he will not be affected by additional market signals during the crop year as long as price is below the loan rate. Given that release and call levels and loan rates are not appropriate agricultural policy instruments, it follows that neither the farmer-owned reserve nor the loan program is an appropriate agricultural policy. In addition, the storage payment and interest rate subsidy associated with the farmer-owned reserve have distorting properties and promote long-run inefficiency.

Two arguments suggest that set-aside and deficiency-payment instruments are also inappropriate. Deficiency payments are motivated by a desire to transfer wealth, and acreage set-asides have become necessary because of a long history of distorted market signals associated with inflexible policies of the past. First, when conditional policies are employed that transmit to the private sector sufficient market signals for adjustment, the need for input or output controls is eliminated. Second, any need for transferring wealth is better met by other noncommodity-specific policies as noted earlier. The use of deficiency payments and acreage set-aside instruments for the transfer of wealth can only promote continued inefficiency and chronic long-term adjustment problems.

These arguments imply that neither the existing loan program nor the farmer-owned reserve should continue to operate; rather, any public stock ownership should be taken directly by government. Of course, government ownership has traditionally suffered from problems of excessive storage cost, quality deterioration, transactions costs associated with the spatial distribution of a commodity, etc., as well as possible costs associated with selling at a price lower than that at which governmental purchases took place. However, as demonstrated below, market institutions exist that can eliminate many of these costs.

The policy proposed here is as follows. Rather than offering to buy all grain at a specific loan rate, the government should specify an easily understandable linear stock purchase/sales rule. For example, the government could specify that, say, 1 million bushels of grain would be purchased for every 1 cent per bushel the actual market price falls below some specified target price level. Similarly, the government could sell 1 million bushels from stocks for every 1 cent per bushel the price is above a specified target price. If these transactions occur at competitive market prices which farmer's grain was actually purchased by the government would make no difference. Note, also, that the actual market price used in determining these transactions is the one that occurs with government intervention rather than in the absence of intervention.

This procedure would provide some stabilizing influence when prices are near equilibrium where stability comes at a very low cost. On the other hand, the policy would not put a "cap" on possible price variations that could lead to high Treasury costs and policy disequilibrium. Thus, the policy is capable of maintaining a balance between the private cost of price uncertainty borne by farmers and the public cost of Treasury uncertainty borne by government. Furthermore, under this type of policy, the stabilizing influence can possibly be provided throughout a marketing season. As price increases, the government could sell stocks to ease the price increases. As price starts downward, the government could buy stocks to ease price declines. Thus, the announced policy of a 1-million-bushel transaction for a 1-cent change in price would be an equilibrium relationship that could be applied continuously in determining governmental stock transactions. Of course, the time interval between transactions should not be too long; otherwise, prices may be too far out of line or cause too much price unsettlement when transactions finally occur.

To make this stabilization policy adjust itself appropriately interseasonally, a conditional policy must be specified for the modification of target prices in a way that is sensitive to changing economic conditions. To do this, a simply understandable linear rule could be specified for modifying the target price from year to year based on accumulated reserves. For example, the target price could be increased (decreased) by 1 cent per bushel for every 3 million bushels actual reserves are below (above) a target reserve stock.[5] Finally, the target reserve stock should be specified as a simple function of real Treasury costs and other concerns that signal policy disequilibrium. With such policies, market signals can still be passed on to producers, albeit in a filtered form, so that the agricultural economy does not stray too far from equilibrium. Also, with specific conditions governing adjustment of policy instruments from year to year, farmers can anticipate governmental policies over the longer run based on their own perception of market prospects. Thus, policy risk is minimized, and inappropriate levels of investment are not encouraged. Furthermore, with this type of adjustment, the policy can be self-correcting with respect to unanticipated Treasury costs, inflation, variation in nominal interest rates and exchange rates,

[5]Alternatively, an appropriate conditional policy for the storage subsidy could be specified to depend upon the accumulated size of the farmer-owned reserve. For example, the new target subsidy could be determined by subtracting (adding) 5 cents per bushel for every million tons the farmer-owned reserve is above (below) some goal level for the reserve size. If this revision rule were known well in advance by producers, the effects of current and expected future market developments could be taken into account; and the uncertainty associated with unanticipated storage-subsidy changes could be avoided. Furthermore, with this type of rule, revisions would be assured so that the reserve would not begin to accumulate indefinitely and lead to excessive Treasury cost. Moreover, the reserve would not be depleted over a period of many years.

technological development, and other changes in conditions of the agricultural and general economies.

Operationally, there are two ways to implement this policy. One is to have the government perform transactions in terms of the physical commodity. The other is to hold the right to buy or sell physical units of the commodity, i.e., to take positions on the futures market. Actual transactions in terms of the physical commodity suffer from significant problems of implementation. Where would the commodity be purchased? Where would the commodity be stored? How would storage costs be paid? What would be the frequency and timing of transactions? Alternatively, the futures market approach solves the problems about where the transaction would take place and where the storage would take place (no physical storage would be involved) and provides much greater flexibility regarding timing and frequency of transactions. Furthermore, the futures market approach eliminates many of the other costs associated with carrying physical stocks such as physical deterioration, payments for storage facilities, transactions costs associated with spatial distribution of a commodity, and most, if not all, of the interest expense. (The margin required for futures market transactions is typically only about 5 to 10 percent of the value of the associate physical commodity and can be held in the form of interest-bearing Treasury bills.) Also, the costs of administering this program through the futures market are orders of magnitude less than with the current program. All of the governmental activity could be accomplished through a small trading office enacting the mechanical rules on one or a few futures trading exchanges as opposed to placing an enormous administrative burden on local Agricultural Stabilization Conservation Service offices all across the country.

Besides the target reserve stock, this policy has the simplicity of requiring a choice of only two parameters: the ratio of stock transactions to the price differential from target (the stock transaction ratio) and the ratio of target price adjustment to the reserve stock differential from target (the target price adjustment ratio). These parameters provide for both short-run and long-run flexibility of the policy in adapting to changing economic conditions. They also allow both short-run and long-run anticipation of farmers in their planning process.

The justification for all of the features of this proposed policy is beyond the scope and space of this paper. However, the success of the policy clearly depends on making the right choices of the above two policy parameters. The choice of the stock transaction ratio determines the trade-off between Treasury cost risk and private price risk. A high ratio is associated with high Treasury cost risk and low private price risk and vice versa. Too high a value of the ratio will lead to policy failure because of excessive Treasury risk, while too low a value will not reduce private price risk. Nevertheless, previous policy experiences give rich evidence for choice of an appropriate level.

The choice of the target price adjustment ratio determines the adaptability of the policy to changing economic conditions. Too low a ratio could lead to eventual government failure because of insufficient adaptability, while too high a ratio could make the policy adapt so quickly to changing conditions that no stabilizing influence is provided or that overshooting occurs. Again, however, previous policy experiences and market information exist so that reasonably good choices should be possible. Furthermore, the information required to select appropriate levels of the ratios, which define these policy equilibrium rules, is no different than that needed to formulate changes in traditional policy instruments. Finally, any particular levels of instruments required to enact the initial policy are not crucial to its success beyond the short run. For example, if the initial target price setting is too high, stocks will begin to accumulate beyond the reserve stock target and automatically draw the target price down.

## Concluding Remarks

The inherent instability and riskiness of the U. S. food and agriculture system is the market-failure justification for U. S. agricultural policy. The implementation of policies to address such market failures is often confronted with government failure. Political-administrative instabilities resulting from government failure can exceed the inherent instabilities of the private sector. This paper argues that an operational approach for dealing with both types of failure is the specification of appropriate conditional policies. Such policies must be designed to balance *Treasury cost uncertainty* with *policy risk* while transmitting sufficient *market signals* for long-term adjustment.

Very recently, economists have begun to realize the potential benefits of policy controls that are determined automatically by market conditions. Theoretical and empirical studies have been done to analyze the type of policies advocated here. In each of these studies, the conclusion has been reached that such policies dominate the fixed policies that have been pursued by and large since the 1930s for U. S. agriculture. Moreover, these studies are short run in nature and, therefore, do not take into account the additional benefits of longer term investment efficiencies that can be obtained from more orderly agricultural policies with built-in self-adjustments that can be well anticipated.

Operationally, the specification of smooth and orderly policy equilibrium rules depends on the risk-sharing arrangement between the public and private sectors. A particular conditional policy design implies a particular level of Treasury cost risk, a particular level of policy risk, and particular levels of output and price risk for the private sector. A fixed loan rate or target price can place much of the risk of fluctuating prices on the government (depending on levels) and is reflected in terms of Treasury cost exposure. If, however, the policy is designed so that both government and the private sector incur some of the losses when prices fall, the risk of market price variation is shared by government and the private sector. Such risk sharing would be preferred by farmers if the fact that government bears all of the risk means that it is more likely to "randomly" change policies. Increased stability for the government means a more stable policy environment.[6]

[6]Once a desirable risk-sharing arrangement has been determined, the frequency by which the conditional policies are allowed to self-adjust can be specified. For example, under the current set of policy instruments, acreage set-asides and target prices must be set on an annual basis well before any resource-allocation decisions are made by farmers, i.e., prior to planting time. All of the remaining instruments (e.g., loan rate, storage payments, interest rate subsidies, call and release prices, and public stock sales) could be allowed to adjust intraseasonally.

## *References*

Babb, E. M. "Commodity Policy Options: Dairy." Paper presented at the Conference on Alternative Agriculture and Food Policies and the 1985 Farm Bill. University of California, Giannini Foundation of Agricultural Economics, and Resources for the Future. Berkeley, June 11 and 12, 1984.

Becker, Gary S. "A Positive Theory of Redistribution of Income and Political Behavior." University of Chicago, CSES Working Paper (October, 1980).

Cochrane, W. W., and Y. Danin. *Reserve Stock Grain Models: The World and the United States, 1975–1985.* University of Minnesota, Minnesota Agricultural Experiment Station, Bulletin No. 305. St. Paul, 1976.

Cochrane, Willard W., and Mary E. Ryan. *American Farm Policy, 1948–1973.* Minneapolis: University of Minnesota Press, 1976.

Danin, Y. "Grain Reserves and Price Stabilization." University of Minnesota, Department of Agricultural and Applied Economics. St. Paul, December, 1975.

Firch, Robert S. "Commodity Policy Options: Cotton." Paper presented at the Conference on Alternative Agriculture and Food Policies and the 1985 Farm Bill. University of California, Giannini Foundation of Agricultural Economics, and Resources for the Future. Berkeley, June 11 and 12, 1984.

Freebairn, John W., Gordon C. Rausser, and Harry de Gorter. "Food and Agricultural Sector Linkages to the International and Domestic Macroeconomies." In *New Directions in Econometric Modeling and Forecasting in U. S. Agriculture,* edited by Gordon C. Rausser. Amsterdam: North-Holland Book Publishing Co., 1982.

Gardner, Bruce L. *The Governing of Agriculture.* Lawrence, Kansas: The Regents Press of Kansas, 1981.

———. "Efficient Redistribution Through Commodity Markets." *American Journal of Agricultural Economics,* Vol. 65, No. 2 (May, 1983):225–234.

Johnson, D. Gale. "Agriculture Policy Alternatives for the 1980s." In *Food and Agriculture Policies for the 1980s,* edited by D. Gale Johnson. Washington, D. C.: American Enterprise Institute, 1981, pp. 183–209.

———. "The Performance of Past Policies: A Critique." Paper presented at the Conference on Alternative Agriculture and Food Policies and the 1985 Farm Bill. University of California, Giannini Foundation of Agricultural Economics, and Resources for the Future. Berkeley, June 11 and 12, 1984.

Just, Richard E. *Farmer-Owned Grain Reserve Program Needs Modification to Improve Effectiveness: Theoretical and Empirical Considerations in Agricultural Buffer Stock Policy Under the Food and Ag-*

*riculture Act of 1977.* U. S. General Accounting Office, CED-81-70, Vol. 3. Washington, D.C., June 26, 1981.

Just, Richard E., Darrell L. Hueth, and Andrew Schmitz. *Applied Welfare Economics and Public Policy.* Englewood Cliffs, N. J.: Prentice-Hall, Inc., 1982.

Just, Richard E., and Andrew Schmitz. "The Instability Storage-Cost Trade off and Nonoptimality of Price Bands in Stabilization Policy." University of California, Department of Agricultural and Resource Economics. Berkeley, February, 1979.

Langley, James A. *Formulation and Policy Implications of Alternative Loan Rates and Target Prices for Major Agricultural Commodities.* U. S. Economic Research Service, National Economics Division, Food and Agricultural Policy Branch. Washington, D. C.: U. S. Government Printing Office, November, 1983.

Meyer, Jack. "The Transformation Approach to Stochastic Dominance: Preliminary Results." Paper presented at the Annual Meeting of the Southern Regional Research Project S-1980. New Orleans, Louisiana, March 25–28, 1984.

Nelson, Fredrick J., and Willard W. Cochrane. "Economic Consequences of Federal Farm Commodity Programs, 1953–72." *Agricultural Economics Research* 28, No. 2 (April, 1976):52–65.

Rausser, Gordon C. "Political Economic Markets: PESTS and PERTS in Food and Agriculture." *American Journal of Agricultural Economics* 64 No. 6 (December, 1982):821–833.

Rausser, Gordon C., and Eithan Hochman. *Dynamic Agricultural Systems: Economic Prediction and Control.* Amsterdam: North-Holland Book Publishing Co., 1979.

Rausser, Gordon C., and D. Peter Stonehouse. "Public Intervention and Producer Supply Response." *American Journal of Agricultural Economics* 60, No. 5 (December, 1978):885–890.

Spitze, Robert G. F. "Revisions of Existing Agricultural and Food Policy as an Alternative for 1985." Presentation at the University of Illinois Texas Agricultural Forum, April 12, 1984.

Wallace, T. D. "Measures of Social Costs of Agricultural Programs." *Journal of Farm Economics* 44, No. 2 (May, 1962):580–594.

Zwart, A. C., and K. D. Mielke. "Economic Implications of International Wheat Reserves." University of Guelph, School of Agriculture, Economics and Extension Education, Discussion Paper No. 1. Guelph, June, 1976.

# 4 • 5 DISCUSSION


## Lynn M. Daft*


In his chapter, Bruce L. Gardner examines several policy options for grains. The treatment of the interplay between supply and demand elasticities, including the elasticity of export demand (short term and long term) in the context of policy options, frames the options nicely. While the author takes a back-of-the-envelope approach, it is highly useful in that it quantifies essential elements of the outcome, thereby permitting comparisons across options.

I am in agreement substantially with many of Gardner's assumptions and observations regarding recent policy and future prospects. For example, though it sounds trite, it *is* important for farmers to look beyond price to income in judging their policy options. I would also agree that target price has had a stimulating effect on production and that the farmer-owned grain reserve has been mismanaged, often being used as a means of supporting price. I would agree, too, that the probability is high that most of the central elements of the existing farm program structure will be retained in the next farm bill, absent a significant further deterioration in the farm economy over the next nine months.

For me, the problems with this chapter begin with the assignment that was given the author, namely, the choice of options to be


*The author is Vice President, Abel, Daft & Early, Washington, D. C.

judged. While there is some utility in looking at policy outliers, there is a great deal more in evaluating options that are more serious candidates for adoption. Unfortunately, that excludes most of those that were discussed in this chapter.

For example, a continuation of the 1981 Act, replicated to the point that it includes mistakes and all, does not strike me as a plausible alternative—not so much because it includes mistakes (which are always a sure bet) but that it assumes they will be the same mistakes. Another option assumes continuation of the deficiency payments, the adoption of low loan rates, and discontinuation of all production controls. While the first two assumptions are probable, the third is not. Likewise, placing primary reliance on production controls alone, given the importance attached to increased farm exports, also appears unlikely.

An option of no deficiency payments, coupled with higher loan rates, seems equally remote for many of the same reasons. Finally, elimination of commodity programs completely is, perhaps, the most unlikely option of all. Thus, while I found the technique of analysis interesting and potentially useful, this particular application was not.

I was disappointed that the author did not devote at least some attention to the division of authority between the Legislative and Executive branches. (Both chapters are negligent in this respect.) Gardner's chapter devotes practically no attention to administrative discretion and the problems and opportunities to which it gives rise. Instead, there is an unstated presumption that the focus should be exclusively on the law and that execution of the law will somehow take care of itself. History, both recent and past, suggests otherwise.

Likewise, I would have preferred to have seen more discussion of the implications of alternative grain policies on interests other than agricultural producers. Although farmers still have a major say in the outcome of food and agricultural policy, there are many others—consumers, taxpayers, input suppliers, and exporters—who have an equally important and growing stake.

While I was not surprised that the author was in favor of a free-market approach, I was surprised at his justification for doing so. To be honest, he did not seem to have his heart in his work at this point. Among the arguments cited in favor of *laissez-faire* were that regulated markets are "dominated by narrow interest group politics," that "hardly any groups seem to like the results," and that "government is tending to get too deeply involved in farmers' decisions on production practices for farmers' own good." Having subjected the other options to analysis that was primarily economic, this justification is rather anemic by comparison.

Gardner concludes with a brief examination of the use of put options as a means of providing price insurance. While this concept might have something to offer down the road and I would encourage the author to further develop the idea, it is not likely to see early application. There are too many unanswered questions regarding the use of put options for them to be incorporated in policy even for day-to-day use as a marketing tool. Here I was disappointed that Gardner did not take the opportunity to examine other income insurance approaches that might offer greater near-term potential. The stabilization program that has operated in Canada for the past few years is an example. A review of the results of this program and its implications for use in the United States would have been most welcome.

The principal strength of the chapter by Richard E. Just and Gordon C. Rausser is its nontraditional view of commodity policy. The authors assume that the reduction of risk and uncertainty is the principal justification for commodity policy. After reviewing the recent track record, they find that commodity policy, itself, has often been a source of instability rather than a cure for it. They attribute much of this policy failure to the attempt by legislators to establish future legislation based on past economic conditions. When future conditions deviate from this assumed state, as they inevitably do, policy failure has resulted. To make matters worse, Congress has responded to these failures by changing policies with increasing frequency. This, of course, has only heightened the uncertainty facing farmers.

Stepping back from this situation, Just and Rausser conclude that legislators should stop trying to anticipate conditions and, instead, adopt policies that would respond automatically to changes in selected economic variables. As an illustration, they suggest that the government agree to purchase 1 million bushels of grain for every 1 cent per bushel that market price falls below a specified target price. Conversely, the government would sell 1 million bushels for every 1 cent the price is above a specified target price. The target price, in turn, would be automatically adjusted in response to changing economic conditions—conditions that might be represented by the difference between a desired stock level and actual stocks. The desired stock level, in turn, might be determined as a function of U. S. Treasury costs among other factors.

The point of departure for this chapter—namely, that the government, itself, has become a source of instability in agriculture—is well documented. The very sharp rise in the Commodity Credit Corporation and reserve grain stocks, the fiscal years 1982 and 1983 budget explosions, the announcement of six separate wheat programs over the past two years, the repeated changes in legislative authority, and the massive and costly payment-in-kind program are but a few examples of the problem. Thus, there is little question that the problem being addressed is both real and serious.

But what about the proposed solution? First, will the proposed approach perform as advertised? The authors argue that the adoption of a set of policy rules of the nature they propose would lessen farmer uncertainty. Yet, it is not clear to this reviewer how replacing unknown human judgments (i.e., future program decisions by the Secretary of Agriculture) with unknown economic conditions (i.e., future real U. S. Treasury costs, the target reserve stock level, etc.) will lessen farmer uncertainty. Can farmers and others with an interest in farm policy foresee these economic conditions with any greater clarity than they can foresee the Secretary's decisions? I doubt it. Furthermore, the Secretary's decisions are usually based in a substantial measure on the same economic variables—farm prices, farm income, stocks, inflation, and budget costs—that the authors propose to represent in their conditional rules.

Second, is the proposed approach feasible? Is it institutionally realistic? In this regard, I believe the proposal suffers from two basic faults. Simply put, it overestimates the potential contribution of economics and underestimates that of the political process. As the authors point out, conditional policies are only as good as the wisdom and foresight of those who define the rules. While I believe economics has an important contribution to make in the specification of such rules and while I applaud the authors' call for defining these rules more explicitly and following them more closely, I believe to set them in law is ill-advised for several reasons. First, those who would define the rules, as proposed here, will be legislators reacting to the usual interplay of interests. Thus, what is adopted ultimately might bear little resemblance to what analysts propose. It is this legislative process that has produced the many conditional policies of the past that the authors justifiably criticize.

Furthermore, it is extremely unlikely that any legislative body is going to be content to establish rules for an industry as politically sensitive and economically volatile as agriculture and then leave them unchanged for any length of time. I would argue that the opportunity for policy based on a relatively stable set of decision rules, primarily grounded in economic reasoning, is substantially greater through the preservation of administrative discretion than the approach proposed here. Commodity program decisions are, after all, political decisions. As long as the federal government maintains its present level of involvement in these markets, the decisions are likely to remain political rather than become programed in such a way that "the role of government is controlled mechanically by the conditioning variables."

Where do we go from here? I would encourage the authors to consider two possibilities. This chapter is, by design, conceptual. If the general proposition of the more widespread use of decision rules is to receive serious consideration, it must be presented in an empirical, real-world form. Assuming various rules had been in place in the past, how would results have differed from those that occurred and what are the implications of those differences? In conducting such an analysis, incidentially, it will be important to avoid treating agriculture as a closed system. The agricultural sector is increasingly driven by policy decisions made outside the food and agricultural policy arena. Though it is questionable whether agricultural policies as they have been framed in the past are capable of effectively dealing with these influences, they will be a part of the policy environment nevertheless.

A second possible direction is to consider the relationship between conditional policies and the policy process. No matter how great the idea, it must be accepted and implemented by this process if it is ever to become a reality. As I have implied above, at least some of our past agricultural policy shortcomings have stemmed more from failure in execution than from design. There is clearly room for improvement. There is also room for accommodation of the proposal to the realities of the policy process.

# 6 A 10-POINT POLICY AGENDA FOR THE FOOD AND AGRICULTURAL SECTOR OF THE UNITED STATES

**Willard W. Cochrane***

## *Introduction*

Farm policy has evolved into a food and agricultural policy over the past 50 years. In this evolution we have added some food programs and some export programs, but the central core of that policy has been and continues to be the commodity programs. We have tinkered with these commodity programs in each piece of new legislation; but, in the 1980s as in the 1930s, they are concerned primarily with managing supplies, supporting prices, and making payments to farmers.

I sense that there is a growing dissatisfaction with the existing food and agricultural policy, but our political system seems unable to effect any fundamental changes in that policy. Agricultural policy in the 1980s is held in the vice-like grip of powerful farm commodity groups. Thus, I am proposing today a 10-point policy agenda to help the body politic and its diverse political leaders to think about some revolutionary changes in food and agricultural policy.

---

*The author is Professor Emeritus, University of Minnesota.

## *Proposal*

That agenda is as follows:

1. Eliminate completely all of the commodity programs in the existing legislation. I would hope that this could be achieved in the 1985 farm legislation; but if that proves impossible, then as quickly as possible. The commodity programs no longer serve their intended purpose, namely, to provide effective economic assistance to small- and medium-sized farmers, as well as large farmers, in periods of farm depression. They have become the mechanism whereby large, already affluent farmers obtain monopoly privileges and gains for themselves—monopoly privileges and gains which become capitalized into land values contributing to the increased wealth of these particular farmers. Further, these programs in the 1980s have become prohibitively expensive.
2. Establish and operate a legitimate grain reserve program with the capacity to stabilize world market prices for grains around the concept of a long-run world equilibrium price. This is necessary to protect both producers and consumers in the domestic market since the food and agricultural sector swings back and forth in an unpredictable fashion between boom and bust. It is also the way to make the United States a reliable supplier in world markets.
3. Institute a system of long-term bilateral trade agreements between the United States and its principal agricultural importing partners. This is necessary to regularize and smooth out the export of agricultural commodities and to dampen the price-income gyrations of the food and agricultural sector. In this connection, we should stop our on-again, off-again practice of using food as a political weapon.
4. Institute a major program of subsidized credit designed to assist only the small- to medium-size farmers—to assist only those full-time farmers grossing less than $100,000 in 1984. These are the vulnerable farmers, these are the farmers who have been going out of business in droves, and these are the farmers who will be gone by 1990 if they do not receive significant economic help.
5. Continue to develop the crop insurance program with only modest government subsidy so that every farmer in the United States who wants to insure his operation against crop losses or failure has the opportunity to do so. This is one important way that eco-

nomic risks to farm operators can be reduced with only a small cost to the government.

6. Maintain a system of food consumption programs with the capacity to protect all persons living in poverty and other vulnerable groups from hunger and malnutrition. This, any rich, civilized society should be prepared to do.

7. Maintain an effective system of food inspection, regulation, and nutrition research and education. A modern, complex, technologically advanced society must depend on its government to protect it from impure, contaminated, and adulterated food. The members of such a society must also depend on their government to assist them in choosing healthful diets and to protect them against fads, false advertising, and special interest promotions.

8. Develop and maintain a soil conservation program with the capacity to protect and conserve the soil and water resources of the nation in perpetuity. In a highly competitive environment, such as the United States farm economy, private enterprisers either cannot or will not undertake the necessary conservation practices. Thus, society, if it is desirous of conserving its soil and water resources, must undertake and bear the costs of an effective conservation program.

9. Develop and maintain an effective farm labor program with minimum wage floors, minimum housing and health standards, educational programs for the children of migrant workers, and control over the importation of foreign labor. This, again, is something that a rich, civilized country, such as the United States, must be prepared to do if it is to be judged a truly civilized country.

10. Overhaul the tax system so that the farm sector is no longer the prime tax shelter target of the economy. This is necessary to help slow down the resource concentration process in farming and to assist legitimate full-time farmers to compete successfully in farming operations.

In conclusion, the purpose of this 10-point program is to provide an effective, long-run economic safety net under participants in the food and agricultural sector—it is not to provide windfall or monopoly gains to any group of participants—and to do all of this at a minimum cost to the government.

# 7 REVIEW AND ASSESSMENT OF ALTERNATIVE AGRICULTURAL POLICY PROPOSALS

**Linda S. Calvin**
**William E. Foster**
**Gordon C. Rausser***

The two central topics that this chapter investigates are (1) the criteria upon which alternative proposals for agricultural policy may be judged and (2) the examination, comparison, and ranking of several alternatives to the present system of farm programs. The analysis of the goals and means of governmental intervention must necessarily begin with an understanding of the underlying problems that policies are designed to solve or mitigate (McCalla, 1984). The choice between policies, whether based on social ideals or political considerations, is a subjective matter of importance but one left too often unattended by agricultural economists.

Without an explicit exposition of the basic objectives of farm policy and the trade-offs between objectives, analyses of alternative programs and schemes are left incomplete. In many studies, only

*The authors are, respectively, Economist, U. S. Department of Agriculture; Research Assistant, Department of Agricultural and Resource Economics, University of California, Berkeley; and Chairman and Professor of Agricultural and Resource Economics and Economist in the Agricultural Experiment Station and on the Giannini Foundation, University of California, Berkeley.

some of the numerous and possibly conflicting objectives are considered; the obvious or advertised objectives of policies are examined; and those lightly weighted are ignored. But more importantly, the ability to judge alternatives is diminished when the analyses are based on differing, implicit criteria. A broad range of policy proposals is presently under consideration in anticipation of the 1985 farm bill and examining them, in the context of specific objectives, would serve to further the evaluation of public action in agricultural markets.

Our evaluation of agricultural policy is divided into five parts. First, we discuss the possible criteria with which the policy alternatives considered may be compared. A distinction is drawn between the normative aspects of public policy and the political (or interest group) aspects. By normative, we mean the "ought to be"—the correction of market failure, or the provision of public goods; by political, we mean the "probably will be"—the allocation of benefits to the politically powerful. Attention is focused on two major concerns: the mitigation of market failure and the avoidance of political failure.[1] The inability of unregulated markets to solve socially intolerable problems often motivates the government's intervention in the economy, but the political system itself is subject to failure. Policies which began to accomplish one task may not be effective or, worse, they may aggravate the original problem.

The second section offers a broad review of the alternatives available for future farm policy. Alternatives are briefly presented by objectives and instruments. A more detailed description of each policy option is given in the third section. We relate each alternative to the social criteria discussed in the first section and compare each to the system of existing programs in terms of the winners and losers. We also present the operational features of the alternatives, noting whether or not they will work, and their susceptibility to

[1]For a thorough discussion of market and political (or nonmarket) failures, see Wolf (1979).

political failure. The policy alternatives are ranked according to the criteria of interest in the fourth section. Finally, our conclusions are presented in the last section.

## *Evaluation Criteria*

### Social Goals

When judging between various policies that a government may institute to influence the agricultural sector, one must distinguish what ought to be the criteria from what will be. Normative standards yield different results than purely positive analyses of the political system. We consider five normative criteria of importance. There are, of course, many more possible social objectives for farm policy, such as to insure the efficient conservation of resources, or to provide variety and choice to the consuming public.[2] We judge the five criteria discussed in this section as those most pertinent for current considerations of future agricultural policy. Once these objectives—along with the objective to avoid political failure (discussed below)—are introduced, policy alternatives can be compared.

**Redistribution of Wealth.**—Traditionally, our society has paid much popular attention to the ideals of a fair distribution of wealth: Equity is good; disparity is bad. The historical "farm problem" has been repeatedly characterized by economists and other interested observers as the disadvantaged economic position of farmers. This notion once served as the single most important justification for state action to redistribute society's wealth to the agricultural sector.[3] Agricultural economists, more often than not, accept the goal of redistribution without comment, focusing their analyses on comparing the various means of attaining this goal (e.g., Wallace, 1962, and Gardner, 1983).

---

[2]For reviews of several rationalizations for farm policy, see Tweeten (1979) and Paarlberg (1980). For an approach closer to our own, see Lee (1981).

[3]For instance, see the discussions of past policies in Brandow (1977) and Penn (1981).

The redistribution argument for farm programs is less relevant today.[4] Perhaps, the more essential criterion in judging policies is the fairness in making the rules of the game—not fairness in final results. Nevertheless, a common implicit assumption is that wealth offers more opportunities to create wealth under the set of social and economic rules now existing. A typical criticism of agricultural programs is that access to benefits depends on farm size which, in turn, leads to increasingly larger farming operations.[5]

**Food Security.**—There is a certain public good aspect to the federal government's participation in commodity storage. Society deems important the preservation of adequate food reserves which the private market may be unwilling to guarantee.[6] First, public welfare may be enhanced not only by assurances of supply for this country but by the availability of grain stocks in the event of foreign crop failures, wars, and other catastrophes. Second, in the case of food price inflation, such stocks provide insurance to U. S. consumers that rapid increases in food prices can be moderated. Private enterprises engaged in the production and allocation of food supplies may lack the incentives necessary to maintain safe levels of stored grain because they are facing a stream of demands over time that are based on the actions of individual agents.

Typically, government programs rely on the manipulation of stocks to affect the price of commodities; but there is, perhaps, an unintended provision of food security when grain storage exceeds what would otherwise result. When debating and structuring farm

[4]Indeed, in farm policy debates, the term "poverty" has evolved into cash poor and economically disadvantaged into financially exposed (see the criterion regarding risk and uncertainty below). While there are still poor farmers, there are many more poor nonfarmers not receiving government support based on their occupations. As a method of transferring wealth to the poor, programs benefiting persons, because they are farmers, seem to offer only a means of keeping those in need tied to their poor farms.

[5]For example, the Interreligious Taskforce on U. S. Food Policy (1983) is one group that contends current programs are hastening the disappearance of the family farm while failing to efficiently meet its primary goal of equity. Economic evidence—that benefits of past programs accrue mainly to large operations—can be found in Lin, Johnson, and Calvin (1981); see theoretical arguments in Rausser, Zilberman, and Just (1983).

[6]For a review of the motivations for a policy of reserve stocks to insure against production shortages, see Cochrane (1974).

policy, some appraisal must be made of the socially optimal level of stored grain; and alternative programs should be judged, in part, by their attention to this question of safe stock levels. The degree of food security may be measured simply by the amount of stored grain in existence at any time. A more subtle and appropriate measure would account also for anticipation of future weather patterns and political and economic conditions throughout the world.

**Risk Sharing.**—The existence of market failures often rationalizes governmental intervention in the marketplace. In particular, the stochastic character of both commodity prices and production is offered to justify public policies aimed at the agricultural sector. A market failure arises from the inability of farmers to adequately trade their risks to other agents in the economy and from the divergence of social interests from actual farmer response to uncertainty. The tendency of a free economy to yield different results than socially optimal has been at least tacitly recognized by policymakers. Farm policies, such as price-stabilization schemes and crop insurance, are designed, in part, to affect directly the ability of the agricultural sector to cope with, and respond to, the capricious nature of its physical and economic environment.

Farming operations have become increasingly reliant on outside sources for their financing and inputs (both material and labor), adding to their sources of risk and uncertainty (see, for instance, Shepard and Collins, 1982, and Harrington *et al.*, 1983). As the nature of agriculture has changed, the exposure to more risks has led to apprehension regarding the cash-flow and debt-asset problems of farmers. The focus of public concern has shifted from agriculture's relative poverty to the difficulties of managing in a risky environment without sufficient means of insurance.

The market failure associated with risk and uncertainty is among the most persuasive rationales for an active farm policy. This, of course, is not a necessary and sufficient condition for governmental intervention in food and agriculture (Randall, 1984). Of the several normative standards, however, by which policy alternatives are judged, the correction of this problem should be weighted heavily. Existing and proposed policies should be considered in light of their potential effect on both income and price variability.

**Flexible Agricultural Sector.**—An often ignored but crucial objective of policy is to insure the ability of the agricultural sector to adapt to changing economic conditions. This is related somewhat to the issues of risk insurance and food security, but another dimension to policymaking may be the desire to avoid future farm policy crises associated with government failure. When production and supply decisions in this country are divorced from the underlying market forces determining demand and production in the rest of the world, conditions arise to pressure U. S. agriculture to become more responsive to the marketplace. Eventually, the market asserts itself, sometimes with spectacular results.[7]

It is most important that alternative programs be judged on their influence on the adaptability of agriculture to signals from the market. An easily observed and reliable measure of flexibility is the deviation of per-unit returns (including government payments and guarantees) from actual world prices. World prices may imperfectly mirror long-run conditions. However, the persistent deviation of farmer returns from market-determined prices would lead to an industry structure and system of resource use that leaves agriculture vulnerable to dislocation and crisis if policies were suddenly unable to isolate it from world market conditions.[8]

**Minimizing Treasury Costs.**—An additional social goal to be considered in the construction of farm policy is the minimization of Treasury outlays. Many of the objectives discussed above can be reached given sufficiently large expenditures of tax dollars. There is, of course, a political limit to the amount of funds that can be spent on addressing agricultural issues. Farm legislation is based on the support of the nonfarm population whose acquiescence depends on the financial cost of programs: A greater proportion of resources channeled to the farm sector means less attention to problems of other social groups. While current programs seem costly, the Treasury's exposure under innovative alternatives is uncertain, perhaps even more costly. The expected dollar value of federal expenditures for agriculture is a simple measure for evaluating policy proposals

---

[7]A recent example is the accumulation of massive stocks during the 1960s and 1970s induced by the federal government's attempt to insulate the domestic feed and food grain markets from world market forces. The heavy subsidization of Soviet grain purchases in 1972 was partially responsible for the large price increases that occurred in 1973.

[8]Using this measure, one could predict, for example, a future series of crises for U. S. sugar policy.

according to this objective of minimizing costs. It may be appropriate to gage proposals by other moments of the distribution of government costs as well (Just and Rausser, 1984).

## Political Failures

In addition to solving market failure, agricultural policy must mitigate the effects of political failure. Political failure is the tendency of the legislative process to produce policies that do not lead to Pareto superior outcomes. Political markets induce politicians to consider personal, not public, benefits and costs. As a result, the existence of market failure is a necessary but not sufficient condition for government intervention. A sufficient condition is that the loss of economic efficiency in the case of the uncorrected market failure is greater than the loss under the government remedy which is influenced by the existence of political failure.

Political failure has two important effects. First, a policy may be selected that does not solve market problems in an efficient manner but, rather, contributes to the short-run goals of politicians. This is the most obvious result of political failure—failure in choice.[9] Economists can do very little to solve this problem other than try to inform the public and politicians about available policy choices. The second result, failure in implementation of a policy, is more subtle. Over time, policies may be modified to serve political concerns.[10] Economists should recognize this potential, as well as the additional costs of political failure in implementation, and design proposals that both alleviate market failures and mitigate and avoid political failures. If economists do not consider the possibility of political failure in the implementation of particular policies, relevant costs are ig-

---

[9]We cite, as an example of failure in choice, Tweeten's (1979, p. 465) remarks concerning policies in the late 1950s:

> "A Democratic Congress and Republican Secretary of Agriculture Benson concocted an unworkable combination: price supports with ineffective controls at an inopportune time when the technological revolution had struck agriculture full force. The result was unconscionable levels of stocks ... A major shift in policies was clearly needed."

[10]The mere existence of a policy that raises income or wealth attracts efforts to increase that transfer; see Rausser (1982).

nored; and the choice between programs may be incorrect. The recent history of the farmer-owned reserve illustrates the consequence of political failure in implementation. Reserve loan rates are intended to stabilize prices, not to raise them; but in 1981–82, loan rates were raised above market-clearing prices in response to political pressures from farmers faced with low returns. Raising the loan rate due to a political crisis led to excessive stock accumulation and eventually contributed to the need for the costly payment-in-kind (PIK) program. In the consideration of farm bill alternatives in 1981, the possibility of increasing reserve loan rates to raise incomes should have been considered a relevant cost. Proposals for the 1985 farm bill should offer policy tools that minimize the potential for political failure in implementation.

## *An Overview of Proposed Policies*

In this section we briefly review the major proposals for agricultural policy. Most of these policy alternatives have been used before in some form. Specific policies are discussed in more detail in the third section. We categorize the current and proposed policies both by type of program and social objectives.

Figure 1 shows the general scheme of possible government intervention.[11] The type of intervention depends on perceived problems which change over time. If the public decides that commodity markets are not functioning correctly, it can introduce programs to correct or compensate for the shortcomings of the marketplace. Since 1933, the government has used agricultural programs to increase farm income. The broad categories of policies that have been used to meet this goal have been supply contraction, demand expansion, and direct income enhancement. Many of the recent policy proposals emphasize reinforcing the operation of the free market. This has coincided with the emergence of the idea that the farm problem is no longer low income but variability of income. These proposed policies aim to improve the ability of farmers to operate within the market environment more effectively. Alternatively, so-

[11]A similar scheme of policies is found in Tweeten (1979).

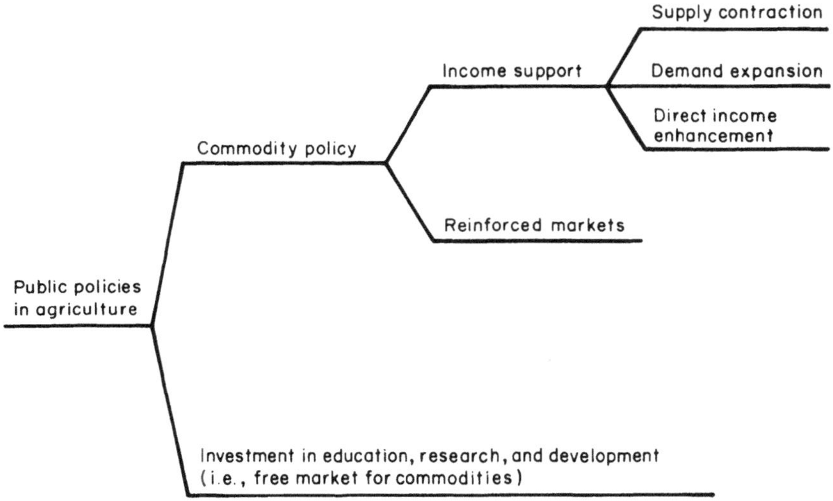


FIGURE 1. Decision Tree Showing General Types of Agricultural Policies

ciety may decide that commodity markets function well. In this case the government may intervene but only to provide traditional public goods, such as education, and research and development. This was essentially the situation in the United States until the 1930s, and current political trends may very well return agriculture to that state.

In Table 1 current and proposed policy tools are arranged both by type and objective. For most objectives, there are several types of policies that could be used; for each type (e.g., supply contraction), there are several possible instruments. (Current instruments are marked with an asterisk.) We now turn to an overview of these broad policy types.

**TABLE 1 Categorization of Proposals by Objective[a]**

| Proposals \ Objectives | Wealth redistribution | Food security | Risk reduction | Flexible agricultural sector | Minimizing Treasury costs | Avoidance of political failure |
|---|---|---|---|---|---|---|
| Free markets | no | no | no | yes | yes | yes |
| *Reinforced free markets* | | | | | | |
| Revenue insurance | no | no | yes | yes | ? | yes |
| Futures markets | no | no | yes | yes | yes | yes |
| Commodity options | no | no | yes | yes | yes | yes |
| Flexible storage | no | yes | yes | yes | yes | yes |
| *Supply contraction* | | | | | | |
| Voluntary set-aside* | yes | no | ? | no | no | no |
| *Mandatory* | | | | | | |
| Tax excess production | yes | no | no | no | yes | no |
| *Demand expansion* | | | | | | |
| Export subsidies* | yes | no | ? | no | no | no |
| CCC loans* | yes | yes | yes | no | no | no |
| Flexible loans | no | yes | yes | yes | yes | yes |
| *Direct income enhancement* | | | | | | |
| Deficiency payments* | yes | no | yes | no | no | no |
| Negative income tax | yes | no | yes | yes | yes | yes |
| Sector income guarantee | yes | | yes | no | yes | yes |

[a] Those proposals that meet an objective are marked "yes" and those that do not are marked "no." When it is uncertain whether a policy meets an objective, it is marked with a question mark.

* Current instruments.

Source: Computed.

The proposed alternative, which offers the most extreme divergence from current policy, is the free market or no programs at all. Prior to the depression of the 1930s, no direct intervention was the rule. Over the years, public dissatisfaction with particular market results has produced a patchwork of government programs to regulate the marketplace. The deregulation of agriculture would eliminate the costs of commodity programs, give farmers complete autonomy of choice, and force the sector to adjust to market signals—enhancing flexibility. A completely free market, however, may exacerbate the problems of equity, market-induced risk, and food security. The uncertainty associated with unpredictable public intervention would be decreased, and the opportunity for political failure of implementation would not exist.

Many proposed alternatives deal with tools to reinforce the operation of markets by improving the farmer's ability to trade risk. They include revenue insurance proposals and suggestions to improve forward-contracting markets such as providing subsidized put options. Flexible storage tools have also been proposed to moderate price and food stock fluctuations without interfering with long-run market conditions. Unlike current and past programs, they would not be as susceptible to political failure due to their reliance on rules and not discretion. All of these proposed policies would reduce income variance and maintain farmer autonomy in production and marketing. Most would be less costly than current programs, although revenue insurance may be an exception. Reinforcing the market would avoid some of the problems of complete *laissez-faire*; but relative to no programs, it would be susceptible to political failure. Also, there is no guarantee that the goals of equity would be achieved.

Under reinforced market policies, farmers would face less risk and uncertainty; but they would have to accept long-run market conditions. Proposed programs to contract supply and expand demand seek to alter both the returns to (and risk profile of) agriculture by manipulating the market. Proposed direct income enhancement would provide a nonmarket means of transferring wealth and could also be used in conjunction with other programs. The remainder of this section reviews these proposals for redistribution.

Different methods of supply contraction, both mandatory and voluntary, have been used since 1933 in efforts to increase farm returns. Restrictions on supply can be used alone to raise market prices or to counter the incentives to increase output provided by production-based income support policies. The voluntary acreage set-aside program is the current policy tool; but due to free-rider and slippage problems, it does not effectively reduce production. It has proved, therefore, a rather expensive program given the limited results.

Alternative proposals to mandatorily restrict supply are introduced in order to reduce government costs and increase effectiveness. These proposals would continue attempts to support agriculture at socially fair levels. Used alone, they would neither reduce price variability nor contribute to food security. They would, however, maintain the barriers to efficient response to world market conditions. It is unlikely that farmers would accept supply contraction without some additional form of income support as an inducement—a feature of past policy. The management of programs restricting supply and raising incomes is subject to intense pressure from segments of the agricultural sector, undermining the social goals these programs are designed to help achieve. Hence, there is little reason to believe that these proposals would reduce the potential for political failure.

Government policies to expand the effective demand for agricultural products were also initiated in 1933 with the introduction of nonrecourse loans. Outward shifts in demand can be accomplished by increases in domestic or foreign consumer demand and by the government acting as a residual demander—targeting price levels below which it acts to absorb supply. State-supported demand has typically been in the form of stock accumulation, or subsidized consumption, as in the case of food stamps and P. L. 480. Government storage to raise prices interferes with the performance of export markets, making U. S. commodities uncompetitive and disrupting trade channels that have taken time to build.

Two alternatives to correct this problem have been proposed: (1) increase the use of export subsidies to make storage programs compatible with world competition[12] or (2) minimize the influence of aspects of storage programs that isolate U. S. prices from world

[12]This proposal was in effect during the 1960s.

prices by making loan programs flexible and responsive to world economic conditions. The first alternative is an example of political failure in choice.[13] It postpones necessary adjustment to market conditions since there is a limit to the government's ability to subsidize exports. Subsidies are more useful when they are clearly a short-term remedy—not a solution to noncompetitive prices. The second alternative is a more long-range solution to the problem of conflict between domestic and export policy goals; domestic policy adjusts to the world market instead of counteracting it at great cost. This program is designed to avoid political failure in implementaton. Export expansion directly benefits farmers in proportion to their production through increases in market prices. Since integration into the world market exposes U. S. farmers to international price shocks, those farmers who have no other sources of income may face increased financial difficulties. Both alternatives provide storage programs which are particularly important when demand varies significantly due to unpredictable foreign needs.

The current direct income enhancement tool—the target price/deficiency payment scheme—was introduced in 1973 as a means to separate price and income support. Farmers who participate in the program are guaranteed a particular price which is frequently above the expected market price, leading to an inefficient allocation of resources. In addition to increasing the returns to farmers, the program eliminates downside risk and reduces the variance of returns. Proposed alternatives attempt to enhance income at the individual and sector level without changing the production incentives of farmers. Ideally, farmers would respond only to market prices and receive payments from the government to compensate for very low incomes. A negative income tax for farmers is an appealing proposal. Whether these alternatives are more expensive than current programs depends on the targeted income level and the cost of any price manipulation policies that might be used in conjunction with an income enhancement policy to mitigate the size of direct payments. The level of current income transfer would probably be accomplished at a lower cost (administrative plus welfare) with direct payments. In addition to reducing the cost of direct

[13]For a discussion of political interests and trade policy, see Hathaway (1983).

payment programs, a storage policy also can be used to assure safe food stocks.

Direct income enhancement tools can be used to foster equity goals. While all farmers can use the production-based programs, larger farmers and those that own land benefit most. Even programs that are production based can be modified to target benefits to farms of certain characteristics.[14] When the benefits are tied to the individual instead of the land, every farmer has an equal chance of benefiting from the program regardless of farm size or tenure. Policies that divorce the marketplace from a farmer's ability to obtain an adequate living prevent the sector from adapting to economic conditions. Preventing natural industry attrition likely hampers the flexibility of agriculture by encouraging outmoded methods and discouraging the adoption of competitive practices.[15] Any income transfer is subject to intense political pressure to increase that transfer. However, if the current programs were eliminated and replaced with only a direct-income enhancement policy, it would be subject to much scrutiny; and the likelihood of political failure would be reduced.

# *Alternative Policy Proposals*

## Free Markets

Historically, the inability of an unfettered market to attain the social objectives discussed above has been the principal justification for federal intervention in agriculture. Dissatisfaction with governmental policies and programs, however, has called into question whether tampering with the market leads to worse problems than those which motivated public action in the first place; and, as political attitudes and goals have changed, there has been a growing

[14]A proposal by the National Grange (Miller, 1983) suggests that the government should buy commodities for stock purposes and food programs from certain categories of farmers, such as beginning farmers, and pay them a higher than market price. Other farmers would have to rely on the market alone.

[15]While farm bankruptcy rates receive wide public attention, the rate of failure for farms is much lower than that of other small businesses (Thompson, 1984).

respect for free-market outcomes. Unsuccessful governmental intervention may be traced either to the technical infeasibility of a public policy or to an adulterated translation of social goals into political reality—the result of political failure. Whether for good or bad, future policy proposals will be more critically judged relative to the expected results of unregulated private enterprise.

Deregulating agriculture would shift the distribution of wealth from one determined in part by the ability of farmers to take advantage of federal programs to one determined primarily by the ability to efficiently manage and produce in a free-market environment.[16] Inefficient producers certainly would be pressured to leave the sector; and their wealth, accumulated as a result of the capitalization of government-sponsored rents into land prices, would deteriorate. Efficient managers would reap most of the benefits of a policy of inaction, and consumers would be relieved of supported prices. If farming exhibits increasing returns to scale, then the tendency to large operations would be accelerated. There is, however, good reason to suspect increasing returns to scale in the production of government benefits; therefore, the structure of agriculture may, in fact, move to smaller units under deregulation.[17]

The most drastic effect of a free market may be on the variability of farm incomes. Present programs isolate agriculture from severe changes in the economy and harvests, and eliminating present policies would result in increased uncertainty for producers and consumers. The uncertainty in agriculture, however, is not due only to

---

[16]The idea that government programs are an input in the management of a successful farming business is illustrated by the recent comment of a farmer and director of Farmland Industries: "I think Farmland's government affairs division is the way to go. People realize that government has an input in most of everything we do these days, and if we want government to work for us then we are going to have to take part in the process" (*Farmland News*, 1984). Similarly, at a recent Western Agricultural Economics Association Council meeting (January 6, 1984), a number of industry leaders supported this perspective. They argued that the greatest returns to commodity group associations could be obtained by direct involvement in the political process.

[17]Smaller operations may have a cost advantage in coping with risk and uncertainty. If this is so, not only do large operations gain more than smaller ones from government programs because benefits are tied to production but there is an additional opportunity afforded to large farms in an environment where the public absorbs much of the risk.

weather and the working of the marketplace; there is also the uncertainty associated with the political system of farm programs. There is a trade to be made between market uncertainty and policy uncertainty.[18]

Whether or not private storage would compensate for the reduction or loss of government-sponsored storage is an empirical question, but the absence of deficiency payments and loan rates would almost certainly create greater instability in supply. This point also relates to the objective of food security. Unregulated markets would not take into account socially optimal stock levels. If private storage does not completely replace government storage, a policy of free enterprise may decrease the provision of adequate, safe food reserves.

Moving to a free market for agriculture would increase the sector's capability to respond to long-run changes in economic conditions. Changes in market conditions—in this country as well as the rest of the world—would be quickly reflected in the returns to farming. Therefore, a more flexible and efficient agricultural sector would be encouraged, avoiding occasional policy crises associated with governmental attempts to shield farming from world market conditions. In addition, the ability of U. S. agriculture to efficiently compete in world markets would be enhanced by the elimination of policies that maintain artificially high prices.

As we have noted above, past policies that attempted to avoid the outcomes associated with free enterprise have been subject to political failures. A policy of free markets would avoid these additional costs. There would no longer be the expenditure of great amounts of human and financial resources for the influence of public policy. Successful producers would be determined by their productive, not political, efficiency. In addition to avoiding political failure of implementation, a large burden on federal expenditures would be eliminated.

## Reinforced Free Markets

The disadvantages associated with eliminating federal involvement in agriculture have led to several proposals designed to correct

---

[18]For a discussion of this trade-off, see Just and Rausser (1984) and Rausser and Stonehouse (1978).

the deficiencies of a free market. The major problem of complete deregulation is the increased volatility of prices and farm incomes that would result. Proposed policies addressing this issue take two forms: governmental sponsorship of risk-trading institutions and direct governmental involvement in dampening shocks through commodity storage. Both forms of policy would maintain the essential advantages of nonintervention while reducing the degree of uncertainty facing producers.

Personalized insurance against adverse crop yields are now available. Insurance against adverse price movements has been provided by government loan-for-storage and deficiency-payments programs. These programs have not only set a floor on prices but have been used as income supports as well, isolating agriculture from the realities of the marketplace. Moreover, past programs have not been designed for individual farmers nor have they attempted to charge producers for society's cost of absorbing risk.[19]

Futures markets do afford farmers hedging opportunities. Nevertheless, futures markets have proven unpopular with farmers due to the short term of existing contracts, imperfect capital markets, and the unattractive degree of exposure to margin calls. The use of futures contracts can be encouraged by lengthening contracts to give price protection for two to three years. (The market, however, would be extremely thin for farm contracts.) In addition, the government could directly intervene in futures markets, reducing the variability of contract prices and, thus, decreasing the exposure of everyone in the market. Put options have been suggested as a better vehicle for farmers wishing to hedge against price decreases (Gardner, 1977, and Schertz and Clayton, 1983). First, purchasing options would remove farmers' exposure to a long string of margin calls. Second, the premiums the farmer would have to pay for price insurance could be subsidized to encourage options use.

A policy of nonintervention in agricultural markets with the reinforcement of market incentives to transfer risk would mitigate the problems of uncertainty facing producers. Supplies and prices, however, would not be directly stabilized. It has been proposed that the government should reduce market shocks by a policy of public stor-

[19]The dairy PIK program is a current exception.

age or by subsidizing private storage, thereby decreasing instability facing producers and consumers. The success of a policy of government-held stocks would depend on the response of private storers. Subsidizing private stocks could be made flexible—the level of subsidy being dependent on the level of stocks, prices, world market conditions, and the like. In addition to the beneficial effects on instability, a storage policy aimed at stabilization would tend to achieve the objective of food security.

Any attempt to reinforce the workings of a free market is necessarily subject to political failures. For example, once in place, a system subsidizing the use of put options for farmers would carry the potential for subsidies to be used to transfer wealth to agriculture. In some sense, overly cheap insurance would induce overly risky production behavior by farmers. As another example, establishing stock subsidies for price stability and food security also leaves open their later political manipulation as a means of enriching storers.

## Revenue Insurance

Revenue insurance is an appealing idea in theory and has attracted much attention, but proposals for this plan are still quite general.[20] A farmer would choose to insure gross revenues at some level—the premiums being based on this level and on farmer and farm characteristics. Revenue insurance would be a more individualized risk management tool than the present system of programs. Using private benefits and costs, a farmer would select the level of insurance desired rather than having to take a package of loan rates, diversion requirements, payment limitations, and other program restrictions.

Farm revenue is difficult to insure. The lack of independence of losses is the most severe problem; if the price is low for one farmer, it is low for all farmers. This exposes an insurance company to great

---

[20]Recent papers (Schuh, 1980; Congressional Budget Office, 1983 and 1984; Schertz and Clayton, 1983) have advocated or examined the idea of revenue insurance. Crop insurance is privately and publicly provided, but efforts to insure prices and yields simultaneously have not succeeded. In 1920 the Hartford Company did provide price and yield insurance although, when prices fell sharply and the company lost $1.7 million, it abandoned the revenue insurance business (Offutt and Lins, 1983).

financial risk and makes large premiums imperative. Moral hazard further complicates the insurance problem: Once a farmer insures his revenue, there is little incentive to allocate past combinations of resources to maintain typical revenues. For example, to avoid some of the problems of moral hazard, premiums could be made contingent on the continuation of historical average yields. Revenue insurance would also have to contend with adverse selection. Insurance would be most attractive to those who are least adept at risk management; and those experienced at using other means to manage risk, such as the futures market, might not purchase revenue insurance. Despite these difficulties, private insurers could offer revenue insurance; but high premiums would severely limit farmer participation.

Those who advocate revenue insurance believe it would be a less expensive and more efficient method of stabilizing revenues than the existing commodity policies, but either the insurers or farmers who buy revenue insurance must be subsidized to make revenue insurance a financially attractive alternative to farmers. The cost of this subsidy has not been estimated and could be quite large. The government could subsidize producers directly (which would entail large administrative cost), subsidize insurance companies, or produce insurance itself.[21] Administrative costs could be reduced if the existing (yield) insurers were used to provide revenue insurance.

Revenue insurance would be a difficult policy to implement. It requires development of better actuarial data on prices and a complicated program structure to minimize the problems of moral hazard and adverse selection. As discussed above, the existence of a subsidy presents the opportunity for political failure. If private insurers were used to provide revenue insurance, political failure might increase as one more interest group becomes involved. We doubt that a revenue insurance scheme could be devised that stabilizes revenues at present levels with a lower cost to the government than current commodity programs.

With revenue insurance, farmers would lose due to lower prices. A decrease in revenue variability, however, would increase farmers'

---

[21]One proposal suggests that private insurance companies could use commodity options as a means to minimize their risk (Schertz and Clayton, 1983). In this case, the government could subsidize the writers of options.

access to capital markets and promote investment and adoption of new technologies which would lower costs. Farmers would also have more freedom to choose the level, if any, of revenue insurance they desired. Rural communities and agriculture-related industries, such as farm machinery dealers, would also benefit from more stability. The effects of farmer income distribution are uncertain, depending on farmers' abilities to benefit from insurance. One attractive advantage of revenue insurance over current commodity programs is that it is possible to target different levels of subsidy to particular farmers. Farmers with certain characteristics, which would be known to the insurer, would pay lower premiums for a particular level of coverage. For example, middle-sized farms that are perceived to have special problems competing with large commercial farms might receive a larger subsidy for the purchase of insurance.

## Flexible Storage Policies

Current commodity storage programs are intended to moderate price fluctuations by the accumulation of government stocks or by the subsidization of private storers. Traditional stabilization tools, however, have been notoriously unresponsive to market signals, burdening farm policy with costly and sometimes embarrassing levels of stocks. In response to this problem, several alternative proposals to the present system of storage programs argue for flexible (that is, market conditioned) storage policies (Miller and Sharples, 1979; Schuh, 1983; Congressional Budget Office, 1984; Just, 1981; Just and Rausser, 1984; and Langley, 1983). Policy tools would be responsive to market signals and vary depending on the economic environment.

When program instruments are left unadjusted as the economic environment changes, a policy disequilibrium develops: The tools are no longer appropriate for, and perhaps contrary to, the original policy objectives. This leads to a policy crisis when dissatisfaction with either program benefits or costs is so widespread that a change in policy becomes inevitable. Proposals for flexible policies, however, introduce program instruments that would change as the economy changed and, therefore, help avoid policy crises. If the 1981 farm bill had included flexible policies, for instance, loan rates could have been conditioned on variables such as stock levels, prices, etc. If loan rates had been a known function of prices and stocks, they would have declined as world price fell in the early 1980s and storage levels increased. The United States would have avoided the buildup in stocks and the disorienting effect of a major change in policy (PIK).

Most of the proposals for flexible storage use current policies tools, such as loan rates, but attempt to reduce the effects of political failure by restructuring the procedure for selecting the setting of tools. Instruments would be specific and well-known functions of commodity price, farm income, and other economic variables. The formulas for determining the setting of policy tools, such as loan rates, would be known to all—farmers, storers, and consumers. If farmers anticipate the adjustment process when making their investment decisions, agricultural production would attain greater economic efficiency with less risk. In fact, with more efficient investment in the farm sector, lower prices could lead to the same levels of income.

Another objective of policy is to avoid excessive government costs because this, too, can lead to abrupt changes in policies (Just and Rausser, 1984). In an open market, there is a trade-off between price stabilization and minimizing government costs. The alternative proposed by Just and Rausser would integrate both of these concerns in a government storage program. A target level of stocks would be determined based on Treasury costs and the need for safe stocks. The government would determine a target price and buy or sell a certain amount of a commodity for every 1 percent decrease or increase in price around the target level. Conditioning the target price on the level of stocks would reflect changes in the economic environment and avoid unmanageable spreads between world prices and target levels. The costs associated with government storage could be reduced, perhaps eliminated, with the use of the futures markets to acquire commitments to buy or sell rather than actually engaging in physical transactions.

When designing a new policy, the natural tendency is to leave ample discretion in case the policies do not respond to changes in the economic environment in the anticipated manner. But discretion is what must be minimized in a flexible policy scenario, or government failure in implementation is likely to result. Devising appropriate adjustment rules would be a formidable task, requiring legislative attention to current economic conditions and also to future exigencies. In addition to large initial setup costs, administrative costs may be high because variables in the economic environment would need to be closely monitored. Needless to say, storage costs would be more closely contained and easier to predict. These, of course, are the major benefits of these proposals.

Farmers will not necessarily gain from flexible policies. Agriculture will become more efficient, but inefficiency at high supported incomes may be preferred by many farmers to efficient production at a lower income. Nevertheless, society as a whole would gain from efficiency in production and storage and the continuity of viable programs. This type of policy makes agriculture much more adaptable than it is currently, and individual farmers would have as much autonomy as they do now.

## Supply Contraction

Past governmental intervention in commodity markets has tended to support excess supplies through the maintenance of higher prices than the market would otherwise allow. Price supports do serve the purpose of raising and stabilizing farmers' incomes, but the government is left with absorbing the cost of storing large amounts of grain. The ability to manage ever-increasing stocks is limited both financially and politically; the inevitable result is a change in policies and an increase in the uncertainty regarding governmental action. Recognition of this tendency to increase commodity supply has led to acreage controls, marketing orders, and pressure to divorce income support from production. Acreage controls are only crudely effective because supply does not correspond exactly to the amount of farmland under production.[22] Marketing orders traditionally have been applied to agricultural commodities where producers are regionally concentrated and easily organized.

---

[22]Love, Rausser, and Freebairn (1984) maintain that slippage is near 100 percent at the historical means of loan rates, target prices, and other instruments.

Orders also tend to weaken farmer independence and are popularly perceived to effect collusion as in the case of milk.

One policy proposal that seeks to control production directly, while avoiding the difficulties associated with previous attempts, is taxation of excess production.[23] This policy has been offered as a more cost-effective means of both supporting income and controlling supplies. In some sense, proposals to tax production over prescribed levels (viz., quotas) are simply a reworking of mandatory controls of the type proposed by Cochrane.[24] The establishment of taxes would merely transfer the enforcement of quotas to tax collection agencies. This alternative is exemplified by recent consideration of programs to reduce the base production in the dairy markets; but, as yet, taxation has been contested as a policy instrument.[25] The basic notion is to set production levels over which farmers participating in commodity programs would be liable to paying a penalty. One variation is to subject all producers to this tax and refund the proceeds to participants in base-reduction programs. The government would still be supporting prices at levels to maintain incomes at levels higher or more stable than an unfettered market would provide, but penalties for increasing production reduces the problems associated with the accumulation of large stocks.

The primary instruments necessary to carry out this proposal are price supports, a level of tax on excess production, and a level of production below which no penalty is assessed. The level of prices, which the government aims to achieve, can be set to accomplish two goals: (1) raise farmers' incomes and (2) stabilize incomes. Where price supports increase incomes above some long-run average (i.e., above a level consistent with market forces), the level of production tax would tend to be high to discourage the investment in more cost-effective methods of production. Farmers will always tend to lower costs and, tax or no tax, supplies will tend to increase as costs per unit fall.

---

[23]For a discussion of this proposal, see Weaver (1983).

[24]For a discussion of Cochrane's proposals, see Tweeten (1979, pp. 466–469).

[25]For a review of the Canadian experience with excess production taxes in the dairy industry, see Stonehouse and Rausser (1981).

The principal rationale for production taxes is the need to make compatible the policy to support farmers' incomes, which is a broad social goal, with the desire to reduce excess supplies which are a symptom of governmental failure. The tendency of traditional programs to raise farm incomes, by giving production incentives greater than market prices, leads to supply levels the market cannot absorb without decreasing prices. In this light, taxing excess production does not deviate from the basic objectives upon which previous policies presumably are based; but it does address directly the major difficulty of public intervention in farming by reversing the incentives to produce overly abundant supplies. Therefore, relative to the status quo, the proposals to tax excess production are best judged by the potential gains in efficiency while pursuing social objectives.

The major advantage of this proposal is the reduction in administrative costs associated with storing large amounts of commodities produced as a result of high price supports and ineffective production controls. Exactly how much the government would save depends on the levels of prescribed yields and the tax on excess production relative to marginal costs. On the surface, this seems to solve the problems associated with setting price supports too high, but the potential for government failure still exists under the proposed policy. The effective price per unit of output under existing programs is not the true measure of producer benefits; rather, it is the total transfers, i.e., effective price multiplied by total output of participants in the programs. In order to sustain the same levels of transfers, the government can resort to increasing deficiency payments (allowing the market to clear whatever supplies are thus generated), to maintaining a storage policy that would restrict supplies available to consumers, or to setting prescribed production levels that would similarly restrict supplies. Whatever course the actual implementation of policy would take—and we suspect a mixture of all three would be attempted, just as in the case of current programs without taxes—there would exist strong, political pressures to manipulate the policy instruments to avoid any decreases in levels of transfers going to farmers.

The effect on farmers' incomes critically depends on the per unit effective prices offered for production within prescribed levels and the ability of taxes to enforce those levels. If present prices are maintained at current levels and taxes are effective at reducing supplies, then farm revenues would decrease. Quotas, however, allow further restrictions of supplies and higher prices without increases in storage costs to the government; therefore, taxing excess production may afford increases in farmers' incomes depending on the elasticity of total demand.

Income uncertainty can be changed in two ways by this proposal. First, by removing much of the storage costs associated with price supports, the probability of drastic changes in policy—due to excessive stocks—is reduced. Second, the absence of large government-controlled stockpiles of commodities gives farmers a chance to take advantage of increases in demand, i.e., the downside risk is still eliminated; but upward shifts in price are not dampened by release of stocks.

## Demand Expansion

Demand expansion is popularly perceived to be an easy solution to the problem of low farm incomes brought about by low prices. Other historically popular policy alternatives for raising incomes are supply restriction, which is unpalatable to farmers who must cut back production, and direct payments to farmers which are costly to the government and too obvious to be politically attractive. Demand expansion, however, does not typically require any unpleasant adjustments by U. S. farmers and is usually inexpensive, unless exports are subsidized. Unfortunately, private domestic demand and government demand are unlikely to expand significantly (particularly after PIK), and exports of major commodities have decreased substantially in the early 1980s. Most proposals for expanding demand concentrate on the foreign market. This is, however, not a perfect solution; as farmers expand into foreign markets, they face more price instability since prices are subject to international shocks in supply and demand.

With high domestic loan prices supporting the world price, demand for U.S. commodities is low. The U. S. market shares have fallen as other countries have expanded production. In a period of sluggish economic growth, the demand curve is unlikely to shift outwards significantly; consequently, U. S. commodities must become more competitive if export performance is to improve. There are two approaches to solving the problem of low foreign demand for U.S. commodities: either lower the domestic cost of commodities or maintain the high domestic price but subsidize the export price at great expense. Lowering the domestic price is not, in general, a popular policy;[26] therefore, subsidized export plans have received much attention.

Subsidizing exports is a costly proposition since other major exporters (in particular, EEC) are also trying to protect their producers by providing high domestic prices. As a result, treasuries compete against each other to subsidize exports.[27] To meet subsidized export competition, direct dollar subsidies, blended credit, expanded P. L. 480 sales, and export PIK programs have been suggested.[28] These proposals are problematic. Direct dollar subsidies are contrary to the General Agreement on Tariffs and Trade (GATT) if they are used to increase market share and if export PIK programs leave the United States open to charges of dumping on the international market which complicates efforts to restrict dumping on the part of other exporters. Also, many charges have been made that P. L. 480 undermines production in developing countries which is contrary to humanitarian goals. Unfortunately, efforts to liberalize trade will probably not succeed until the costs of subsidizing exports become greater than the political costs of forcing adjustment on the domestic agricultural sector.

The alternative is to restructure U. S. agricultural policy so that it does not artificially increase prices to levels that impede export market performance. Much of the enthusiasm for this type of program results from a fear that U. S. market isolation will cause irreparable damage to U. S. export potential. During 1980–1982, the value of the dollar appreciated about 20 percent while loan rates for wheat and corn were steady or increased. Longmire and Morey (1983) estimate that 20 million tons of grain entered farm program

---

[26]Agricultural input industries gain from policies that expand output, and at least one representative from the farm equipment industry has favored such program (McKee, 1983).

[27]The United States has accused the EEC of using export subsidies and import restrictions opposed by GATT to protect their farmers. The United States has retaliated in kind by subsidizing wheat flour sales to Egypt, a traditional French market (see, for example, the *Wall Street Journal* (1983).

[28]For more information on export subsidies, see Mayer (1983).

stocks as a result of the loan rate not declining as the value of the dollar increased. This represented about a quarter of the stocks carried over into the 1983 crop year. A flexible loan policy, where the loan and target prices would be a function of the value of the U. S. dollar, has been proposed by Longmire and Morey (1983); Schuh (1980); and McKee (1983). Such a policy would leave less room for discretion and so reduce political failure. Just and Rausser argue that flexible policies should be based on the factors that indicate crises or major policy disequilibria. They contend that there is simply too much noise in the linkages between the policy instruments and the underlying causal factors (e.g., exchange rates and the real rate of interest). For example, instead of basing the loan rate on the value of the dollar, the level of stock (reflecting, in part, the dollar's value) would be a more appropriate variable.

If demand could be expanded without any change in domestic policy or use of export subsidies, price would increase; farmers would unequivocally gain; and consumers would pay higher prices but would gain with cheaper program costs. If demand is expanded by resorting to export subsidies, farmers still would gain; but taxpayers and consumers lose as they pay higher taxes and prices. If commodity programs are revised to bring domestic prices more in line with world prices, program costs would fall benefiting consumers and taxpayers. Farmers, however, might not fare as well under this scheme since they would receive a lower price for their output. Also, although exports would increase, it would be, at least partially, at the expense of sales to the government leaving total quantity uncertain. In addition, farmers would also face more price variability.

## Direct Income Enhancement

When supply contraction and demand expansion policies are used to raise income, market distortions can be quite costly. In particular, high domestic market prices distort economic signals to farmers, restrict export sales, and lead to high consumer prices. These policies can be replaced or supplemented by income enhancement policies that make direct payments to farmers and reduce the need for market-distorting actions. The costs, however, of income support could be excessive in the absence of a price-support program.[29]

Current direct-payment programs are based on the volume of production, allowing those with large output to gain more than smaller producers. There is no income guarantee to small or poor producers. Payment limitations to reduce excessive transfers to larger producers have been ineffective (Lin, Johnson, and Calvin, 1981), and the perception remains that these programs use taxpayer dollars to aid wealthier farmers. Direct payments based on production also alter production incentives and lead to excess production.

Dissatisfaction with the performance of existing direct income-enhancement programs has led to alternative proposals that aim to reduce or eliminate incorrect market signals and meet equity goals. One proposal—a minimum income target plan—would provide an income floor for individual farmers and so better address questions of equity (Congressional Budget Office, 1984, and Council for Agricultural Science and Technology, 1983). Another alternative, minimum sector income target plan, would guarantee the agricultural sector a minimum aggregate income (Schertz and Clayton, 1983).

The minimum income plan for individuals is, in reality, a negative income tax plan. Individual farmers with inadequate income would receive payments to bring their income up to a selected target. A severe problem with this type of plan is that administration would be extremely awkward. Calculating net income would present considerable difficulty. In contrast, with a production-based direct payment scheme, only total output and deficiency payment per unit of output need be determined.

While a negative income tax scheme would not interfere with production decisions, it may, nevertheless, interfere with management decisions. If benefits are based on some definition of a farmer instead of on production, there would be an incentive for nonfarmers to try to qualify as farmers by buying small parcels of land and for farmers to divide up land among family members to increase their family benefits. A policy would have to be carefully designed

[29]There is strong incentive for the government to raise loan prices (aside from pressure from producers) when market prices are low to minimize deficiency payments costs, i.e., to minimize its Treasury exposure risk.

to discourage this type of activity. The guaranteed income level would also have to be sufficiently low not to interfere with the incentive to produce efficiently.

A guaranteed income plan for individuals would be equitable in that it would transfer income to those who most need publicly provided support, not those who have a large volume of production. Also, any farmer could benefit in a bad year regardless of size or crop produced. However, concern over providing aid to low-income farmers must be weighed against the idea that farmers, as anyone else, should be allowed to succeed or fail on their own merits. Other citizens who pay for this program, but are not entitled to such an income guarantee themselves, may find such a plan unfair and unsupportable. It may be difficult for a nonfarmer to understand that the complex system of existing agricultural programs already provides large wealth transfers to farmers; a negative income tax, however, is quite obvious. In general, farmers would be expected to lobby diligently to increase the level of the guaranteed income, but nonfarmers would probably be equally intent on avoiding any increase. As a result, the very directness of the proposal may guarantee its greater immunity to political failure in implementation.

If a direct income-enhancement policy were to be used alone without policies to manipulate price, the market would be free to adjust to economic signals just as in the free-market case. Price risk would not be reduced, and the variability of income would only be truncated in the lower range. The vast majority of farmers would rarely benefit from a minimum-income program and would still face some income uncertainty. Storage programs could be implemented to reduce price risk and the cost of the income-enhancement program. The expense of a negative income tax program, of course, depends on the level of guaranteed income. In 1983 commodity programs (including PIK) cost $28.3 billion. In a worst case scenario, with a zero net income for all farmers, each one could have received a guaranteed income of $11,791.66 in 1983 with the same expenditure.[30]

With a sector-based income policy, the agricultural sector as a whole is guaranteed a certain net income; but individual farmers

---

[30]That is, assuming there were 2.4 million farmers in 1983 (U. S. Department of Agriculture, 1983).

receive no income guarantee of any sort. The objective, as before, is to transfer income to a "disadvantaged sector" without distorting production incentives. The income guaranteed to the sector is apportioned based on the value of farm marketing, suggesting that those with high value of output would benefit most whether or not they need public aid. Placing a ceiling on federal expenditures to agricultural sector would reduce the risk of high Treasury outlays. The government would spend up to a maximum guaranteed sector income and no more. One disadvantage of this proposal, relative to a negative income tax for farmers, is that it does not eliminate production distortions completely. It would be in a farmer's best interest to increase the value of his marketing—expanding production and changing his crop mix accordingly.

## *Assessment of Alternative Proposals*

We have explicitly outlined what we consider to be the relevant goals of agricultural policy and have reviewed some of the major proposed alternatives to the current system of farm programs. In this section, we present a ranking of these proposals alone and in combination.[31] Because of the general nature of these alternatives and the difficulty in assessing exactly their effects on the various criteria of interest, we turn to a lexicographic ranking of several proposals.

There are three basic means to evaluate proposed policy. The one most often used by the profession is to determine the variety of possible effects that changes in policy would have on the economy. It entails no explicit, subjective judgments regarding social goals; the results of such analyses are presented to the public and politicians.[32] Alternatively, the goals could be embedded in some criterion function where the trade-offs between the goals would be re-

---

[31]It is possible to implement any combination of instruments as well as any single policy. We, however, limit our discussion to combinations that are complementary and consistent with respect to the underlying criteria. We ignore combinations that have significant overlapping of policies (e.g., flexible loans and flexible futures trading) or where policies counter each other (e.g., free market and any intervention).

[32]For review and illustration of this method, see Johnson and Rausser (1977).

flected in assigned (subjective) weights. This requires considerably more information regarding the specifics of each proposed policy, their costs, and the effects of each objective.[33] In what follows, we employ a third method that ranks alternative proposal without considering the exact settings of relative weights between particular objectives. Our treatment of this method is only illustrative of how a lexicographic ranking of the alternatives might be conducted.[34]

The objectives are ordered according to their subjective importance; and a threshold, or satisficing level, is established for each to determine whether policies meet these goals. This simplifies the evaluation by allowing the examination of each policy's performance one objective at a time. We present a ranking of objectives based on our subjective view of agricultural problems and past attempts to solve those problems. The subjective ranking of social goals is the crucial determinant of optimal policies: Different rankings may lead to different appraisals of the several proposals considered.

We view the most important objective as the avoidance of political failure. Programs are evaluated according to their intended effects on social problems; but if a policy is subject to political failure in implementation, the actual consequences are uncertain and not necessarily in the public's interest. What society gets may not be what it expected. To insure a correct ranking of policies, the primary criterion is that the only policies to be considered are those where desired and realized policy instrument settings are not likely to diverge. The only proposals meeting the first objective are those where this potential is considered less than the threshold level; others are rejected. The threshold level is specified in terms of the expected divergence between the desired and realized instrument settings—specifically, 20 percent of the desired settings.

The second ranked objective is that of reducing the level of risk and uncertainty in agriculture. As discussed above, we consider this the most serious problem now facing the farm sector. To satisfy this goal, policies must reduce the degree of uncertainty below current levels. Third in importance is what we rank the objective of assuring

---

[33]For an overview and several examples of this approach, see Rausser and Hochman (1979).

[34]For a motivation of this method of policy evaluation and a discussion of satisficing, see Simon (1979).

a flexible agricultural economy. Acceptable policies equate the returns to farming with long-run market conditions. The fourth ranked objective is the maintenance of adequate food stocks for humanitarian reasons and to insure U. S. consumers against rapid food price inflation. Policies meeting this requirement are those assuring a level of stocks greater than that which would exist, on average, in free market environment.[35]

The fifth ranked objective is that of farm-income enhancement. The threshold for acceptability is some positive level of wealth transfer to small and mid-size family farming operations. Our final goal is to maximize Treasury costs. If more than one policy meets all of the above goals, the least expensive proposal would be preferred.

In Table 1 we show whether various proposals meet the satisficing level for our criteria. We judge that only four policies have acceptably low probabilities for political failure in implementation. Free markets offer no opportunity for such a problem arising since there is no government intervention and, therefore, no rents to be pursued through the bureaucratic process. Flexible storage and flexible policies aimed at expanding demand also satisfy these criteria. Proposals to establish fixed rules for program instruments do involve federal action, but they are particularly structured to avoid discretionary modification.

The potential for political failure in program implementation is greater, and still acceptable, with reinforced markets proposals. Policies encouraging the use of forward-contracting markets, or providing revenue insurance, seem to allow small leeway for program changes to benefit specific interest groups. Nevertheless, their susceptibility to failure does depend on whether or not subsidies are used and, if so, the level of subsidization.

Negative income tax and guaranteed sector income programs are also judged to satisfy the criteria of political failure. These programs

[35]Many consider that the maintenance of safe stocks for the United States has rarely been a problem, and few proposals we consider mention stock programs although they were frequently implied. In our evaluation, however, we attempt to account for world conditions. It must be remembered that, during consideration of the 1981 farm bill, a great deal of political concern was expressed (perhaps mistakenly) regarding the future of world food supplies; see Lesher (1984).

would be sufficiently straightforward in their intent and final payment levels that it would be politically difficult to influence the mandated process of income transfer. Program costs would be more simply computed than present programs; the beneficiaries less hidden to public scrutiny. The effective opposition to any manipulation of these programs once in place would leave them open to change only after a great deal of public debate.

The remaining policy proposals are similar to existing programs. We judge them to have too great a potential for political failure to be acceptable. An excess production tax plan, for instance, would retain the fundamental structure of present policy and add an instrument for taxation. The likelihood remains that instrument settings, such as loan rates and tax levels, would be subject to the same failures in implementation that current programs face. Such policies would maintain the complicated means for income transfers, leaving them subject to interest group pressure and change. By contrast, proposals to reinforce the market are simply ill-prepared as vehicles for large tax transfers. Similarly, enacted flexible policies would protect program administration from pressure by establishing rules to adjust various instruments. Finally, the uncomplicated and blatent costs and benefits of direct payment schemes would aid more effective opposition to interest groups attempting to alter program settings.

The number of acceptable policies is further reduced when the second objective of risk reduction is considered. The proposal to eliminate intervention exposes farmers to the full variability of revenue. It also is likely to lead to lower average revenues due, in part, to farmers paying for risk reduction to replace absent federal programs. Attempts to expand exports with flexible loan policies would also increase farmers' market risk through greater susceptibility to international price shocks. In addition, these proposals would trim income supports from uncompetitive loan rates, contributing to lower average revenues and, at the same time, increasing foreign demand uncertainty. Nevertheless, we consider that all flexible policy proposals (i.e., storage programs, futures markets trading rules, etc.) are acceptable. They would aid in risk management by absorbing price shocks, regulating commodity stock fluctuations, and, more importantly, reducing the uncertainty due to policy crises. Reinforced market proposals also reduce risk sufficiently through market institutions to meet this criterion. Lastly, negative income tax and guaranteed sector income programs reduce the risk of exceptionally low incomes.

Since reinforced market and negative income tax proposals do not intervene to hold prices above long-run levels, they allow market adjustments necessary for a flexible agricultural sector. Flexible policies might temporarily distort market signals, but their design would allow eventual gravitation of prices to free market levels. Hence, agriculture's adaptability to more durable economic conditions is not impaired. The guaranteed sector income plan, on the other hand, introduces incentives to expand marketings in order to receive an increasing share of government largesse. Since such a proposal interferes with efficient resource allocation in a manner similar to the present system, we judge it an unacceptable program.

Flexible storage is the only proposal closely addressing the food security criterion although storage programs could be made compatible with other programs. A policy to maintain safe food stocks is possible with all other proposed means of governmental intervention. A flexible storage plan could be used with other programs although the resulting complexity (relative to simply one program) would increase vulnerability to political failure in implemention.[36] However, considering each policy separately, of the proposals meeting the criteria to this point, only flexible storage rules meet the criterion of food security.

The objective of income maintenance for family farming is met by several proposals: negative income tax, sector income guarantees, supply contraction, and government demand expansion. These policies, however, separately do not fulfill all higher ranked criteria. Although a program of flexible storage is the only policy considered that does meet higher ranked objectives, it does not satisfy the criterion of income maintenance. Hence, to this point, only a combination of policies would be acceptable. We rule out the use of sector income guarantees with other policies because such a combination would not satisfy the criterion of an adaptable agricultural sector. Similarly, a proposal combination with supply contraction, or

---

[36]For example, there would exist at times considerable pressure to relieve the burden of income maintenance from, say, a negative income tax and place it on the price-increasing means of a storage policy.

government demand expansion, maintains the potential for political failure at unacceptable levels and hinders the sector's adaptability. However, the use of negative income taxes with a flexible storage policy would satisfy income maintenance and all higher ranked criteria. A second acceptable combination of policies is comprised of negative income taxes, flexible storage, and other reinforced market policies such as establishing put options on commodities. The latter combination would further reduce farmers' risk.

The final criterion to consider is that of minimizing Treasury costs. The costs of a negative income tax program would depend upon the target income level, the distribution of farmer incomes below this level, and the number of recipients. Under this scheme, Treasury costs associated with present income transfer levels may be large but less than current programs. Flexible storage policies also would be less expensive and more predictable at accomplishing their objectives than current storage programs. Similarly, the costs of other reinforced market policies to aid in risk management are not likely to be large although they would depend upon chosen levels of subsidies. Clearly, the composite policy with negative income taxes and flexible storage would be the least expensive of the two acceptable proposal combinations since the addition of any reinforced market policy would increase Treasury costs regardless of any gains in farmer welfare it would yield. In any event, based on our method of evaluation, the combination of proposals for negative income taxes and flexible storage is best relative to the other single proposals or proposal combinations considered.

## *Conclusions*

Agricultural economists do not presume to have an exclusive knowledge of the problems facing agriculture and possible solutions nor do they attempt to shepherd the public's consideration of policies. But the profession does have an informed perspective from which it can contribute to the subjective evaluation of policies and not merely confine itself to technical analyses.

In our evaluation of proposals for future farm policy, we stress the need to examine the underlying objectives of such policy in order to better compare alternatives. In addition, policies should be assessed in terms of their susceptibility to political failures as well as their ability to mitigate market failures. We argue that, in the design of policy, ignoring the potential for political failures in implementation increases the likelihood of an alteration of program instrument settings. Such an alteration may be counter to the original ranking (or weights) on social objectives that determined the enactment of policy.

We have presented a brief review of the social objectives we consider most relevant to the current debate on future agricultural policy. Once these social goals and the goal to avoid political failure are ranked or given relative weights in some criterion function, a formal evaluation can be made. We have discussed several of the proposed alternatives presently under consideration, categorized them by their means of intervention, and noted their effect on the various objectives. These alternatives cover a broad range—from no intervention (free markets) to a number of means of governmental manipulation of prices and stocks.

Our assessment of alternative proposals made use of a lexicographic method of ranking with satisficing levels subjectively chosen for each criterion (except minimizing Treasury costs, the last ordered objective). To summarize, flexible storage policies satisfy the criteria of avoiding political failure of implementation, risk reduction, sector adaptability, and safe food stocks. This program alone, however, would not contribute to the objective of income maintenance. The negative income tax proposal satisfies the criteria of avoiding political failure, risk reduction, sector adaptability, and income enhancement, although it does not contribute to food security. However, a composite policy of negative income taxes and flexible storage, with or without other reinforced market programs, would meet these criteria. A combination of proposals for negative income taxes and flexible storage would offer the least Treasury cost; therefore, we conclude that it would be the best set of programs based on our means of evaluation.

An unqualified recommendation of this composite policy would depend on the evaluation of proposed alternatives by the other two methods of analysis. First, a detailed quantitative assessment of the economic effects of several proposed alternatives is needed. Second, the objectives could be embedded in a criterion function representing the weights society places on the various goals and the criterion function maximized over the set of proposals and proposal combinations determining some optimal choice of program(s).

To be sure, there are other problems that must be addressed prior to implementing the proposed combination of agricultural policies. First, a program of negative income taxes must define exactly the characteristics of persons eligible. Second, the means of determining farmers' incomes and the income target level must be established. Third, for flexible storage policies, rules must be determined by which instruments adjust to prices, stock levels, and other factors.

## References

Brandow, G. E. "Policy for Commercial Agriculture, 1945–1971." In *A Survey of Agricultural Economics Literature*, Vol. 1, edited by Lee Martin. Minneapolis: University of Minnesota Press, 1977.

Cochrane, Willard. *Feast or Famine: The Uncertain World of Food and Agriculture and Its Policy Implications for the United States.* Washington, D. C.: National Planning Association, 1974.

Congressional Budget Office. *Farm Revenue Insurance: An Alternative Risk-Management Option for Crop Farmers.* Washington, D. C.: U. S. Government Printing Office, August, 1983.

———. *Crop Price-Support Programs: Policy Options For Contemporary Agriculture.* Washington, D. C.: U. S. Government Printing Office, February, 1984.

Council for Agricultural Science and Technology. *The Emerging Economics of Agriculture: Review and Policy Options.* Report No. 98. Ames, Iowa: Council for Agricultural Science and Technology, September, 1983.

*Farmland News.* May 31, 1984, p. 12.

Gardner, Bruce L. "Commodity Options for Agriculture." *American Journal of Agricultural Economics* 59(1977):986–992.

———. "Efficient Redistribution Through Commodity Markets." *American Journal of Agricultural Economics* 65(1983):225–234.

Harrington, David, Donn Reimund, Kenneth Baum, and Neal Peterson. *U.S. Farming in the Early 1980's: Production and Financial Structure.* U.S. Economic Research Service, Agricultural Economic Report No. 504. Washington, D. C.: U. S. Government Printing Office, 1983.

Hathaway, Dale E. "U. S. Agricultural Policy and World Trade." *Increasing Understanding of Public Problems and Policies—1983.* Oak Brook, Illinois: Farm Foundation, 1983.

Interreligious Taskforce on U. S. Food Policy. "Target Prices: Target, Don't Freeze." Legislative Proposal. Washington, D. C., July, 1983.

Johnson, Stanley R., and Gordon C. Rausser. "Systems Analysis and Simulation: A Survey of Applications in Agricultural and Resource Economics." In *A Survey of Agricultural Economics Literature*, Vol. 2, edited by Lee Martin. Minneapolis: University of Minnesota Press, 1977.

Just, Richard E. "Farmer-Owned Grain Reserve Program Needs Modification to Improve Effectiveness: Theoretical and Empirical Considerations in Agricultural Buffer Stock Policy Under the Food and Agriculture Act of 1977." U. S. General Accounting Office, CED-81-70, Vol. 3. Washington, D. C., June 26, 1981.

Just, Richard E., and Gordon C. Rausser. "Uncertain Economic Environments and Conditional Policies." In *Alternative Agricultural and Food Policies and the 1985 Farm Bill,* edited by Gordon C. Rausser and Kenneth R. Farrell. Giannini Foundation of Agricultural Economics, Division of Agriculture and Natural Resources. Berkeley: University of California, 1984.

Langley, James A. "Formulation and Policy Implications of Alternative Loan Rates and Target Prices for Major Agricultural Commodities." National Economics Division, Food and Agricultural Policy Branch, U. S. Economic Research Service, Washington, D. C., November, 1983.

Lee, John E. "Food and Agricultural Policy: A Suggested Approach." *Agricultural-Food Policy Review: Perspectives for the 1980s,* AFPR-4, U.S. Department of Agriculture, Economics and Statistics Service, April, 1981, pp. 136–148.

Lesher, William G. "Impact of the Agriculture and Food Act of 1981 and the Current Situation in U. S. Agriculture." In *Alternative Agricultural and Food Policies and the 1985 Farm Bill,* edited by Gordon C. Rausser and Kenneth R. Farrell. Giannini Foundation of Agricultural Economics, Division of Agriculture and Natural Resources. Berkeley: University of California, 1984.

Lin, William, James Johnson, and Linda Calvin. *Farm Commodity Programs: Who Participates and Who Benefits?* U. S. Economic Research Service. Report No. 474. Washington D. C.: U. S. Government Printing Office, 1981.

Longmire, Jim, and Art Morey. *Strong Dollar Dampens Demand for U.S. Farm Exports.* U. S. Economic Research Service, Foreign Agricultural Economic Report No. 193. Washington, D. C.: U. S. Government Printing Office, December, 1983.

Love, Alan, Gordon C. Rauser, and John Freebairn. "The Effectiveness of Government Policy in Controlling Agricultural Output." Paper presented at the summer meeting of the American Agricultural Economics Association, Ithaca, New York, August, 1984.

Mayer, Leo V. "Export Options and Current Farm Programs: Alternatives and Consequences." Proceedings for the National Agricultural Policy Symposium, Kansas City, March 27–29, 1983.

McCalla, Alex F. "Consistencies, Inconsistencies, and Other Views on Commodity Policies: A Review and Critique of Preceding Papers." In *Alternative Agricultural and Food Policies and the 1985 Farm Bill*, edited by Gordon C. Rausser and Kenneth R. Farrell. Giannini Foundation of Agricultural Economics, Division of Agriculture and Natural Resources. Berkeley: University of California, 1984.

McKee, Dean E. Deere and Company. Letter to R. W. Boeke, June 15, 1983.

Miller, James C. National Grange Legislative Department. Letter to NC-169 Project Participants, August 2, 1983.

Miller, Thomas A., and Jerry A. Sharples. "Issues Concerning the Level of Price and Income Supports." In *Structural Issues of American Agriculture*. U. S. Department of Agriculture, Economics, Statistics, and Cooperative Service, Agricultural Economics Report No. 438. Washington, D. C.: U. S. Government Printing Office, November, 1979.

Offutt, Susan E., and David A. Lins. "Income Insurance for Commodity Producers: Program Issues and Design Alternatives." Unpublished manuscript. University of Illinois, Urbana, 1983.

Paarlberg, Don. *Farm and Food Policy: Issues of the 1980s*. Lincoln: University of Nebraska Press, 1980.

Penn, J. B. "The Changing Farm Sector and Future Public Policy: An Economic Perspective." *Agricultural-Food Policy Review: Perspectives for the 1980s*, AFPR-4. U. S. Department of Agriculture, Economics and Statistics Service. Washington, D. C.: U. S. Government Printing Office, April, 1981, pp. 28–58.

Randall, Alan J. "General Discussion." In *Alternative Agricultural and Food Policies and the 1985 Farm Bill*, edited by Gordon C. Rausser and Kenneth R. Farrell. Giannini Foundation of Agricultural Economics, Division of Agriculture and Natural Resources. Berkeley: University of California, 1984.

Rausser, Gordon C. "Political Economic Markets: PERTs and PESTs in Food and Agriculture." *American Journal of Agricultural Economics* 64(1982):821–838.

Rausser, Gordon C., and Eithan Hochman. *Dynamic Agricultural Systems: Economic Prediction and Control*. New York: North Holland Press, 1979.

Rausser, Gordon C., and D. Peter Stonehouse. "Public Intervention and Producer Supply Response." *American Journal of Agricultural Economics* 60(1978): 885–890.

Rausser, Gordon C., David Zilberman, and Richard E. Just. "The Distributional Effects of Land Controls in Agriculture: Theoretical Implications of Programming Models." University of California, Department of Agricultural and Resource Economics, Working Paper No. 80 (Berkeley, 1983).

Schertz, Lyle P., and Kenneth C. Clayton. "Alternatives to Current Commodity Programs." Paper presented at the 1983 annual meeting of the American Association for the Advancement of Science, Detroit, Michigan, May 28, 1983.

Schuh, Edward G. "U. S. Agriculture in an Interdependent World Economy: Policy Alternatives for the 1980's." Paper presented at the American Enterprise Institute Conference on Food and Agricultural Policy, Washington, D. C., October 2 and 3, 1980.

______. "Future Directions for Food and Agricultural Trade Policy." Paper presented at the Meetings of Allied Social Science Association, San Francisco, December 27–30, 1983.

Shepard, L. E., and R. A. Collins. "Why Do Farmers Fail? Farm Bankruptcies 1910–1978." *American Journal of Agricultural Economics* 64(1982):609–615.

Simon, Herbert. "Rational Decision Making in Business Organizations." *American Economic Review* 69(1979):493–513.

Stonehouse, D. Peter, and Gordon C. Rausser. "A Quarterly Model of the Canadian Dairy Industry with Endogenous Policy Variables." University of Guelph, Working Paper AEEE/81/7, Guelph, Ontario, April, 1981.

Thompson, Robert. Speech on Future Farm Policy, Berkeley, California, June, 1984.

Tweeten, Luther G. *Foundations of Farm Policy.* 2nd ed., rev. Lincoln: University of Nebraska Press, 1979.

U. S. Department of Agriculture. *Agricultural Statistics, 1983.* Washington, D. C.: U. S. Government Printing Office, 1983.

*Wall Street Journal.* January 19, 1983.

Wallace, T. D. "Measures of Social Costs of Agricultural Programs." *Journal of Farm Economics* 44(1962):580–594.

Weaver, Robert D. "Tax Excess Production: A New Production Control/Price Support Strategy." Cooperative Extension Service Publication, Pennsylvania State University, November/December, 1983.

Wolf, Charles, Jr. "A Theory of Nonmarket Failure: Framework for Implementation Analysis." *Journal of Law and Economics,* XXI(1979): 107–139.

# INDIVIDUAL COMMODITY POLICY OPTIONS

# 8 COTTON

## Robert S. Firch*

In a paper I wrote in 1973, I began: "During the last 20 years the U. S. cotton production industry has endured such sweeping changes in government policy that many observers of the scene may have failed to notice that it has been what the economist may reasonably classify as a declining industry" (Firch, 1973). The bad news is that, since that time, the domestic purchase of raw cotton has continued on a downward trend. The good news is that over the last 10 years foreign purchase of U. S.-grown cotton seems to have had a growth rate somewhat greater than the decline of the domestic market.

The plan of this paper is, first, to review the condition and trends of the markets for U. S. cotton. The performance of the government programs and the cotton industry for the past 20 years will be reviewed with emphasis on the last 10 years. I will argue that the payment-in-kind (PIK) program of 1983 was an extraordinary program that was needed to solve the problems resulting from extraordinary circumstances in 1981–1983 and should not have any particular relevance for future programs, except when similar extraordinary circumstances exist. Then I will argue that the 1973–1981 legislation for cotton has generally performed well and should

*The author is Professor of Agricultural Economics, University of Arizona.

be continued. Some potential problems with this loan price-target price program will be discussed with proposed solutions.

## *Market Trends for U. S. Cotton*

The reason that market trends are so very important to the U. S. cotton production industry is that cotton is continuously threatened by man-made fibers which were displacing cotton at an alarming rate during the 1960s. Ill-conceived government programs could speed this decline of cotton markets as, undoubtedly, they did during the high price-support/acreage allotment program of the 1953–1963 period. The decline of both domestic and foreign markets for U. S. cotton during 1967–1972 was greater than generally perceived because it occurred with slowly declining quantities but substantially reduced real prices.

Table 1 presents some data on domestic and foreign markets for U.S. cotton for the period 1964–1983. The deflated foreign price can be considered only an approximation as it is the Liverpool price of SLM 1 1/16 cotton in British pounds deflated by the British consumer price index. A Far East price for U. S. cotton would be more representative of the foreign price, but no such price series is available. Over the entire 20-year period, there is obviously no trend in either the deflated U. S. mill price or the deflated foreign price. The first half of the period covered in Table 1 had government programs preceding the loan price-target price program of the last 10 years to the present. Focusing on the 1974–1983 period, there have been statistically significant negative time trends in both U. S. and foreign prices. The average decline in the deflated U. S. price was 3.43 percent per year; the deflated foreign price declined 6.39 percent per year.

There were a statistically significant negative time trend in U. S. mill consumption and a nearly significant positive time trend in exports of U. S. cotton in the 1974–1983 period. The average decline in domestic consumption was 2.05 percent per year; the average increase in exports was 5.5 percent per year. With negative trends in both real domestic price and domestic mill consumption, it is obvious that the demand of domestic mills has continued to shift to the left as it had done during the preceding 20 years.

**TABLE 1. Deflated U.S. Mill and Foreign Mill Prices (1972 = 100) and U.S. Mill Consumption and Exports, 1964–1983[a]**

| Crop year | Deflated mill price per pound | | United States Consumption | United States Exports |
|---|---|---|---|---|
| | U. S. cents | British pence | 1,000 bales | |
| 1964 | 38.5 | 16.0 | 9,107 | 4,174 |
| 1965 | 36.8 | 14.8 | 9,454 | 3,029 |
| 1966 | 34.7 | 14.1 | 9,438 | 4,819 |
| 1967 | 40.4 | 19.1 | 8,948 | 4,316 |
| 1968 | 34.1 | 16.2 | 8,204 | 2,816 |
| 1969 | 30.4 | 14.8 | 8,001 | 2,863 |
| 1970 | 31.2 | 15.1 | 8,105 | 3,885 |
| 1971 | 39.8 | 15.5 | 8,163 | 3,376 |
| 1972 | 36.2 | 18.0 | 7,670 | 5,306 |
| 1973 | 54.8 | 31.0 | 7,384 | 6,111 |
| 1974 | 52.1 | 18.2 | 5,797 | 3,914 |
| 1975 | 39.1 | 21.2 | 7,160 | 3,300 |
| 1976 | 54.5 | 25.4 | 6,595 | 4,779 |
| 1977 | 47.0 | 16.3 | 6,416 | 5,459 |
| 1978 | 42.8 | 16.2 | 6,286 | 6,150 |
| 1979 | 42.2 | 14.3 | 6,441 | 9,177 |
| 1980 | 49.3 | 13.3 | 5,828 | 5,893 |
| 1981 | 41.0 | 11.2 | 5,216 | 6,555 |
| 1982 | 32.9 | 12.4 | 5,457 | 5,194 |
| 1983 | 36.2 | 15.0 | 5,750 | 6,750 |

[a] Broken lines identify the years covered by each agricultural act, and asterisks separate the loan price-target price years from the previous years.

Source: U. S. Economic Research Service (various issues).

In the earlier years the leftward shift of domestic demand for cotton was primarily the result of man-made fibers displacing cotton in domestic consumption. With the cotton share of domestic consumption fairly well stabilized in the last 10 years, it appears that the decline in domestic mill demand for cotton was caused primarily by increasing imports of garments and fabrics and the growing comparative advantage of foreign mill-fabric-garment industries. This results in a perplexing problem for U. S. cotton farmers. To maintain good relationships with domestic mills, they feel compelled to support the mills in their efforts to establish "voluntary" restrictions on imports of garments and fabrics even though their pure economic interest may be better served by having no restrictions on imports.

In recent years, the U. S. Department of Agriculture has put considerable effort into researching and reporting that substantial amounts of the cotton imported in garments and fabrics is produced in other countries (Glade, 1983). I believe that it is in the best interests of U. S. farmers to keep the spread between farm price and retail price as small as possible and that this would be facilitated by having no restrictions on imports. The consequence of no restrictions on imports would be further declines in domestic mill purchases and a more than compensating increase in exports. In other words, I believe that the United States has a strong comparative advantage in cotton production but a comparative disadvantage in activities between the cotton gin and the retail shelf. The comparative advantage in production is in part cancelled by forcing too large a proportion of the U. S. crop through domestic mills.

## *A Review of the 1964–1983 Cotton Programs*

Some perspective is needed for evaluation of the loan price-target price program that began with the 1974 crop and has had 10 complete years of performance. That perspective will be gained by reference to the preceding 10 years of government programs that gradually changed toward the structure of the rather stable programs of the last 10 years.

The era of high level loan price with acreage allotments to keep the Commodity Credit Corporation (CCC) stocks from unlimited accumulation spanned the period from the Korean War through 1965. The Agricultural Act of 1964 set the nonrecourse loan price at 30 cents per pound and, under discretion allowed in the Act, the

Secretary of Agriculture set the 1965 loan price at 29 cents (Evans, 1980). Growers planting within their allotments qualified for price-support payments of 3.50 cents per pound in 1964 and 4.35 cents in 1965. The 1964 Act also established subsidy payments for domestic manufacturers of cotton who believed they were disadvantaged because foreign mills benefited from U. S. export subsidies. Stocks owned by the CCC grew in every crop year from 1961 to 1965. By the end of the 1965 marketing year, carry-over had reached 16.7 million bales—more than the quantity of domestic consumption plus exports in any of the previous 25 years (Appendix Table 1).

Domestic users of cotton were anxious for a change in the cotton program because their large subsidy checks drew unwanted publicity. Drastic changes in policy seemed to be needed to stop the rapid accumulation of CCC stocks and set a course toward normal carry-overs. The Agricultural Act of 1965, effective for the 1966 through 1970 cotton crops, began acreage diversion programs with associated payments; lowered the loan price to 90 percent of the estimated world price; and, in effect, paid the subsidies that had gone to domestic users to cotton producers as price-support payments (McArthur, 1977). The 1965 Act substantially reduced planted acres, and lower per acre yields assisted in producing unexpectedly rapid depletion of stocks.

The 1965 Act should be considered the precursor of the loan price-target price program that first applied to the 1974 cotton crop. The 1965 Act moved the program away from high-level price supports (as the principal device for assisting agriculture) to payments directly to farmers with loan prices low enough so as not to affect substantially the domestic or foreign market prices after the huge surpluses, existing at the end of the 1965 crop, had disappeared. Price-support payments ranged from 9.5 cents per pound in 1966 to 16.8 cents per pound, while loan prices ranged from 20.9 to 21.8 cents per pound. In 1967, the combined total of price-support and diversion payments made directly to cotton farmers was $935 million while the receipts from market sales were only $938 million (Appendix Table 2).

The Agricultural Act of 1970, effective on the 1971–1973 cotton crops, continued the loan price at 90 percent of world price; price-support payments were to equal the difference between the average spot price in the first five months of the marketing year and 35 cents, with a minimum of 15 cents. A set-aside of 20 percent of the domestic allotment was required for eligibility for price-support payments, but there were no longer any penalties for planting more than the farm allotment. The 1970 Act imposed a $55,000 limit on price-support payments per program crop to an individual producer.

The Agriculture and Consumer Production Act of 1973 established the target price concept in its current form. For the first time, a deficiency payment was determined as the difference between a predetermined price—the target price—and the average upland cotton farm price for the calendar year, but the payment could not exceed the difference between the loan price and the target price. The 1973 Act set the target price at 38 cents for 1974 and 1975; for 1976 and 1977, target prices would be adjusted by cotton yields and changes in an aggregate index of prices paid by farmers. The resulting target prices for the last two years of the program were 43.2 and 47.8 cents. Because the computed average market prices rose above the target price during the four years of the 1973 Act, no deficiency payments were made.

The Food and Agriculture Act of 1977 continued the basic features of the loan price-target price program established in the preceding act. However, a potentially very significant change was the elimination of acreage allotments that had tied program benefits to geographic planting patterns in the past. This change—basing program benefits on current plantings—combined with the freedom to plant beyond allotments or even without allotments during the previous program allowed benefits to move to low-cost producers and away from producers whose claim was based on planting patterns of the past (Firch, 1973). The formula for setting the target price was changed slightly to reflect individual commodity average production costs. The resulting target prices for 1978, 1979, 1980, and 1981 crops under the 1977 Act were 52, 57.7, 58.4, and 70.87 cents, respectively. The maximum total deficiency payments per person for all crops were $45,000 for 1979 and $50,000 for 1980.

The Food and Agriculture Act of 1981 is effective for the 1982–1985 cotton crops. It generally continues the policies of the preceding act. A major break with the past was that the legislation set a minimum target price for each year without any tie to inflation—contrasted with the two previous acts which had made the target price a function of the cost of production. The minimum target prices were to be 71, 76, 81, and 86 cents per pound, respectively, for 1982, 1983, 1984, and 1985. This increase might be characterized as a legislative prediction of the future rate of inflation. The fact that Congress lowered the rate to 81 cents for the final year of the Act (by amendment signed into law on April 10, 1984) could be interpreted as an admission that it had predicted too high a rate of inflation in 1981, an understandable if not entirely forgivable error.

The loan price is continued at 90 percent of the world price but with a 55-cent minimum. The 1984 program requires a 25 percent reduction from average acres planted in 1982–1983 (counting acres in the 1983 PIK program as planted) for eligibility for the 55-cent loan price and deficiency payments toward the 81-cent target price.

## *Evaluation of 1964–1983 Cotton Programs*

The consequences of the major features of the cotton programs of the last 20 years with emphasis on the loan price-target price program of the last 10 years will be presented in this section. The section will conclude with a brief discussion of the PIK program of 1983.

### Loan Price and Market Price

The Agricultural Act of 1964, which concerned the 1964 and 1965 cotton crops, was the last in which loan prices for cotton were set substantially above the market price that would have prevailed without any government program. Table 2 (adjusted for inflation) shows the substantial change imposed by the 1965 Act as the loan price was lowered almost 30 percent for the 1966 crop. The 1965 Act and later acts set the loan price at 90 percent of the "world" price using contemporaneous world price data for the computations. The deflated loan price has remained in the range of 45 percent to 80 percent of its 1964 level.

The high loan price of the 1964 Act and earlier acts effectively set the U. S. and world prices of cotton, and the United States continued as a residual supplier and world pricesetter until the huge

**TABLE 2 Deflated Cents per Pound of Government Cotton Program Benefits and Farmer Receipts per Pound of Production (1972 = 100), 1964–1983[a]**

| Crop year | Program benefits | | | Farmer receipts | | |
|---|---|---|---|---|---|---|
| | Loan Price | Support payments | Total support | Market value | Government payments | Total receipts |
| | deflated cents per pound | | | | | |
| 1964 | 41.23 | 4.81 | 46.04 | 42.49 | .75 | 43.24 |
| 1965 | 39.00 | 5.85 | 44.85 | 39.36 | 1.31 | 40.67 |
| 1966 | 27.36 | 12.27 | 39.63 | 28.08 | 22.22 | 50.30 |
| 1967 | 25.61 | 14.58 | 40.19 | 33.52 | 33.41 | 66.93 |
| 1968 | 24.53 | 14.83 | 39.36 | 27.78 | 18.23 | 46.01 |
| 1969 | 23.33 | 16.97 | 40.30 | 25.18 | 19.93 | 45.11 |
| 1970 | 22.14 | 18.37 | 40.51 | 24.95 | 20.56 | 45.51 |
| 1971 | 20.31 | 15.62 | 35.93 | 29.25 | 17.11 | 46.36 |
| 1972 | 19.50 | 15.00 | 34.50 | 27.22 | 12.36 | 39.58 |
| 1973 | 18.44 | 14.18 | 32.62 | 41.97 | 10.76 | 52.73 |
| ******** | ******** | ******** | ******** | ******** | ******** | ******** |
| 1974 | 23.51 | 9.51 | 33.02 | 37.09 | 2.02 | 39.11 |
| 1975 | 28.71 | 1.50 | 30.21 | 40.62 | 2.36 | 42.98 |
| 1976 | 29.41 | 3.23 | 32.64 | 48.24 | 1.47 | 49.71 |
| 1977 | 31.87 | 2.26 | 34.13 | 37.18 | .72 | 37.90 |
| 1978 | 31.91 | 2.66 | 34.57 | 38.66 | 2.94 | 41.60 |
| 1979 | 30.74 | 4.57 | 35.31 | 38.11 | .94 | 39.05 |
| 1980 | 26.90 | 5.83 | 32.73 | 41.68 | 3.20 | 44.88 |
| 1981 | 26.88 | 9.44 | 36.32 | 27.69 | 3.76 | 31.45 |
| 1982 | 27.59 | 6.73 | 34.32 | 28.20 | 5.56 | 33.76 |
| 1983 | 25.51 | 9.74 | 35.25 | [c] | | |
| 1984[b] | 24.06 | 11.38 | 35.44 | | | |
| 1985[b] | | | 33.43 | | | |

[a] Broken lines identify the years covered by each agricultural act, and asterisks separate the loan price-target price years from the previous years.

[b] Projected inflation, 6 percent.

[c] Blanks indicate no data available.

Source: Computed from data in Appendix tables 1 and 2, Table 4, and from U. S. Department of Agriculture, Agricultural Stabilization and Conservation Service (1983).

CCC stocks of those programs disappeared in the 1967 crop year (Firch, 1973). However, the average market value of cotton sold by U. S. farmers was substantially above the loan price for the 1967–1980 crops. The market value, an average of all cotton sold by farmers, is consistently slightly below the price of SLM 1 1/16 cotton, the cotton of the loan price in Table 2. In November, 1982, the spot price of SLM 1 1/16 cotton dropped to a very small margin above its loan price. In December, 1982, when people began to believe that there would be a PIK program for the 1983 crop, the spot price began to widen its margin over the loan price. The data strongly suggest that the loan price has had very little direct effect on the U. S. spot price and probably even less effect on world prices during the last 17 years.

The data in Table 3 further support the contention of minimal effect of the loan price on the market price of cotton. The year 1969 was the last year there was any substantial failure of farmers to repay their nonrecourse loans on cotton. It appears that, since that time, cotton farmers generally have found the market price to be enough above the loan price plus storage and interest charges that they have had clear incentives to repay the nonrecourse loans.

As the loan price apparently has not substantially affected the market price, perhaps the loan program no longer plays an important role in the system. However, Table 3 shows that, since 1969, farmers have placed from 9 percent to 42 percent of their cotton crops under loan with an average of almost 20 percent. This suggests that the loan program has been an important part of cotton farmers' marketing programs in spite of the high repayment rate. Only in recent years has the interest rate on nonrecourse loans (paid only if the loan is repaid) been closer to market interest rates. The relatively high use of the loan program suggests that farmers have found the interest rates and/or the price protection offered by the loan program attractive relative to commercial loans. The use rate of the loan program, also, is relatively high considering that many farmers during those years planted more acres of cotton than allowed for qualification for the loan program.

It is very likely that there would have been substantial nonrepayment of nonrecourse loans in 1983 and that market prices would have been at loan price levels if the PIK program or some other major change in the program had not been instituted. In another

**TABLE 3. Cotton Placed Under Loan, Unredeemed, and Purchased by the Commodity Credit Corporation, 1964–1983[a]**

| Crop year | Placed under loan | Unredeemed from loan | Placed under loan | Unredeemed from loan |
|---|---|---|---|---|
| | million bales | | percent of production | |
| 1964 | 7.3 | 4.84 | 49 | 32 |
| 1965 | 7.0 | 5.33 | 47 | 36 |
| ---------- | | | | |
| 1966 | 3.1 | 1.43 | 33 | 15 |
| 1967 | 1.5 | .31 | 20 | 4 |
| 1968 | 4.4 | 2.78 | 41 | 26 |
| 1969 | 3.7 | 1.07 | 37 | 22 |
| 1970 | 2.4 | x[b] | 24 | y[c] |
| ---------- | | | | |
| 1971 | 1.2 | x | 12 | y |
| 1972 | 1.9 | x | 14 | y |
| 1973 | 1.7 | x | 13 | y |
| ********** | | | | |
| 1974 | 2.4 | x | 21 | y |
| 1975 | .7 | x | 9 | y |
| 1976 | .9 | x | 9 | y |
| 1977 | 4.5 | x | 31 | y |
| ---------- | | | | |
| 1978 | 1.5 | x | 14 | y |
| 1979 | 1.7 | x | 12 | y |
| 1980 | 2.3 | z[d] | 21 | y |
| 1981 | 6.0 | z | 38 | y |
| ---------- | | | | |
| 1982 | 5.0 | z | 42 | y |
| 1983 | 1.3 | x | 17 | y |

[a] Broken lines identify the years covered by each agricultural act, and asterisks separate the loan price-target price years from the previous years.

[b] x Indicates less than 50,000 bales.

[c] y Indicates less than 1 percent.

[d] z Indicates more than 1.6 million bales of the 1980–1982 cotton crops under loan were purchased by the Commodity Credit Corporation at above-loan price to satisfy the needs of the 1983 PIK program.

Source: Unpublished data supplied by the U. S. Department of Agriculture, Commidity Credit Corporation.

section of this paper, I will argue that the conditions of 1983 were sufficiently unique that the prospect of substantial loan forfeiture and market price at the loan price should not be attributed to a loan price that was too high.

## Potential Payments and Actual Payments

The support payments column under the benefits heading in Table 2 shows the maximum price-support payments that farmers could have been paid each year if they had participated fully in the acreage reduction programs; diversion payments are not included. The 1965 and 1970 acts had relatively high potential price-support payments for the 1966–1973 cotton crops. The 1973 Act dramatically reduced the potential price-support payments as the real loan price rose for the 1974 and 1975 cotton crops. The potential support payments rose somewhat in the 1981–1984 crops as the loan price trended downward and the deflated target price rose substantially.

The government payments column under the receipts section of Table 2 shows the average payments that were actually made. This differs from the support payments column in two ways: It includes payments made for diversion and disaster programs on cotton, and it reflects farmers' actual participation in acreage reduction programs with average payments declining as participation declines. Under the 1965 Act, the burden of disposing of the huge stock accumulations of previous programs led to large diversion payments in addition to price-support payments. The 1967 crop situation is in the same infamous class with the 1983 PIK program crop: Government payments nearly equaled the market value of the crop. With the exception of relatively small diversion payments on the 1978 crop, the actual payments on the 1974–1980 cotton crops were composed entirely of disaster payments and were rather small relative to the payments under the 1965 and 1970 acts.

Government payments per pound of actual production during the loan price-target price period (1974–1983) were small and composed primarily of disaster payments until the 1983 crop situation created a need for the PIK program. Does the loan price-target price program deserve all the credit for the dramatic reduction in government payments relative to earlier programs? The data in Table 4 provide some supplementary/alternative explanations for the absence of large payments to cotton farmers during the loan price-target price program period. The rapid rate of inflation in the United States in the 1973–1975 period lifted market prices above and away from target prices, eliminating deficiency payments. The large decline in the trade-weighted value of the dollar in 1972 and 1973 also contributed to market price being consistently above the target price. However, in 1981, when the trade-weighted value of the dollar rose dramatically and the rate of inflation began to drop at a rather rapid pace, the market value dropped. Thus, it is likely that deficiency payments would have been made on at least some of the cotton crops in the 1974–1980 period except for rapid inflation and the declining value of the dollar.

## Total Benefits and Total Receipts

The total support column in Table 2 shows the deflated minimum price that farmers could have expected for each cotton crop with full participation in government cotton programs not including diversion payments. Because this price was known each year before the crop was planted, it may be considered the expected minimum price farmers used in making planting decisions. However, in some years, available diversion payments and concerns about maintaining program per acre yields may also have played a role in the decision on how many acres to plant. Before the 1974 crop, support payments were at least partially independent of total benefits; therefore, total benefits are basically the sum of the first two columns. In the 1974–1985 period, total benefits (target prices) are set independently with support payments (deficiency payments) as the difference between support price (column 1) and target price (column 3).

Total potential benefits dropped substantially between the 1964 and 1965 acts. Deflated total benefits were remarkably stable and trendless during the 1966–1970 crops of the 1965 Act even though inflation accelerated from 2.18 percent to 5.37 percent per year. Deflated total benefits had a significant downward trend for the 1971–1973 crops of the 1970 Act. During the loan price-target

**TABLE 4. Implicit Price Deflator of Gross National Product and the Trade-Weighted Value of the U. S. Dollar, 1964–1983[a]**

| Calendar year | Implicit price deflator 1972 = 100 | Trade-weighted value of dollar March, 1973 = 100 | Implicit price deflator | Trade-weighted value of dollar |
|---|---|---|---|---|
| | | | percent change | |
| 1964 | 72.77 | [b] | 1.53 | |
| 1965 | 74.36 | | 2.18 | |
| ------- | | | | |
| 1966 | 76.76 | | 3.23 | |
| 1967 | 79.06 | 120.0 | 3.00 | |
| 1968 | 82.54 | 122.1 | 4.40 | 1.75 |
| 1969 | 86.79 | 122.4 | 5.15 | .25 |
| 1970 | 91.45 | 121.1 | 5.37 | − 1.06 |
| ------- | | | | |
| 1971 | 96.01 | 117.8 | 4.99 | − 2.73 |
| 1972 | 100.00 | 109.1 | 4.16 | − 7.39 |
| 1973 | 105.75 | 99.1 | 5.75 | − 9.17 |
| ******* | | | | |
| 1974 | 115.08 | 101.4 | 8.82 | 2.32 |
| 1975 | 125.79 | 98.5 | 9.31 | − 2.86 |
| 1976 | 132.34 | 105.6 | 5.21 | 7.21 |
| 1977 | 140.05 | 103.3 | 5.83 | − 2.18 |
| ------- | | | | |
| 1978 | 150.42 | 92.4 | 7.40 | −10.55 |
| 1979 | 163.42 | 88.1 | 8.64 | − 4.65 |
| 1980 | 178.42 | 87.4 | 9.18 | − .79 |
| 1981 | 195.14 | 102.9 | 9.37 | 17.73 |
| ------- | | | | |
| 1982 | 206.88 | 116.6 | 6.02 | 13.31 |
| 1983[c] | 215.63 | 125.3 | 4.23 | 7.46 |

[a] Broken lines identify the years covered by each agricultural act, and asterisks separate the loan price-target price years from the previous years.

[b] Blanks indicate no data available.

[c] Preliminary.

Source: U. S. Council of Economic Advisers (1984).

price period of the 1974–1985 crops, deflated total benefits have been trendless but with more year-to-year variability. If Congress had not lowered the target price of cotton for 1985 from 86 to 81 cents, there might have been some basis for concluding that total benefits were trending upward during the loan price-target price period. For the 1974–1980 crops, market prices were substantially above target prices; therefore, the instability of the target price was not of great consequence. Unfortunately, the target price rose substantially in 1981 to its highest level since 1970—just as the 1981 crop with unusually high per acre yields produced the largest U. S. cotton crop since 1953.

The lack of trend in total benefits under the loan price-target price program might be considered a desirable feature, but there is substantial evidence that the normal course of real commodity prices is to decline slowly over time as technology lowers the per unit cost of production. After allowance for inflation, unless the target price is programmed to decline slowly over time, it will rise continuously relative to the free-market price and will progressively become more important in farmers' decisions on how much cotton to produce. The consequences of this growing discrepancy between the target price trend and the natural downward trend in the free-market price of cotton will be discussed in a later section of this paper. (Estimation of the natural rate of decline of cotton prices would make an interesting research project.)

The ideal program might be one that would move the real target price down each year by the natural rate of decline of the free-market price. (This could not be done by simply adjusting the target price by some general price index.) A justification for the real target price falling at its natural rate from some original level at which it did not overly stimulate production could be made in terms of the more efficient resource utilization generally attributed to stable prices.

Another dimension of the problem usually overlooked by policymakers is that a guaranteed price, such as a high loan or target price, will induce greater production than the same price would if it were only an average with great variability. This response to a certain price would need to be factored into the design of an optimum target price.

The real total receipts per pound of actual production in the 1964–1982 period (Table 2) reached their highest level in the crops of the 1965 Act. This occurred largely because of the lowering of support prices to world price levels combined with increased diversion payments to cut production drastically in order to reduce the huge stocks accumulated during the era of high price supports and increased support payments.

The unusually high total receipts per pound of actual production in 1967 reflect very large price-support and diversion payments and the smallest crop of cotton harvested in at least 40 years. There was no obvious trend in total receipts in the first seven years of the loan price-target price program that operated under free-market conditions. The sharp drop in market price in the 1981 and 1982 crops could not be offset by deficiency payments because either (1) many farmers were ineligible for payments because they had not complied with acreage reduction programs when the market price was above the target price for several years or (2) the $50,000 payment limitation curtailed payments for certain farmers. The payment limitation was probably the lesser factor as all but the largest farm operations have generally found ways to evade the limit. (Farmer receipts for 1983 are not shown in Table 2 because the 1983 crop marketing year had not been completed and dependable estimates of the value of the cotton PIK program payments to farmers were not available at the time of writing.)

## Regional Redistribution of Production

The 1970 Act allowed planting cotton beyond or even without acreage allotments, but the relatively high price-support payments tied to acreage allotments held the Southeast region in cotton production. Production from these southeastern acres contributed toward lower market prices discouraging farmers who would have planted beyond allotments at somewhat more favorable prices. As a consequence, the shift of the share of total planted acres toward the West did not gain momentum until the 1973 crop year, as shown in Table 5, when market price began a strong advance prior to and during planting time. The higher market price made the price-support payments tied to acreage allotments less important in determining regional shares of planted acres. The sharp decline in po-

**TABLE 5. Total Planted Acres of Cotton in Each of Four Regions, 1964–1983[a]**

| Year | West[b] | Southwest[c] | Delta[d] | Southeast[e] |
|---|---|---|---|---|
| | percent | | | |
| 1964 | 9.0 | 46.1 | 28.2 | 16.7 |
| 1965 | 9.0 | 45.5 | 28.9 | 16.6 |
| ----- | | | | |
| 1966 | 10.0 | 45.5 | 28.9 | 15.6 |
| 1967 | 10.3 | 46.4 | 28.8 | 14.5 |
| 1968 | 10.6 | 44.7 | 30.6 | 14.1 |
| 1969 | 9.9 | 47.8 | 29.4 | 12.9 |
| 1970 | 9.2 | 48.4 | 29.8 | 12.6 |
| ----- | | | | |
| 1971 | 9.8 | 46.2 | 31.1 | 12.9 |
| 1972 | 9.6 | 44.0 | 34.3 | 12.1 |
| 1973 | 11.3 | 47.9 | 29.2 | 11.6 |
| ***** | | | | |
| 1974 | 13.5 | 42.4 | 33.2 | 10.9 |
| 1975 | 13.8 | 49.9 | 28.6 | 7.7 |
| 1976 | 13.5 | 44.3 | 33.9 | 8.3 |
| 1977 | 15.3 | 52.6 | 25.4 | 6.7 |
| ----- | | | | |
| 1978 | 16.5 | 56.8 | 22.2 | 4.5 |
| 1979 | 17.5 | 59.7 | 18.2 | 4.6 |
| 1980 | 15.8 | 59.2 | 20.3 | 4.7 |
| 1981 | 16.2 | 56.7 | 21.7 | 5.4 |
| ----- | | | | |
| 1982 | 17.4 | 55.6 | 21.4 | 5.6 |
| 1983 | 17.2 | 54.5 | 22.3 | 6.0 |

[a] Broken lines identify the years covered by each agricultural act and asterisks separate the loan price-target price years from the previous years.

[b] Includes California, Arizona, and New Mexico.

[c] Includes Texas and Oklahoma.

[d] Includes Missouri, Arkansas, Tennessee, Mississippi, Louisiana, and Kentucky.

[e] Includes North Carolina, South Carolina, Georgia, and Alabama.

Source: U. S. Economic Research Service (various issues).

tential and then actual price-support and diversion payments under the 1973 Act combined with the earlier relaxation on planting beyond allotments allowed the share of planted acres to increase substantially in the West and Southwest regions and to decline in the Delta and Southeast regions in the 1974–1979 period. During this period of free-market prices, the redistribution of planted acres is presumably a function of relative regional comparative advantage in cotton production. The 1977 Act solidified this regional shift in production by dissolving the historically based acreage allotments. Benefits of recent years have been based on acres planted in the one to three years preceding the year of payment. In this way, program benefits are generally paid to the more efficient producers rather than to the least efficient producers as was the case under the acreage allotment-based programs. The shift of production is even more dramatic than the shift in planted acres since the West has much higher yields than in the Southeast or Delta regions.

The loan price-target price program should be applauded for allowing cotton production to make dramatic shifts to areas with lower resource costs. This interregional shift of cotton production was apparently impossible with the high loan rate acreage allotment program. More efficient production is good not only on purely economic grounds of resource utilization but it also allows the United States to exploit more effectively its natural comparative advantage in cotton production relative to other areas of the world.

## The 1983 Payment-in-Kind Program

The need for a PIK program to reduce substantially the planted acres and, thus, the total cotton production and carry-over can be attributed entirely to unusually high yields of cotton in 1981 and 1982. The 1981 and 1982 yields were, respectively, 15 percent and 25 percent above the average of the preceding five years (Appendix Table 1). Prior to 1981–82, the 1964–65 period held the record for the highest cotton yield during a two-year period; the 1981–82 yield was nearly 10 percent above that record.

If the carry-over from the 1981 and 1982 crops is computed using the average yield of the 1976–1980 period—holding acres, domestic consumption, and exports constant—the ending stocks after the

1982 crop would have been 3.3 million bales rather than the actual 7.8 million bales. A 3.3-million bale carry-over is a level at which cotton prices are still fairly responsive. (Note that this tends to understate the lower yield carry-over because, with lower carry-over, anticipated and actual prices would have been higher, and both domestic consumption and exports would have been somewhat lower.)

The unusually high yields of 1981 and 1982 coincided with a rather rapid decline in inflation, a very large increase in the trade-weighted value of the dollar, and the world economy in recession. The combination and magnitude of all of these forces, simultaneously directed toward lowering prices for cotton in 1981–1983, should be considered truly extraordinary and well beyond the conditions that the loan price-target price program could or should be expected to control in terms of supply and income. The PIK program should be considered an extraordinary program that was needed to cope with the extraordinary situation that existed at the end of the harvest of the 1982 cotton crop.

The PIK program has been criticized severely for its "excessive" cost and "seat of the pants" administration, but the critics either do not understand or do not acknowledge the conditions that prevailed at the time the decision was made to implement the program. With substantial cooperation from the weather that lowered yields considerably on the 1983 crop, the PIK program solved the excess supply problem in cotton in just one year.

It is clear that the PIK program for cotton was more generous than hindsight tells us it needed to be. The participation rate was much higher than the administration expected it would be making it difficult for the CCC to gain control of enough cotton to satisfy its responsibility to deliver cotton to participants. One of the reasons the participation rate was much higher than it had been for recent cotton programs was that the payments in cotton under the PIK program were exempt from the $50,000 payment limitation.

## *Some Potential Problems for Loan Price-Target Price Programs*

This section will discuss some potential problems for loan price-target price programs that may arise because of the level at which

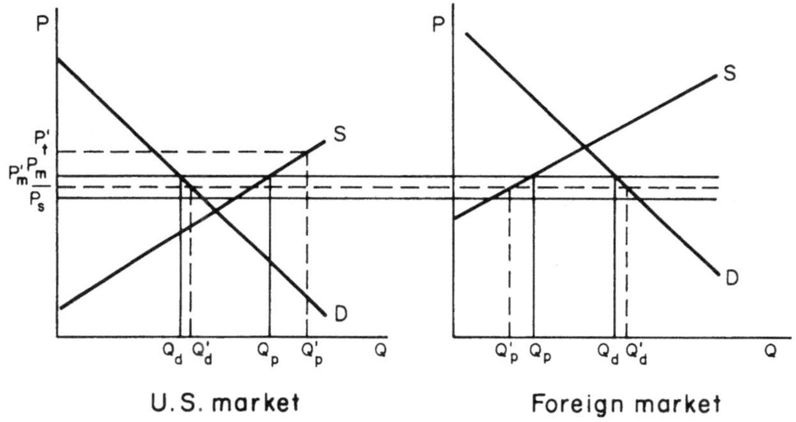


FIGURE 1. U.S. and Foreign Markets for Cotton

target or support prices are set or because of other features of the programs. In general, these problems have been of minimal importance because, to date, prices have been set low.

## Target Price Set Too High

Figure 1 provides the basis for discussing the consequences of a target price set too high. $P_m$ is the price determined by supply and demand in the U. S. and foreign markets; $P_s$ is the loan price available on nonrecourse loans; $Q_d$ is the quantity demanded in each market; and $Q_p$ is the quantity produced in each area. The supply curve in each market represents farmer response to a price available with certainty. In the figure, $P_s$ has been set below $P_m$ reflecting setting the support price at 90 percent of the world price and the fact that no cotton has been forfeited on the loan program during the loan price-target price program. If the target price, $P_t$, is set at or below $P_m$, the program will not have any effect as was the case during much of the time after it began with the 1974 crop. $P_t'$ is a target price set above $P_m$. Now, U. S. cotton farmers will be induced to produce $Q_p'$, a quantity that will force the market price down to $P_m'$ where the quantity consumed in the United States ($Q_d'$) plus imports by the foreign market ($Q_d' - Q_p'$) equals $Q_p'$.

A target price even higher than $P_t'$ could force the market price down to the loan price, $P_s$, with further increases in domestic consumption and exports. Because $P_m$ cannot drop below $P_s$, a further increase in the target price would result in CCC accumulations of cotton. ($Q_p$ would exceed $Q_d$ in the domestic market plus $Q_d$ minus $Q_p$ in the foreign market.)

Total deficiency payments would be equal to $Q_p'$ multiplied by the difference between $P_t'$ and $P_m'$. The loan price constrains the market price and, therefore, also constrains the deficiency payment per pound; but the total deficiency payment could continue to rise in direct proportion to increases in $Q_p'$ resulting from higher $P_t'$.

Voluntary acreage reduction programs could reduce CCC accumulations, raise the market price, reduce deficiency payments per pound of cotton, and reduce total payments. An acreage reduction program would also reduce domestic consumption, reduce foreign consumption, increase foreign production, and give some of the cotton market share to man-made fibers. If the payment for reduction of planted acres is a flat amount per pound, the total payment to produce price $P_m$ would have to be $P_t'$ minus $P_m$ multiplied by $Q_p'$ minus $Q_p'$. If, instead, the payment is based on farmer bids with the lowest bid price per pound accepted, the payment could be reduced. Acreage reduction programs imposed on either a high target price or a high loan price program require that someone attempt to set both price and quantity to clear the markets. The likelihood of success in setting both price and acreage reduction programs in perfect coordination is nil. Problems in structuring an acreage reduction program to ensure a particular reduction in acres are well known; in addition, there is substantial variability in yields per acre. Although it implies lower average prices, the loan price-target price program would function much more efficiently without acreage reduction programs. If the combined domestic and foreign demand for U. S. farm output is elastic, as some now believe, output restriction would not seem to be a wise program applied to U. S. cotton production which seems to have a rather strong comparative advantage (Johnson, 1984).

The target price should be set low enough that deficiency payments would average relatively small amounts, and the loan price should be set low enough that average loan forfeitures would be small. The long-term trends of declining real prices of commodities imply that supply is shifting to the right at a relatively faster rate than the demand. If the $P_t$ and $P_s$ are not allowed to drop at the rate of the long-term decline in $P_m$, the consequences eventually will be similar to those already discussed for a target price set above the market price.

What justifications can be given for continuing a government program with relatively low loan and target prices and no production restrictions such as for the 1974–1980 crops? One justification would be that a reduced variability in expected price would result in more efficient utilization of resources and a somewhat larger average output which would allow U. S. agriculture to utilize more effectively its comparative advantage. The continual use of the loan program, even when the loan price is far below the market price, says that cotton farmers have found the program valuable. The value probably comes in terms of slightly subsidized interest rates, easy availability of loans, and some price protection.

Another possible justification is that agriculture has great difficulty in capturing the benefits of new technologies because the competitive system passes these on so quickly to consumers and other segments of the economy. Perhaps, this service that agriculture provides should earn some reward in income transfers when its income would otherwise be depressed severely. However, this argument is weak because the deficiency payment system is a very ineffective way of making transfers to those in agriculture affected adversely by the adoption of new technologies.

## The Time Period for Computing Average Market Price

The 1970 Act required the computation of an average price received by farmers as one of the determinants of payments to cotton farmers. For the 1970–1973 crops, the computation was based on the first five months of the marketing year (August through December). The cotton harvest with months ranked by volume of harvest would typically be October, November, December, September, and January; thus, the 1970 Act included the four heaviest harvest months and then substituted August (a nonharvest month) for January (the fifth heaviest harvest month), providing a reasonable compromise between accuracy in reflecting actual prices received by farmers and allowing fairly early payments to them.

The 1973 Act and later acts changed the period for computing the average cotton price received by farmers to the calendar year in which the crop was grown, but no change was made in the earliest date payments could be made to farmers. Also, cotton sales were added to the computations, but sales have little relevance in determining what should be the appropriate deficiency payment on a given crop. For example, if farmers delay selling a large portion of one year's crop until the first few months of the following year, payments would be low for the low-sales year and high in the year they sell the most. As a result, the deficiency payment system has considerable potential for destabilizing gross income from cotton and even more for destabilizing net income. This destabilizing effect could be reduced substantially by changing the computation back to the time period specified in the 1970 Act.

## *Summary and Conclusions*

The loan price-target price program for the 1974 through 1985 cotton crops has generally performed well to this point. This performance was especally good for the 1974–1980 cotton crops when no price-support payments and very little in diversion payments were made for cotton (Appendix Table 2). Substantial disaster payments were made in some of those years, but I do not consider disaster payments a direct part of the loan price-target price program. The disaster payments probably had a great effect in stabilizing incomes of individual farmers when they experienced unusually adverse growing conditions in certain years.

It is reasonable to wonder whether the very low cost of the loan price-target price program during the 1974–1980 period was caused merely by unusually favorable conditions outside the agricultural sector. Generally, accelerating inflation and the declining value of the dollar in foreign exchange both contributed to market prices of cotton rising faster and generally remaining above the target price with the loan price far below the market price throughout the period. In spite of the large gap between the loan price and the market price, farmers used the loan program extensively but almost universally repaid their loans plus interest and storage costs.

The 1981–1983 period was basically the reverse of the 1974–1980 period in terms of outside influences. The rate of inflation dropped substantially, the value of the dollar in foreign exchange strengthened dramatically, and real interest rates were at historically high levels discouraging the ownership of stocks of cotton. Elsewhere in the paper I have argued that the unusually high cotton yields in both 1981 and 1982 combined with these outside influences resulted in extraordinarily adverse conditions that were neither caused by nor within the reasonable scope of the loan price-target price program to solve.

How well would the loan price-target price program have functioned under "normal" conditions? The precise answer to that question for the 1974–1983 period will never be known. The decline in the foreign exchange value of the dollar in the 1972–1980 period was a combination of the move away from fixed exchange rates with the dollar overvalued and generally accelerating inflation in the United States (Johnson, 1984). I believe that the move to flexible exchange rates with no inflation in the 1974–1983 period would have resulted in some deficiency payments in the 1974–1980 period. The unusually high yields of the 1981–82 crops would have resulted in large deficiency payments and some strain on the program; but the loan price-target price program could have solved the problem without mobilization of the extraordinary PIK program of 1983.

The organizers of this conference have asked the authors to respond specifically to each of the following four "policy options." These options are necessarily mutually exclusive.

**Continuation of the Food and Agriculture Act of 1981.**—I do not favor a simple extension of the 1981 Act for cotton because I believe it contains one serious flaw and another potentially very serious flaw. The actual flaw is the use of the calendar year for computing the average farm price in determining deficiency payments that I have discussed above. The potential flaw is that there is no mechanism in the current legislation to assure that the real loan and target prices will move down over time at the natural rate of decline of commodity prices. Without this device for moving the real program prices down over time, it is possible that the program prices will rise relative to free-market prices and that the program could be a major drain on the U. S. Treasury and a major distorter of prices and resource allocation.

I believe that a continuation of the 1981 Act, with the modifications alluded to here, would make a very desirable program for cotton. The 1985 crop loan price and target price levels (especially since the 1985 target price was lowered to 81 cents) seem to be very satisfactory levels from which to begin to reduce these prices by their natural real rate of decline. The average costs per year of the program will probably be more than for the 1974–1980 period but less than during the 1981–82 period.

**Adjustment of the Food and Agriculture Act of 1981 to Make Policy Instruments Sensitive to Key Variables in the Domestic and International Economic Environment.**—This option implies that the program would be made more complex and that program prices and other policy instruments would be less predictable. Under free-market conditions, farmers face an almost overwhelming level of uncertainty; this uncertainty is one of the major justifications for having government commodity programs. Farmers do not need government commodity programs that increase rather than reduce their uncertainty. However, the adjustment of the period for computing the average farmer price for cotton is logically a matter of making the deficiency payment more "sensitive to a key variable": the actual market price received for a given cotton crop. The mechanism for reducing the real program prices over time would necessarily involve the rate of inflation unless inflation held at a zero rate.

**A Radical Alternative to the Food and Agriculture Act of 1981 with No Specific Commodity or Land Management Policies.**—From the perspective of my assignment to review policy options for cotton, it is not logical for me to react to this policy option. Other writers with broader assignments can deal more effectively with this policy option.

**No Government Intervention.**—I believe that a modified price-support, target price program would be superior to no government intervention. The principal justification for government intervention would be that assured minimum loan and target prices, even substantially below recent past market prices, would have favorable effects on resource utilization. A secondary, but less easily defended, justification would be that agriculture deserves some income transfer when its income from the market is unusually low because society, rather than farmers, receives most of the benefits from the adoption of new technology by farmers. However, commodity price programs have proven to be unsatisfactory in distributing benefits to the farmers most adversely affected by new technology.

**APPENDIX TABLE 1. Upland Cotton Production and Ending Carry-Over, 1964–1983[a]**

| Crop year | Planted acres | Yield per harvested acre | Production | Ending carry-over |
|---|---|---|---|---|
| | thousand acreas | pounds per acre | thousand bales | |
| 1964 | 14,725 | 517 | 15,025 | 13,980 |
| 1965 | 14,075 | 527 | 14,850 | 16,734 |
| ---------- | | | | |
| 1966 | 10,269 | 480 | 9,484 | 12,081 |
| 1967 | 9,381 | 446 | 7,374 | 6,379 |
| 1968 | 10,844 | 516 | 10,847 | 6,377 |
| 1969 | 11,805 | 433 | 9,913 | 5,727 |
| 1970 | 11,869 | 439 | 10,135 | 4,134 |
| ---------- | | | | |
| 1971 | 12,253 | 438 | 10,379 | 3,182 |
| 1972 | 13,903 | 507 | 13,608 | 4,153 |
| 1973 | 12,395 | 521 | 12,896 | 3,753 |
| ********** | | | | |
| 1974 | 13,596 | 441 | 11,450 | 5,649 |
| 1975 | 9,408 | 453 | 8,247 | 3,615 |
| 1976 | 11,590 | 464 | 10,517 | 2,879 |
| 1977 | 13,604 | 519 | 14,277 | 5,278 |
| ---------- | | | | |
| 1978 | 13,298 | 419 | 10,762 | 3,905 |
| 1979 | 13,887 | 547 | 14,531 | 2,962 |
| 1980 | 14,461 | 402 | 11,018 | 2,614 |
| 1981 | 14,272 | 542 | 15,566 | 6,567 |
| ---------- | | | | |
| 1982 | 11,269 | 590 | 11,864 | 7,844 |
| 1983[b] | 7,899 | 504 | 7,634 | 3,200 |

[a] Broken lines identify the years covered by each agricultural act, and asterisks separate the loan price-target price years from the previous years.

[b] Preliminary.

Source: U. S. Department of Agriculture, Agricultural Stabilization and Conservation Service (1983); and U. S. Economic Research Service (various issues).

**APPENDIX TABLE 2. Government Payments to Cotton Farmers and Total Value of Cotton Sold by Farmers, 1964–1982[a]**

| Crop year | Payment programs | | | Total payments | Value of production |
|---|---|---|---|---|---|
| | Price support | Diversion | Disaster | | |
| | 1,000,000 current dollars | | | | |
| 1964 | 39.3 | [b] | | 39.3 | 2,230 |
| 1965 | 69.3 | | | 69.3 | 2,086 |
| ------ | | | | | |
| 1966 | 489.4 | 284.7 | 2.2 | 776.3 | 981 |
| 1967 | 611.7 | 323.3 | | 935.0 | 938 |
| 1968 | 638.4 | 144.7 | | 783.5 | 1,194 |
| 1969 | 796.8 | 26.4 | | 823.2 | 1,040 |
| 1970 | 889.9 | 24.9 | | 914.8 | 1,110 |
| ------ | | | | | |
| 1971 | | 818.3 | | 818.3 | 1,399 |
| 1972 | | 807.3 | | 807.3 | 1,778 |
| 1973 | | 704.6 | | 704.6 | 2,747 |
| ****** | | | | | |
| 1974 | | | 127.8 | 127.8 | 2,346 |
| 1975 | | | 117.6 | 117.6 | 2,023 |
| 1976 | | | 98.0 | 98.0 | 3,223 |
| ------ | | | | | |
| 1978 | | 40.5 | 187.8 | 228.3 | 3,004 |
| 1979 | | | 107.5 | 107.5 | 4,344 |
| 1980 | | | 302.0 | 302.0 | 3,933 |
| 1981 | 467.4 | | 81.3 | 548.6 | 4,038 |
| ------ | | | | | |
| 1982 | 550.0 | | 105.0 | 655.0 | 3,322 |

[a] Broken lines identify the years covered by each agricultural act, and asterisks separate the loan price-target price years from the previous years.

[b] Blanks indicate zero value.

Source: U. S. Department of Agriculture, Agricultural Stabilization and Conservation Service (1983).

## *References*

Evans, Sam. "Analysis of Upland Cotton Farm Programs Since 1965." *Cotton and Wool Situation.* U. S. Department of Agriculture, Economics, Statistics, and Cooperative Service. Washington, D. C.: U. S. Government Printing Office, May, 1980.

Firch, Robert S. "Adjustments in a Slowly Declining U. S. Cotton Production Industry." *American Journal of Agricultural Economics* 55, No. 5 (December 1973):892–902.

Glade, Edward H., Jr. "Analysis of U. S. Raw Cotton Content of Textile Imports." *Proceedings: Beltwide Cotton Conference.* National Cotton Council, January 1983.

Johnson, D. Gale. "United States Agricultural Policy—National and International Contexts." University of Chicago, Office of Agricultural Economics Research, Paper No. 84:3, February 1984.

McArthur, W. C. *Agricultural Programs Affecting Cotton.* U. S. Department of Agriculture, Commodity Economics Division Working Paper. Washington, D. C.: U. S. Government Printing Office, September 1977.

U. S. Council of Economic Advisers. *Economic Report of the President.* Washington, D. C.: U. S. Government Printing Office, February 1984.

U. S. Department of Agriculture, Agricultural Stabilization and Conservation Service. *ASCS Fact Sheet: Upland Cotton.* Washington, D. C.: U. S. Government Printing Office, May 1983.

U. S. Economic Research Service. *The Cotton Situation.* Washington, D. C.: U. S. Government Printing Office, various issues.

# 9 TOBACCO AND PEANUTS

**Dale M. Hoover**
**Daniel A. Sumner***

Tobacco and peanuts are frequently linked in policy discussions: They are grown in the South; they have been labor intensive; production units have been small in land area relative to most field crops; and they are localized crops with a considerable history of quota and allotment systems and apparent producer power in Congress. However, there are major policy and production differences as well. Consequently, we will handle the two crops separately, putting most of our emphasis on tobacco because it is economically more important and because the potential for, and impact of, policy changes are much greater.

## *Tobacco*

Tobacco policy is different from that of most other commodities in that it does not have to be rewritten periodically. While the basic tobacco program has not changed, there have been major alterations

*The authors are, respectively, Professor and Head, Department of Economics and Business, and Associate Professor, Department of Economics and Business, North Carolina State University.

of program features in each of the last three years. Further, the current law has provisions in place that imply further changes in the next few years including movements toward a more competitive price in the world market. Therefore, we will spend most of our efforts providing background and examining recent policy changes. The effect on producers of a possible discontinuation of the program is analyzed to conclude the section on tobacco policy. For a further description of the tobacco industry and policy and an extensive list of references, see Sumner and Alston (1984).

## Summary of Previous Tobacco Policy and Industry Characteristics

The program operates under the Agricultural Adjustment Act (AAA) of 1938, which has been amended many times, but the basic points have not changed. They provide for a national quota and price-support mechanism for the various types of tobacco, each subject to the approval of two-thirds of the producers in periodic referenda. The two major types are burley and flue-cured, accounting for approximately 30 percent and 60 percent of average annual tobacco production, respectively. North Carolina is the dominant flue-cured tobacco state, while Kentucky is the major burley tobacco producer (Table 1). Most other types have allotment and price-support programs; however, Maryland tobacco (ranking third in value) has had no program for the last few years. Burley and flue-cured tobacco quotas have been in force since 1934. The date for the next flue-cured tobacco referendum is January, 1986; for burley, it is February, 1986.

## Price Supports

Price supports have been set by a formula that has been changed a number of times—in 1940 to raise the real price level and then to lower it in 1960 and 1982. Concern over the decline in the U. S. share in world trade has been the cause of each price adjustment. The export market has taken a constant amount of flue-cured tobacco, but the U. S. share in the volume of trade and in world production has fallen dramatically (Table 2). Lately, imports of to-

**TABLE 1. Distribution of Tobacco Production Rights by State, 1982**

| State | Burley | | Flue-cured | |
|---|---|---|---|---|
| | Number of allotments | Tobacco production | Number of allotments | Tobacco production |
| | | thousand pounds | | thousand pounds |
| Florida | 0 | 0 | 7,307 | 21,322 |
| Georgia | a | a | 25,767 | 103,940 |
| Indiana | 10,259 | 19,381 | 0 | 0 |
| Kentucky | 148,238 | 488,066 | 0 | 0 |
| North Carolina | 18,866 | 35,833 | 117,102 | 642,489 |
| Ohio | 11,021 | 27,927 | 0 | 0 |
| South Carolina | a | a | 24,173 | 123,546 |
| Tennessee | 95,915 | 157,521 | 0 | 0 |
| Virginia | 16,640 | 33,290 | 20,873 | 84,723 |
| Others | 5,879[b] | 15,822[b] | 279[c] | 807[c] |
| Total | 306,818 | 777,840 | 195,501 | 976,827 |

[a] Minor amounts included with "others."

[b] States include Alabama, Arkansas, Georgia, Illinois, Kansas, Missouri, South Carolina, and West Virginia.

[c] For Alabama only.

Source: U. S. Agricultural Marketing Service (1982).

bacco into the United States have risen rapidly as well—up from 10 percent to about 30 percent of the content of a U. S. cigarette.

Operationally, the price-support formula establishes the mean price for a type of tobacco. Then an array of supports by grade is established which attempts to shadow-price the commercial market. An overestimate in a relative grade support price leads to relatively more of that grade being acquired by the support agency. It is the average support price that counts for overall program effects because relative grade prices are readjusted each year.

**TABLE 2. Average U. S. Share of World Tobacco Production and Export**

| Year | Burley | | Flue-cured | | Total[a] | |
|---|---|---|---|---|---|---|
| | Production | Export | Production | Export | Production | Export |
| | percent | | | | | |
| 1955–1959 | 82 | 60 | 41 | 60 | 23 | 35 |
| 1960–1964 | 80 | 57 | 40 | 52 | 25 | 30 |
| 1970 | 62 | 33 | 30 | 46 | 19 | 28 |
| 1975 | 52 | 27 | 28 | 33 | 18 | 20 |
| 1980 | 45 | 27 | 20 | 29 | 15 | 21 |
| 1981 | 52 | 21 | 18 | 26 | 16 | 18 |
| 1982 | 49 | 25 | 13 | 25 | 13 | 18 |

[a] Includes Puerto Rico.

Source: U. S. Department of Agriculture (1983).

With the exception of a small amount of subsidized tobacco under the Food for Peace Program, tobacco purchased by producer cooperatives with nonrecourse loans from the Commodity Credit Corporation (CCC) has been resold to the trade. In years past, profits were paid out as a patronage dividend to producers, while losses were passed on to the CCC. This has been changed by the No-Net Cost Program of 1982 which will be discussed later. Losses were few; and stocks did not increase without bound as they did for cotton, wheat, and feedgrains during several periods. Tobacco was different because quota determination is tied to total stocks including those held by the cooperatives in conjunction with the maintenance of price supports. It is these features that historically kept the U. S. Treasury losses low and gave tobacco policy its good name.

## Quota Ownership and Distribution

At several points, important changes have been made in the supply restriction policy either by legislation or by introducing the change in the referendum submitted to and approved by producers.

These include intracounty lease and farm-to-farm transfer of allotments for flue cured in 1961 and burley in 1971; the shift from acreage allotments to individual farm quotas for flue cured in 1965 and burley in 1971; and loose-leaf marketing of flue-cured tobacco to accord with new mechanical harvesting and bulk-cured tobacco technologies in 1966. A prominent effect of flue-cured and burley tobacco policy has been to freeze the location of production. Movement of quotas across county lines has been virtually prohibited, but quota has been movable among producers within counties.

## Costs of the Program

The cost to the U. S. Treasury for tobacco programs has been relatively low. Most tobacco purchased by producer cooperatives in connection with price-support operations has been sold back to the trade without significant nominal loss because quota has been reduced whenever stocks have risen.

Even though the U. S. Treasury losses on price supports and administrative costs have been low, there has been a general misconception that government policy has stimulated production and encouraged smoking. Of course, the reverse is true. The quota programs restrain output and raise tobacco prices above the level that would otherwise prevail. In fact, the quota program operates just as excise taxes do. The "tax" collected by quota owners is significant but only a fraction of the amount collected by the federal, state, and local governments on cigarettes.

## Farm Size and Farmer Characteristics

Production consolidation has proceeded very rapidly for flue-cured tobacco and somewhat more moderately for burley (Table 3). Between 1954 and 1982, the number of flue-cured tobacco farms in the United States fell more than 6 percent per year. In 1954 there were six times as many farms reporting sales of flue-cured tobacco as in 1982. At this rate of decline, there would be approximately 20,000 flue-cured tobacco producers by 1992. The pattern for North Carolina flue-cured tobacco farms was similar, and the

**TABLE 3. Number of Farms Growing Tobacco**

| Year | Tobacco | | | | All tobacco | All farms | | |
|---|---|---|---|---|---|---|---|---|
| | Burley | | Flue-cured | | | | | |
| | United States | Kentucky[a] | United States | North Carolina[b] | United States | North Carolina | Kentucky | United States |
| | thousands | | | | | | | |
| 1954 | 240 | 116 | 227 | 135 | 512 | 268 | 193 | 4,782 |
| 1959 | 213 | 103 | 164 | 98 | 417 | 191 | 151 | 3,711 |
| 1964 | 168 | 89 | 129 | 79 | 331 | 148 | 130 | 3,158 |
| 1969 | 157 | 85 | 92 | 58 | 276 | 119 | 125 | 2,730 |
| 1974 | 119 | 67 | 60 | 37 | 198 | 91 | 102 | 2,314 |
| 1978 | 124 | 72 | 49 | 33 | 203 | 89 | 110 | 2,479 |
| 1982[c] | 123 | 69 | 37 | 24 | 179 | 73 | 102 | 2,241 |

[a] Has been adjusted to exclude farms growing flue-cured and other types of tobacco.

[b] Has been adjusted to exclude farms growing burley tobacco.

[c] Preliminary.

Source: U. S. Bureau of the Census (various years).

decline was more rapid for flue-cured tobacco farms than for all farms in the state.

Burley farm numbers fell by 50 percent from 1954 to 1974 but have remained roughly constant since that time. This pattern also holds for Kentucky burley farms and all Kentucky farms. (The U. S. farm numbers also declined by over one-half in the same period.) Technological changes have occurred less rapidly in the burley crop than in flue-cured tobacco. Technology and returns to scale led to the introduction of lease and transfer of quota 10 years earlier in flue-cured tobacco which, in turn, probably helped to speed farm consolidation. For the last decade, burley and flue-cured tobacco programs have been very similar; and, except for limiting the counties in which tobacco is produced, tobacco programs do not seem to have greatly affected the growth in farm size.

Recent data (U. S. Bureau of the Census, various years) show that tobacco enterprises typically have fewer acres than those for most other field crops. In 1978, the (predominantly flue-cured) tobacco farms of North Carolina averaged tobacco sales of about $26,700 and total (average) farm sales of about $32,000. Kentucky tobacco farms sold an average of about $6,900 of tobacco and about $9,000 in total sales. Average total sales of all farms in these "tobacco" states were about $34,000 for North Carolina and about $17,000 for Kentucky, compared to a national average of $43,600. Flue-cured tobacco farms are closer to the state and national average measured by sales, while burley farms are smaller than Kentucky or national average farms. For flue-cured tobacco farms, only about one-quarter of the farms sold less than $10,000 while, for burley farms, almost two-thirds were in this category in 1978. Preliminary acreage figures indicate that there was little change in these distributions between 1978 and 1982. Again, flue-cured tobacco farms tend to be much larger. Note that there have been no significant program differences to account for these size distribution differences.

Census data for 1974 and 1978 show that the age distribution of tobacco farmers is very similar between the two types and is also quite similar to all U. S. farmers. Flue-cured tobacco farmers were slightly more middle aged than burley farmers. The change in the tobacco age distributions reflected a national trend toward younger farmers between 1974 and 1978. Schooling levels for tobacco farmers are available from sample survey data in the mid-1970s. These indicated that there were no differences between burley and flue-cured tobacco farmers; about 40 percent of tobacco farms have 12 years of schooling or more. Off-farm employment patterns are a further indication of farmer characteristics. Farmers producing flue-cured tobacco are somewhat less involved in off-farm work than the national average, while burley farmers do more off-farm work. These numbers are consistent with the size relationships noted earlier. It seems that there is no particular policy influence in any of these patterns. Further, it seems that tobacco farmers are comparable to all U. S. farmers in basic characteristics which may indicate abilities to adjust to economic and policy change.

## Producer-Quota Owner Policy Differences

When output is restricted, the inelastic resource claims the residual between total revenue and cost of production. The value of the tobacco program resides in the transferable acreage allotments and quotas. The onset of the lease and transfer program produced a handy index of the value of the program. Rent per pound in the more productive counties has averaged about 30 percent of the market price for both burley and flue-cured tobacco. There is a clear separation between the interests of producers (1) who owned little of the quota for tobacco they produced and (2) quota owners who produced little or no tobacco. Producers sell their managerial services and benefit from a greater volume of the crop; quota owners wish to maximize rental income (volume times rent). Growers who lease most of their quota have the same interest in greater volume as farm suppliers, the marketing subsector, and manufacturers. This conflict among tobacco interests is an additional divisive element in the legislative process.

## Recent Restrictions on Quota Allocation

Some provisions of the 1982 legislation are directed toward the conflict between quota owners and lessees. In addition to requiring nonfarm corporations and certain other institutions to sell their flue-cured tobacco quotas before December 31, 1984, it allowed the sale of flue-cured tobacco quotas (without land) within counties. Burley quotas have to be sold by institutions but cannot be sold by individuals, reflecting less sharply drawn differences between owners and lessees in burley areas as well as less pressure to expand production unit size. The 1983 legislation includes a provision for elimination of lease and transfer of quotas by the end of 1986. Nonproducing quota owners could again become producers or rent their tobacco land. Both of these options have clear costs so this legislation creates pressure for a transfer of quota ownership to active tobacco producers. If this reduction in the proportion of quotas owned by nonproducers occurs, it will decrease dissension but, also, it will reduce the number of interested voters and could weaken future congressional responsiveness. A move to allow the relocation of burley or flue-cured tobacco to the high rent (low cost) regions would reduce the number of congressional districts in which tobacco would be important, but this move has not been actively pursued by those in the industry.

## Real Price-Support Reductions

The 1982 tobacco legislation set the price-support adjustment equal to only two-thirds rather than 100 percent of the advance in the producer price index. In 1983 Congress froze the nominal flue-cured tobacco support price for 1983, 1984, and 1985 crops. Burley price support in these years is required to maintain its historic relationship with flue-cured tobacco; it has been effectively frozen as well.

The decline of the nation's share of world tobacco export trade and booming imports has caused many to conclude that the demand faced by U. S. producers is more elastic than had been expected in setting the degree of production restriction and the level of recent price supports. Several schemes have been suggested to deal with the more elastic demand. One is to assess quota owners for export subsidization. This would, of course, be a two-price plan and would be subject to retaliation and challenge under the General Agreement on Tariffs and Trade (GATT). Another plan has been the request that quotas or tariffs be placed on imports. This proposal has been presented to the last two presidents and may yet lead to action even though it runs against the long-run movement toward freer trade in tobacco.

A third scheme would allow U. S. production outside of the quota system for selected lower priced grades which would then not be eligible for price supports. The implicit assumption is that the elasticity of demand is greater for lower priced grades that compete directly with tobacco produced abroad. If U. S. producers are allowed to compete for these markets, imports might fall and, perhaps, exports would increase.

The price-support freeze has considerable impact on quota owners, producers, and other tobacco sector interests. Assuming no nominal price-support increases between 1982 and 1985 and a 4 percent per year increase in all other prices, the real price of tobacco will have fallen by 12 percent by the end of this period.

One impact of a fall in real price, assuming no further reductions in quota, would be in the lease paid per pound. A 12 percent decline in the real price support would be about 20 cents. Since quota is an inelastic "input" used in fixed proportions (in 1984 dollars), the quota lease rate would fall by about this same amount. In the highest lease-rate counties, this would be equal to about one-third of the lease rate. In the highest cost of production counties, the lease rate would fall to zero and quota would not be fully produced. This could bring about the first major shift in the relative importance of the various geographic regions in the last 40 years.

Consumption would be expected to increase some at home (primarily by displacing imports) and, even more abroad (after a time), displacing foreign competitors. The aggregate price elasticity of demand in the intermediate run is probably greater than one and, perhaps, greater than two (Sumner and Alston, 1984). Consequently, given time for storage stocks to be worked down, aggregate quota might increase 20 percent or more relative to the alternative of no freeze. In regions of medium to high lease rate, quotas could expand even more as some other regions reduced production.

Given more quota but a lower lease rate, the aggregate lease income of quota will probably decline, and the net income of producers who do not own quotas might increase. This hypothesis is very tentative because the working off of current stock may defer or disguise the effects of the freeze. In addition, it is possible that the underlying demand and foreign supply conditions have shifted more than has yet been detected. In this case the income of producers and quota owners may decline, and quotas may be cut still further. In such a case the effects traced above represent directions of impact, not necessarily observed results.

## The "No Net Cost" Provisions

The No-Net Cost Act of 1982 provides that an assessment shall be made on quota holders and producers by the cooperatives sufficient to pay CCC interest on the stocks they acquire in price-support operations. In each crop year the cooperatives estimate the portion of the crop they will acquire, the average number of years the purchase will remain in storage, and the interest rate they will have to pay. The proposed assessment is submitted to the U. S. Department of Agriculture (USDA) for final approval. If any one of the four expected quantities in the formula is incorrect (portion of crop acquired, interest rate, average storage period, and size of the crop that will be marketed), assessments will be too high or too low. This procedure implicitly assumes that the price of stocks will rise enough to cover noninterest costs of storage. No provision for absorbing losses has been made should discounting prices be required to move the stocks into trade channels.

In the first three years of operation (1982, 1983, and 1984), the per pound assessments were for burley, 1 cent, 5 cents, and 9 cents, respectively; for flue-cured, 3 cents, 7 cents marketed plus 7 cents leased, and 7 cents, respectively. The "double" assessment for flue-cured tobacco quota in 1983 meant that, for each pound of tobacco quota that was leased and transferred, the quota owner was also required to pay a 7 cent assessment. But this disincentive to lease or transfer was not effective in significantly reducing the practice and was discontinued with the expectation that overall assessment would rise. It now appears that the USDA acquiesced in an assessment rate that will fall short of costs for 1984.

In the last decade or so, cigarette manufacturers and dealers have maintained storage stocks about three times annual usage. Typically, tobacco is stored at least a year before being used in cigarettes. Further storage probably leads to positive quality changes as well as the flexibility to blend different vintages to retain tastes and other consumption attributes from year to year. However, cooperative stocks may not advance enough to cover a sizable portion of costs at the present time because (1) government and private stocks are large (about 3.4 times historic usage), and quota cuts have not been great enough to bring stocks back to what most analysts consider consistent with current demand; (2) usage has declined recently due to an increase in the federal excise tax and more international competition leading to a further decline in demand for U. S. tobacco; and (3) there has been a temporary fall in demand because the nominal price support is expected to be unchanged through the 1985 crop year, while the price level is expected to rise at least 4 percent a year. Faced with a lower real price for new crop tobacco, the demand for stored tobacco has fallen. Buyers may also expect the cooperatives to reduce the price to move the old crop out of storage as they occasionally have in the past.

The leaf sector is faced with a dilemma: cut quota and output sharply in order to move tobacco out of cooperative storage or do not cut quotas and continue to produce, maintaining current gross revenue but increasing the probability of significantly higher assessments. If past assessments do not prove to have been sufficient and a large assessment is levied some year to catch up, it is conceivable that the next referendum could fail, perhaps leaving the cooperatives insolvent. Some in the industry have suggested an elimination of the 1985 price freeze as a solution to the cooperative storage problem.

## Tobacco Production and Income Without a Program

Thus far, the analysis has been directed to current legislation and its developing impact on producers and others. No legislative action is necessary in 1985 which makes it difficult to establish the range of possible future policy actions. Of course, one striking policy change would be for the quota and price-support programs to be discontinued. This could occur if the referenda for burley and flue-cured tobacco were to receive less than the two-thirds vote required or if the program were to be ended legislatively, either gradually or abruptly.

## Price, Quantities, and Incomes

A recent study by Sumner and Alston (1984) projects the new prices and quantities that would occur if the quota and price-support programs were ended and sufficient time were to elapse to allow adjustment. This is a "long run" projection in the sense that the time required for current stocks to be worked off is assumed to pass. Any initial shock to price and incomes is also assumed to be dissipated.

With the end of the program, quota rent would disappear and depress the incomes of quota owners substantially in the range of $800 million per year. Although producers would gain from increased demand for managerial services, most producers also own some quota so they are affected in both ways.

Estimates of supply response are based on the reasoning that (1) tobacco would be produced in the lowest cost production regions; (2) the cost of production in those areas would be somewhat greater than the current price less the current quota rental rate, reflecting some increase in marginal costs as output expands; and (3) there may be a relative shift toward lower cost grades as production expands.

The elasticity of supply assumed in this analysis is high because land and managerial talent are in relatively plentiful supply. Quota occupies less than 10 percent of the cropland in the regions of lowest cost. There is also a large number of experienced tobacco farmers in prime low-cost areas who are not presently producing. As noted briefly above, the supply curve of the lowest cost regions represents the supply curve of unrestricted production. Counties with high costs would no longer produce tobacco.

On the demand side, expansion would occur in four areas: (1) major increased demand by U. S. cigarette manufacturers due to substitution of tobacco produced in the United States for imported leaf, (2) small increased demand by U. S. cigarette manufacturers due to substitution of tobacco for other inputs used in cigarette production, (3) small increased demand by U. S. cigarette manufacturers due to increased demand for U. S. cigarettes in the United States and for export, and (4) major increased tobacco exports due to substitution of U. S. tobacco (for tobacco produced outside the United States) by foreign cigarette manufacturers. All told, these increases could amount to a 50-percent-or-more expansion in consumption from a 25 percent decline in price.

Consumption of U. S. tobacco would increase as imports of low-quality tobacco were displaced. The share of domestic production in U. S. cigarettes would be expected to rebound from its current level of 70 percent to almost 85 or 90 percent. This could amount to about 200 million additional pounds of tobacco or 15 percent of current production. Domestic consumption of cigarettes would expand some, perhaps as much as 1 percent or 10 million pounds in response to a 3–4 percent decline in wholesale prices and a low (-.30) elasticity of demand for cigarettes. The U. S. cigarette sales abroad could be as much as 10 percent greater, in response to a demand elasticity of –3.0, which would yield an expansion of 20 million pounds in tobacco sales. Export of leaf is, perhaps, the most difficult to estimate. At an elasticity of –4.0, exports would double, increasing sales by 800 million pounds. The sum of these forces amounts to 1 billion pounds or 50 percent of current production arising from an aggregate elasticity of demand (weighted) of –2 and a 25 percent decline in price. Consumers at home and abroad would gain from a lower product price. Table 4, using a range of elasticity estimates, summarizes some of the basic results of eliminating the program.

## Other Considerations

In the absence of programs, farm prices would vary more than they have in the recent past, but it is difficult to estimate how great the variance would be. Price variability is caused by shifts in the supply and demand curves and in the slopes of the curves. On the supply side, evidence on yields by regions suggests that year-to-year yield variance, usually ranging within 10 percent of the mean, would remain unaffected by removal of the program. Year-to-year variation in planting could increase, but the specialized capital used for producing and curing would limit this source of variability. The question of price variability hinges mostly on demand shifts and the slope of the aggregate demand curve. The price-support process made the demand curve infinitely elastic for a given crop year but more responsive over a several-year period. In the absence of price

**TABLE 4. Effects of Deregulation on Prices, Quantities, and Value of U. S. Tobacco for a Range of Supply and Demand Elasticities[a]**

| Elasticity: Supply (S) | Elasticity: Demand (D) | Change in U. S. tobacco price | Increase in U. S. tobacco output | Increase in tobacco revenue |
|---|---|---|---|---|
| | | percent | | |
| 5 | −2 | −23.4 | 46.9 | 14 |
| | −3 | −21.1 | 63.4 | 30 |
| | −4 | −19.2 | 76.9 | 47 |
| | −5 | −17.7 | 88.3 | 55 |
| 10 | −2 | −26.3 | 52.6 | 13 |
| | −3 | −24.8 | 74.4 | 30 |
| | −4 | −23.4 | 93.8 | 48 |
| | −5 | −22.2 | 111.1 | 64 |
| infinity | −2 | −30.0 | 60.0 | 12 |
| | −3 | −30.0 | 90.0 | 33 |
| | −4 | −30.0 | 120.0 | 54 |
| | −5 | −30.0 | 150.0 | 75 |

[a] Estimates are based on the assumption that removal of supply control and price supports would reduce the marginal cost of current U. S. tobacco output by 30 percent.

The estimates are calculated as:

$$\text{Percent change in tobacco price} = \frac{0.3S}{0.7D - S}$$

$$\text{Percent change in tobacco output} = \frac{0.3SD}{0.7D - S}$$

where

D = percentage change in the quanitity of U. S. tobacco demanded per percentage change in price over the range of a price fall of 20 percent to 30 percent, i.e., an arc elasticity over the relevant range and

S = percentage change in the quantity of U. S. tobacco output per percentage change in the price after removal of supply control and price supports.

Percent change in tobacco revenue = % change in price + % change in output + [(% change in price) × (% change in output)].

Source: Sumner and Alston (1984).

supports, storability of the crop would dampen price variability associated with year-to-year acreage shifts and yield changes arising from good and bad weather. Shifts in the demand in response, say, to an unanticipated shift in excise taxes or import tariffs might produce some additional price variability. Even so, given the high capital intensity of production, elastic supply in the long run, and the presence of market outlook services, there is no reason to assume a wild cobweb price and production cycle would develop in the absence of the current programs.

Mean size of tobacco enterprises would grow rapidly for a short time as tobacco production shifted to low-cost production areas which currently have the larger production units. But within a low-cost production county, farm size growth probably would not change much since the programs do not currently prevent mechanization and quota leasing. Direct marketing might develop, but there is every reason to expect auction marketing of tobacco in warehouses to continue to be the predominant marketing method in the near term. Over the long term, all kinds of adjustments are possible; but most of the discussion about loss of the crop to a completely new region for which no special varieties have been developed or management skills established is unrealistic. The location of specific marketing and distribution services also suggests that a major advantage in production costs would be required to shift the crop dramatically.

## *Peanuts*

Unlike tobacco, the peanut program is not covered by permanent legislation. The current policy has been evolving since the 1977 farm bill, and continuation on the present course requires enactment of renewal legislation in 1985. Title VII of the Food and Agriculture Act of 1981 suspends for the 1982–1985 crops the marketing quota and acreage allotment provisions of the AAA of 1938. In the absence of new legislation, the peanut program will revert to 1.4 million acres of allotment subject to price supports between 75 percent and 90 percent of parity. This would mark an abrupt reversal of the policy trend for the last eight years during which time the price-support level has been reduced from 75 to 55 percent of parity. Under the last two acts, the national quota will have been reduced by 46 percent for edible peanuts by 1985. The allotment program was first modified by creating a farm-level quota in 1977 and a separate market for export and domestic nonedible peanuts. In this market for "additionals" (nonquota peanuts), supply and demand forces determine prices and quantities except for the occasional impact of a price-support mechanism. In order to provide some background, we will review production, consumption, prices, and the recent program history briefly before discussing 1985 policy alternatives. For further description of the peanut industry and for an extensive list of references, see McArthur *et al.* (1982).

## Production and Consumption

There are three major production belts scattered geographically in seven states from Virginia across the south to Texas and Oklahoma (Table 5). In the aggregate, peanuts use only 1.5 million acres (0.5 percent of the U. S. total of harvested cropland) and rank about 18th in gross receipts from crops. The United States accounts for about 10 percent of world production and between 25 percent and 50 percent of world trade in peanuts.

Peanuts can be consumed as whole kernels (in shell, salted, and in candy), as peanut butter, as oil, and as meal. The U. S. exports of edible peanuts have been growing; most exports are for the whole kernel uses. Peanut oil has a small portion of the total U. S. cooking oil market—a share that has been declining as sunflower, corn, and other similar oils have increased in use. Most of the production of peanuts outside the United States is for domestic oil consumption.

In 1978 there were about 28,000 peanut producers averaging 52 acres each. The average size of peanut enterprises grew rapidly over the last three decades as mechanization of production and forced hot air drying were developed and adopted. Allotment and quota leasing between quota owners and producers has been possible for a number of years; but leasing for peanuts has been less important than for tobacco because there are fewer small quotas, the quota system being a more recent development in the peanut belts. Quotas may now be sold but, like tobacco, only in restricted geographic areas. In addition, since 1983, underplanting of quotas on a farm can lead to loss of its quota.

**TABLE 5. Geographic Distribution of Peanut Production and Prices, 1983**

| | Virginia and North Carolina | Georgia and Florida (Southeast) | Texas and Oklahoma (Southwest) | United States total[a] |
|---|---|---|---|---|
| *Production* | | | | |
| 1983 (million pounds) | 517 | 2,189 | 539 | 3,296 |
| Change since 1954 (percent) | 20.3 | 414.9 | 216.9 | 224.7 |
| Percent of U. S., 1983 | 15.7 | 66.4 | 16.4 | 100.0 |
| *Area harvested* | | | | |
| 1983 (thousand acres) | 242.0 | 802.0 | 306.0 | 1,373.5 |
| Change since 1954 (percent) | –14.5 | +15.1 | –18.6 | –0.8 |
| *Yield* | | | | |
| 1983 (pounds per acre) | 2,112 | 2,719 | 1,758 | 2,380 |
| Increase since 1954 (percent) | +40.2 | +343.6 | +353.1 | +227.4 |
| *Average real price* | | | | |
| Cents per pound ÷ producers' price index | 8.4 | 7.7 | 8.3 | 8.0 |
| Type of nut | Virginia | Runner | Spanish | |
| Primary edible uses | In shell<br>Salted | Peanut butter<br>Salted<br>Candy | Peanut butter<br>Salted<br>Candy | |

[a] The U. S. total includes Mississippi, New Mexico, South Carolina, and other minor peanut-producing states.

Source: U. S. Department of Agriculture (1983 and 1984) and McArthur *et al.* (1984)

The different production belts grow slightly different peanuts and market them through separate growers' cooperative marketing associations. The southwestern region has had a relatively stable share of the market for the last three decades. It has had low but increasing yields and slightly declining acreage harvested (Table 5).

Peanuts produced in the other two belts (Virginia-North Carolina and the Southeast) are botanically identical, but varieties differ. In the past, the southeastern peanuts have been somewhat smaller on the average and of less regular shape. However, a variety introduced in about 1970—the Florunner—has contributed to the more rapidly increasing yields in the Southeast belt as well as to a larger kernel. Currently, screenings from the larger of this runner variety compete with Virginia-type peanuts for some of the whole-kernel market.

Yields and acreage in the Southeast belt have grown over time. The southeastern market share increased dramatically from the early 1950s to the mid-1970s and has seen some increase since that time, especially in the last two years. The Virginia-North Carolina belt has had little yield increase and has been the loser of the market share.

## Recent Program Developments

In 1977 producer referenda were suspended, and individual farm marketing quotas were introduced and assigned to acreage allotment holders. The aggregate quota for 1978 was set at 80 percent of the 1977 national marketing target. These quota peanuts were supported at $420 per ton, approximately the effective 1977 level (Table 6). An additional 20 percent of the 1977 national marketing target was assigned to these same allotment holders and tagged additionals. A lower price-support level of $250 per ton was allowed for these peanuts. In 1979, 1980, and 1981, the quota level was adjusted downward. The price-support level for quota peanuts was raised in 1980. For additionals, the price support was first raised in 1979 and then lowered in 1980. In summary, the 1977 Act reduced the quota, reduced CCC acquisitions and losses on storage stocks, and established a two-price program in which the edible market was separated from the oil and export market.

The 1981 legislation continued the major innovations of the 1977 Act, but some changes were introduced. Acreage allotments were

**TABLE 6. Peanut Production and Prices, 1972–1984**

| Year | Production | National average support price | | Average price to farmers | Farm value |
|---|---|---|---|---|---|
| | | Quota | Nonquota | | |
| | million pounds | dollars per ton | | dollars per ton | million dollars |
| 1972 | 3,274.8 | 285.00 | —[a] | 290 | 475.4 |
| 1973 | 3,473.8 | 328.50[b] | — | 324 | 562.9 |
| 1974 | 3,667.6 | 366.00[b] | — | 358 | 657.6 |
| 1975 | 3,857.1 | 394.50[b] | — | 392 | 756.5 |
| 1976 | 3,750.9 | 414.00[b] | — | 400 | 750.3 |
| 1977 | 3,726.0 | 430.50[b] | — | 420 | 783.3 |
| 1978 | 3,952.4 | 420.00 | 250 | 422 | 841.6 |
| 1979 | 3,968.5 | 420.00 | 300 | 412 | 819.3 |
| 1980 | 2,301.3 | 455.00 | 250 | 502 | 578.3 |
| 1981 | 3,981.8 | 455.00 | 250 | 538 | 1,069.5 |
| 1982 | 3,438.3 | 550.00 | 200 | 502 | 862.2 |
| 1983 | 3,273.5 | 550.00 | 185 | 480 | 785.6 |
| 1984 | — | 550.00 | 185 | — | — |

[a] Dashes indicate not applicable.

[b] Before deduction for storage, handling, and inspection costs.

Source: U. S. Department of Agriculture, Agricultural Stabilization and Conservation Service (1984).

abolished, but quotas continued to be assigned to individual farms. The price support on quota peanuts was raised to $550 per ton in 1982, and provision was made for its advance after 1983. However, the 1982 level is still in force for 1984.

What had been additionals in the sense of being tied to an allotment allocation can now be produced in unlimited quantities by anyone. The price support on nonquota peanuts was reduced to $200 per ton in 1982 and further reduced to $185 per ton in 1983.

This price support for nonquota peanuts was equal to 71 percent of the level of quota peanuts in 1979 but only 34 percent in 1983 and 1984. The level of price supports for nonquota peanuts is set at a level that will normally be below the price in the oil and export market so that the government will not incur losses for nonquota peanuts. In fact, the market price has been well above the support price of $185 for the last several years.

If nonquota peanuts are not contracted for oil use or export by April 15, they must be delivered to cooperative pools that are operated independently in the various belts. They may be sold by the cooperative for exports, be crushed for oil, or, if demand is sufficiently strong for domestic edible peanuts, nonquota peanuts may legally be sold on the domestic market at or above the price-support level for quota peanuts. The sale of nonquota peanuts on the domestic edible market by the growers' association is known as the "buyback" provision. The differential between the loan rate and the sales price is retained by the cooperative. If the cooperative has a financial surplus after offsetting any losses arising from the sale of peanuts crushed for oil, it is returned to producers. As the demand for in-shell and large salted peanuts has grown, buybacks in Virginia-North Carolina and the Southeast have provided a way to increase consumption of edible peanuts from these belts. Historically, quota adjustments have been uniform across belts and have not reflected changes in demand. Further cuts in the quota for 1984 and those programmed for 1985 will reduce CCC losses on quota acquisitions which have been incurred on Southwest belt quota peanuts crushed for oil. Thus, a differential shift in production will have been brought about by a general reduction in quotas and the use of the buyback provision.

## 1985 Program Alternatives

The effects of three possible policy alternatives will be considered: (1) extension of the present program, (2) reversion to the pre-1977 acreage allotment program, and (3) no program.

The present program can be characterized as a two-price program. The aggregate quota and the limitation on marketing of nonquota peanuts mean that quantities in the domestic market will be above the quota amount only if the domestic market price is above the support price. If the buyback provision is maintained, production will have expanded in the belts of increased demand, and the price-support level will be just slightly below the market price. In this situation the price-support level becomes very important. Currently, it is scheduled to be adjusted as much as 6 percent per year to reflect changes in the national average cost of production, excluding any increase in land cost. In the nonquota market, production is free to move to the least-cost areas. Price and quantity will be set by production and consumption conditions if the policy is unchanged. In 1978 and 1979, U. S. exports increased to about 50 percent of world trade; they fell back due to the 1980 drought and have been increasing some since then in response to current policy. Depending on exports, the low-price market for nonquota peanuts could become an even more important source of income for peanut producers. If the buyback provision is ended and government costs are kept low, the quota will have to be kept sufficiently low so that the market price is at the price-support level in the belt which has experienced the least growth in demand. Alternatively, differential quota adjustments would be required.

Reversion to pre-1977 legislation would provide for a minimum of 1.4 million acres of peanuts and price supports at 75 percent of parity. The CCC losses from buying peanuts at high support prices and selling at oil prices would be very high. In the 1972–73 crop year, losses ran very high, about $100 million as compared to $500 million–$600 million for farm-level production values. If large losses were to be incurred again, it seems unlikely the program would continue to exist without major modification.

The third alternative is to have no federal program. Production, consumption, price, and income with unrestricted production and no price supports are difficult to estimate. However, it is possible to conclude that the farmer price for domestic edible peanuts would be the same as for exports and peanuts for oil production. Given the availability of land and other resources, the supply curve is probably very elastic. This would lead to the price being determined primarily by supply and quantity being determined by demand. The supply price under no program can be approximated by looking at the price in the unrestricted, nonedible market. An alternative would be to subtract the current quota lease rate from the price-support level for quota peanuts. These two methods suggest that the unrestricted market price would be $80 to $140 less per ton per year on 1.1 million tons.

Using an estimate of $120 per ton, the aggregate reduction in income for quota owners would be $130 million per year—about 14–16 percent of the total value of production in 1982 and 1983. This estimate overstates losses that quota-owning producers would experience because production would expand if farm prices for edible peanuts were to fall to the level currently observed for non-quota peanuts. We have no reliable estimate of the elasticity of demand on which to base estimates of this expansion.

Like tobacco, there are other implications of the no-program alternative. With unrestricted production and no price supports, market prices for edible peanuts would be more variable than they have been under the present program. There would be some relocation of production among belts; but, probably, there would not be large-scale shifting. This projection is based on the observation that non-quota peanuts did not shift location dramatically in 1982 or 1983 when adjustments were allowed. It is also reinforced by the similarity of quota lease rates in the various belts suggesting that the cost of production is similar in the various belts.

The major changes in the peanut program begun in 1977 and continued in 1981 have provided for an industry that is more competitive on world markets. If this policy continues after 1985, the operation of the buyback provision may allow expansion and a geographic distribution of production that is more efficient.

## References

McArthur, W. C., Verner M. Grise, Harry O. Doty, Jr., and Duane Hacklander. *U. S. Peanut Industry.* U. S. Economic Research Service, Agricultural Economics Report No. 493. Washington, D. C.: U. S. Government Printing Office, November 1982.

Sumner, Daniel A., and Julian M. Alston. *Consequences of Elimination of the Tobacco Program.* North Carolina State University, North Carolina Agricultural Research Service, Bulletin No. 469, March, 1984.

U. S. Agricultural Marketing Service. *Annual Report on Tobacco Statistics, 1982.* Washington, D. C.: U. S. Government Printing Office, June, 1982.

U. S. Department of Agriculture, Agricultural Stabilization and Conservation Service. *Peanuts: Summary of 1984 Support Program and Related Information.* Washington, D. C.: U. S. Government Printing Office, March, 1984.

———. *Tobacco Outlook and Situation.* Washington, D. C.: U. S. Government Printing Office, March, 1983, and December, 1983.

U. S. Bureau of the Census. *The Census of Agriculture: 1982 Preliminary Report.* Washington, D. C.: U. S. Government Printing Office, 1978 and previous years.

# 10 THE U. S. SUGAR PROGRAM AND ITS EFFECTS

**Andrew Schmitz**
**Roy Allen**
**Gwo-Jiun M. Leu***

Sugar programs have existed around the world for quite some time. The essence of the programs is that they are protectionist in nature; such policies began with Napoleon. There was a large trade between the Americas and Europe in the late 18th Century, but it was vulnerable to naval blockade. Napoleon established a domestic source through protectionist measures and built 40 sugar beet factories between 1811 and 1815.

In the United States, after the adoption of the Constitution, a tariff was imposed on sugar in an act passed on July 4, 1789. The object of the tariff was to raise money since no sugar was produced in the United States at that time. Generally, duties have been in place since then except from 1890 to 1894 when there was no duty on raw

*The authors are, respectively, Professor of Agricultural and Resource Economics and Economist in the Agricultural Experiment Station and on the Giannini Foundation, University of California, Berkeley; Senior Economist, Recon Research Corporation, San Francisco, California; and Postgraduate Research Agricultural Economist, Department of Agricultural and Resource Economics, University of California, Berkeley. The authors would like to thank Hai-Yen-Sung for helpful comments.

sugar. For those years, the domestic producing industry received a 2-cent per pound subsidy and the refining industry was protected by a tariff of 0.5 cent per pound on refined sugar. Thus, tariffs have always been in effect; but another major protective device (quotas) has not always been used.

The purpose of this chapter is to describe the changing nature of U. S. sugar policy and provide an economic assessment of the costs and benefits of the present program with some suggestions on how it should be modified. The topic is extremely interesting partly because of the changing nature of sugar programs through time.

## *The U. S. Sugar Program*

Between 1934 and 1974, U. S. sugar policy was expressed in a series of sugar acts. Through this legislation, Congress tried to achieve several objectives: (1) guarantee a price that would provide an adequate return to farmers, (2) maintain a level of self-sufficiency that would produce a reasonable degree of supply security and price stability, and (3) allow enough foreign imports to provide reasonable prices to consumers. A mixture of policy instruments—including domestic and foreign marketing quotas, direct payments to producers, and tariffs and excise taxes—was used to achieve these objectives.

When the U. S. Sugar Act expired in 1974, U. S. sugar producers were placed in direct competition with foreign producers for the years 1975 and 1976 except for minor trade restrictions. However, in view of the sharp price rise in 1975 (Figure 1), protection for U. S. sugar producers was clearly not needed.

In May, 1977, under authority of the Agricultural Act of 1949, the President instructed the Secretary of Agriculture to institute an interim price-support payment program. The payment program began September 15 for the 1977 sugar crop and ended on November 7, 1977.

The Food and Agriculture Act of 1977, which amended the Agricultural Act of 1949, contained sugar provisions for the 1977–78 and 1978–79 crops. Sugar beets and sugarcane were listed under Title II as "designated nonbasic agricultural commodities" and, under Section 201, were to be supported through loans or purchases

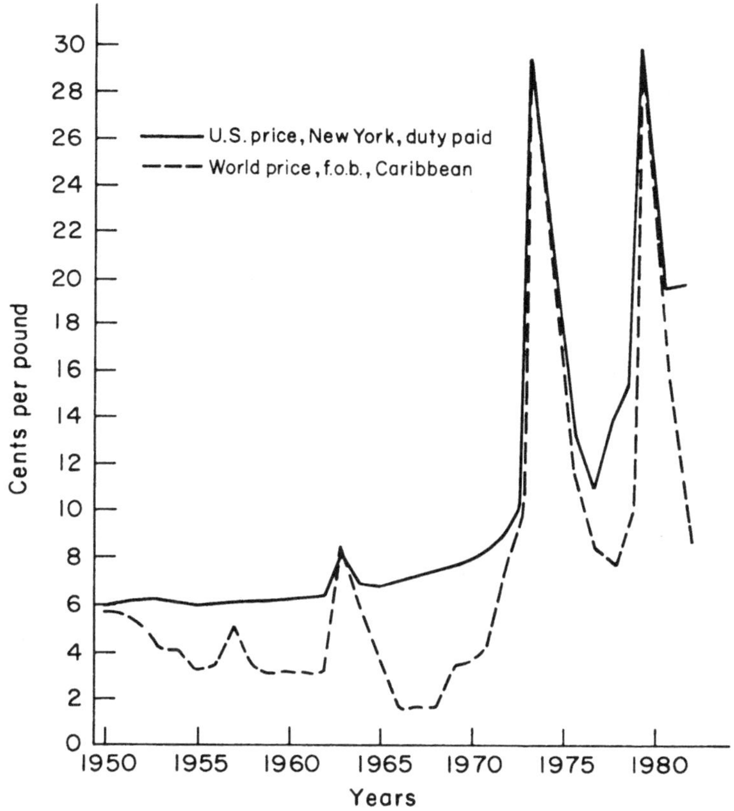


FIGURE 1. Raw Sugar Prices, 1950-1982

at 52.5 percent to 65 percent of the parity price (but in no event less than 13.5 cents a pound raw-sugar equivalent). The 1977 crop price-support loan program began November 8, 1977. Loan rates for the 1977 and 1978 crops were established at 13.5 cents and 14.73 cents a point, raw value, respectively.

Because the 1979–80 and 1980–81 sugar crops were not designated to receive price support, price-support authority reverted to Title III, Section 301 of the 1949 Act, which gives the Secretary of Agriculture discretionary authority to establish a sugar program. The price-support loan program was available for part of the 1977 sugar crop and all of the 1978 and 1979 crops.

A purchase program was available for sugar processed between December 22, 1981, and March 31, 1982, at a national average purchase price.

On May 5, 1982, partly in response to the dramatic price drop (Figure 1), the President proclaimed an emergency quota program to limit sugar imports into the U. S. market. The Secretary of Agriculture was given authority to set future minimum quota levels, set quota periods besides calendar quarters, and permit the carrying forward of unused quota allotments into subsequent quota periods.

A loan program for processors became effective October 1, 1982, for the 1982 refined sugar beet crop, raw cane sugar, cane sirup, and edible molasses. Sugar processed between April 1, 1982, and June 30, 1983, was eligible for the 1982 crop loan.

The sugar program which existed from 1934 to 1974 had as its main features (Johnson, 1974):

1. Total annual U. S. consumption requirements established yearly by the Secretary of Agriculture.
2. Allocation of total consumption requirements among domestic areas (cane- and beet-producing areas) and foreign countries.
3. Establishment of import quotas to control shipments by foreign countries of both raw and refined sugar to the United States.
4. Establishment of limits on quantities of direct-consumption sugar from Hawaii and Puerto Rico shipped to the continental United States.
5. Benefit payments made to domestic producers (including producers in Puerto Rico) for abiding by the terms of the Sugar Act.
6. A tax of 50 cents per hundredweight of raw sugar on all sugar used in the United States.
7. A tariff of 62.5 cents per hundredweight on raw sugar.

8. Provisions for the "fair division of the benefits" of the sugar program.

The key element in the program, as far as protection to U. S. producers was concerned, was the quota instrument. Total U. S. consumption requirements were allocated among domestic areas and foreign countries. Quotas applied to both raw and refined sugar imports. As shown later, the effects of existing tariffs were minor compared to the quota effects. In addition, the Commodity Credit Corporation (CCC) was not actively involved in sugar since producers received conditional payments if they abided by their proportionate share. Conditional payments to producers were graduated, declining as the amount of sugar produced on farms increased.

In 1975, 1976, and part of 1977, largely because of the dramatic increase in world sugar prices, the market was allowed to operate freely. The only protection afforded U. S. producers was a relatively small tariff; domestic and foreign supplier quotas were no longer in effect.

From 1977–78, until the reintroduction of quotas in 1982, producers received both tariff protection via a variable levy and price supports derived by producer negotiations with processors and/or sugar refiners who could sell processed and refined sugar to CCC at the government-established loan rate. In this period, CCC became a major buyer of sugar.

When quotas were reintroduced in 1982, they, once again, became the major protective measure for producers. Due to ceilings on the variable tariff, its magnitude did not have the desired effect; hence, the government reintroduced quotas. Unlike from 1934 to 1974, however, quotas were not placed on domestic production. In addition, CCC would still purchase sugar at the loan rate; but since the U. S. market price exceeded the loan rate, this has not been of any significance.

## Production

Sugar is derived from sugarcane in tropical and semitropical regions of the world and from sugar beets in temperate regions. Because of this duality of sources, sugar is produced in about 125 countries in the world. By 1981, about 39 percent of all centrifugal sugar came from beets. In 1981, world production of centrifugal sugar was 91.9 million metric tons, consumption was 87.7 million metric tons, and stocks were 42.1 million metric tons or 48 percent of consumption. Noncentrifugal sugar, produced in less-developed countries, is ordinarily not traded but is consumed where produced. Reliable production data on noncentrifugal sugar do not exist. What data there are suggest that it is an inferior product rapidly replaced by centrifugal sugar as income rises.

Close to 50 percent of the world's sugar is produced in developing countries, while another 30 percent comes fom developed countries and the remainder from centrally planned economies. The EEC is the world's leading sugar producer, accounting for about 15.3 percent of total world production. The USSR, Brazil, Cuba, India, and the United States are also important producers which, together with the EEC, produce over half the world's sugar. The USSR and the United States consume most of their production, while Brazil, Cuba, and the EEC export significant portions (Table 1).

## Consumption

The leading consumers of sugar are the USSR, the EEC, the United States, Brazil, the People's Republic of China (PRC), India, and Japan—together consuming over 55 percent of the world's sugar. Per-capita consumption of sugar for the world averages about 44.2 pounds per year; for developed countries, 100 to 120 pounds. For developing countries that export sugar, consumption is about 75 to 100 pounds per year while, for developing countries that import sugar, it is much less.

## Trade

Leading exporters of sugar have been Cuba, the EEC, Australia, Brazil, and the Philippines; leading importers have been the USSR, the United States, Japan, the EEC, the PRC, and Canada (Table 1). Of the leading exporting and importing countries, only Canada and India (since August, 1978), allow sugar to be produced and traded nearly free of government intervention. In most other countries, governments have established policies, control devices, and exclu-

**TABLE 1. World Sugar Statistics, 1981**

| Country or region | Production | Consumption | Net exports | Net imports |
|---|---|---|---|---|
| | million metric tons | | | |
| EEC | 15.48 | 10.59 | 3.98 | 0 |
| Brazil | 8.73 | 5.87 | 2.67 | 0 |
| Cuba | 7.93 | 0.55 | 7.07 | 0 |
| USSR | 6.20 | 12.50 | 0 | 5.02 |
| United States | 5.63 | 8.90 | 0 | 3.70 |
| Australia | 3.51 | 0.79 | 2.98 | 0 |
| People's Republic of China | 3.45 | 4.10 | 0 | 1.04 |
| Philippines | 2.38 | 1.13 | 1.28 | 0 |
| Japan | 0.81 | 2.75 | 0 | 1.60 |
| Canada | 0.10 | 0.94 | 0 | 0.77 |
| Other Central America | 5.75 | 4.56 | 1.80 | 0.72 |
| Other South America | 4.91 | 4.46 | 1.24 | 0.79 |
| Other Europe | 8.95 | 9.44 | 0.19 | 1.85 |
| Other Asia | 11.27 | 13.63 | 1.45 | 5.15 |
| Africa | 2.00 | 7.29 | 1.96 | 3.33 |
| Other Oceania | 0.42 | 0.24 | 0.42 | 0.16 |
| TOTAL | 91.93 | 87.75 | 25.04 | 24.13 |

Source: International Sugar Organization (1981).

sive trade arrangements which allow them to insulate themselves from the free market. Consequently, the internal price of sugar in many countries does not reflect the price of sugar on the world market.

The so-called free market for sugar sold in nonpreferential international markets is a very thin market accounting for only about one-sixth of world sugar production. When sugar is in abundant supply, this market becomes a distress market for subsidized exports or for surplus sugar from countries that normally sell part of their exports in controlled markets. Consequently, the world price of sugar in this free market is often below the cost of production. Chief exporters to the free market have been the EEC, Australia, Brazil, Cuba, and the Philippines. The chief importers from the free market have been the United States, Japan, and the USSR (Table 1). The United States and many of its leading suppliers were exposed to the free market after the expiration of the U. S. Sugar Act in 1974.

As a consequence of increased self-sufficiency of the developed countries, the share of sugar production that enters international trade has declined continuously. In the mid-1950s, about 40 percent of world production was exported whereas in the mid-1970s only about 25 percent was exported. However, a few countries, notably the United States, USSR, and Japan, rely heavily on imports which are provided by relatively few countries, predominately Cuba, Brazil, and Australia.

## Market Controls and Trade Agreements

To talk of a "world sugar market" is somewhat misleading. There are really three kinds of markets. First, there are the domestic markets within producing countries which accounted for 69 percent of all sugar in 1981. Second, there are international agreements between some of the largest importers and their suppliers (Table 2). Such agreements may cover price, quantity, or both price and quantity. Examples are the current agreements covering imports to the EEC and USSR, or expired agreements covering imports to the United States and United Kingdom. Third, there is a residual "free market" in sugar which has from time to time been regulated by International Sugar Agreements. The 1977 International Sugar Agreement, a five-year agreement which became effective January 1, 1978, was the latest attempt to stabilize the world sugar market.

In 1973, before the expiration of the U. S. Sugar Act, just over half of traded sugar entered the free market: The free market sugar rep-

**TABLE 2. International Trade by Type of Market, 1973 and 1981**

| Type of market | Exporter | Importer | 1973 | | 1981 | |
|---|---|---|---|---|---|---|
| | | | 1,000 metric tons | percent | 1,000 metric tons | percent |
| Under U. S. Sugar Act | Philippines<br>Dominican Republic<br>Brazil<br>Mexico<br>Others | United States | | | | |
| TOTAL | | | 4,835 | 21.8 | 0 | 0 |
| Under Common-wealth Sugar Agreement | West India<br>Mauritius<br>Australia<br>Guyana<br>Others | United Kingdom | | | | |
| TOTAL | | | 1,770 | 8.0 | 0 | 0 |
| Under Lome Agreement | 47 ACP countries | EEC | | | | |
| TOTAL | | | 0 | 0 | 1,365 | 4.71 |
| Under Bilateral Cuban Agreement | Cuba | USSR<br>China<br>East Germany<br>Bulgaria<br>Others | | | | |
| TOTAL | | | 2,996 | 13.5 | 4,827 | 16.67 |
| Free Market Exporter | Brazil<br>Cuba<br>Australia<br>EEC<br>South Africa<br>Taiwan<br>Poland<br>Dominica<br>Argentina<br>Others | | | | | |
| TOTAL | | | 11,665 | 52.7 | 20,479 | 70.73 |
| Free Market Importer | | Japan<br>Iran<br>USSR<br>Canada<br>Iraq<br>Yugoslavia<br>Malaysia<br>Indonesia<br>Others | | | | |
| TOTAL | | | 11,803 | 53.2 | 19,688 | 68.00 |
| Gross export in all market | | | 22,415 | 100.0 | 28,952 | 100.00 |
| Domestically consumed | | | 55,950 | | 62,980 | |
| Total production | | | 78,095 | | 91,932 | |

Source: International Sugar Organization (1973 and 1981).

resented 14 percent of total production. The U. S. Sugar Act covered 22 percent of trade, Cuban Agreements covered 14 percent, and the Commonwealth Sugar Agreement covered 8 percent. In contrast, after the expiration of the U. S. Sugar Act, about 70 percent of traded sugar entered the free market in 1981 with that free-traded sugar accounting for 22 percent of total production. Cuban Agreements covered 16.7 percent of trade, and the Lome Agreement covered 4.7 percent.

## *Literature Review*

Various research efforts have been made in the areas of modeling world sugar trade and studying policy effects. Snape (1963) studied the effect of protection on trade and consumption in the free-world sugar market. He compared the estimated average receipts from home-produced raw sugar with import or export parity prices as they were in 1959 and concluded that protection shifted production away from exporting countries to importing countries. Bates and Schmitz (1970) applied the spatial-price-equilibrium approach to modeling the world sugar market in order to assess both the impact of the Cuban embargo on the United States and the long-run efficiency of current U. S. sugar policies.

Johnson (1967) calculated net protection from international competition afforded by sugar programs as the sum of the quota premium, the tariff, and the direct payment. He then estimated the cost of the sugar program and its distribution. Snape (1969) calculated the cost of protection and taxation in terms of changes in customs revenue and producers' and consumers' surplus and then studied the effects of protection and/or taxation on price and net imports in the actual situation and in a hypothetical world-free trade situation. Flores (1972) used both the transportation model and the spatial-price-equilibrium model with linear demand and supply functions to predict U. S. sugar production, consumption, and trade-flow patterns under alternative sugar policies. Johnson (1974) again studied the costs of the sugar program to consumers and taxpayers and the net benefits that have been derived by domestic and foreign producers. He concluded that the sugar program should be replaced with at least a deficiency payment scheme because the cost was much greater than benefits to producers and was inconsistent with a liberal trade policy. Gemmill (1976) estimated very sophisticated nonlinear demand and supply functions for 75 consuming and 68 producing regions in the world and applied reactive programming to solve for spatial-price-equilibrium under alternative international and national policies.

Jesse and Zepp (1977) used a simplified two-country econometric model to project sugar price levels and market conditions under free trade. Stocks were included in the model but were treated as an exogenous variable. Jesse (1977) modified this model using revised data and included sugar substitutes as part of sugar supply, but stocks were treated as an "identity." Conway *et al.* (1978) applied spatial equilibrium models to represent the world sugar market. Sugar demand and supply functions were estimated by two-stage least squares. Demand and supply stock functions were constructed to simulate activity by the International Sugar Organization to maintain a stabilization price band for sugar.

## *Costs and Benefits*

### Pre-1975

Excellent studies are available on the costs and benefits of the sugar program which ended in December, 1974. The most notable of these is Johnson's (1974) analysis. His results of the costs of the program are given in Table 3. They are based on actual U. S. sugar prices and consumption levels in 1972. The price received by foreign producers was greater: Caribbean ports for sugar sold to the United States was 7.99 cents per pound. Total utilization of raw sugar was 11.4 million short tons. The excise tax was 0.50 cents per pound and the tariff rate was 0.625 cents per pound, both in terms of raw sugar.

Consumer cost was estimated to be $826.5 million at the maximum; $498.5 million at the minimum. The elements are quota premium cost, tariff cost, and the excise tax cost. The government collected tariff revenues and excise taxes; but from this, they paid producers conditional payments. The net cost to consumers and

**TABLE 3. Alternative Estimates of the Cost of the U. S. Sugar Program, 1972**

| | Quota premium (cents per pound) | | |
|---|---|---|---|
| | 2.5 | 2.0 | 1.5 |
| | million dollars | | |
| *Consumer cost* | | | |
| Quota premium cost | 570.0 | 456.0 | 342.0 |
| Tariff cost | 142.5 | 142.5 | 142.5 |
| Excise tax cost | 114.0 | 114.0 | 114.0 |
| TOTAL | 826.5 | 712.5 | 598.5 |
| *Taxpayer cost* | | | |
| Recipts | | | |
| Tariff | 64.1 | 64.1 | 64.1 |
| Excise tax | 114.0 | 114.0 | 114.0 |
| Total receipts | 178.1 | 178.1 | 178.1 |
| Less payments | 82.0 | 82.0 | 82.0 |
| Net cost | –96.1 | –96.1 | –96.1 |
| Net cost | 730.4 | 616.4 | 502.4 |

Source: Johnson (1974, p. 50).

taxpayers ranged from $502.4 million to $730.4 million. It is clear the largest cost component, by far, was created by the quota.

According to Johnson's estimates, of the total value of the quota premium, $200 million went to foreign suppliers of raw sugar and $256 million to domestic suppliers. (The 1972 sugar supply consisted of 5 million tons of imported sugar and 6.4 million tons of domestic sugar.) In addition, domestic producers of cane and beets benefited from the tariff effect and the direct payments. The total gain to producers was $418 million. However, this is overestimated since some went to sugar processors because of the virtual prohibition on the importation of refined sugar.

Johnson gives some interesting numbers of the distribution of transfers among farms of different sizes (Table 4). The range is enormous: from $9,495 in Puerto Rico to $506,385 in Florida. Large farms clearly dominate sugar production, and it is these farms that derive the greatest benefits from the sugar program. In terms of concentration, in 1971 at least one-third of all U. S. sugar was produced by 224 farms in Hawaii, Louisiana, Florida, and Puerto Rico.

Net benefits of farmer-owned resources are only a small fraction of gross benefits because the United States is a high-cost producer of sugar and many resources used could find employment elsewhere. Johnson's estimate of the net benefit is, at most, $128 million or only 30 percent of the gross transfer ( $418 million). This compares to the middle cost estimate for consumers and taxpayers of $616 million.

## The 1977–78 to 1980–81 Period

As pointed out, the program which existed in the late 1970s was of a different nature than the one analyzed by Johnson. Quotas were no longer in effect; tariffs were larger; and the CCC bought significant amounts of sugar since, for part of the period, the U. S. market price was below the loan rate. It is clear, based on Allen, Dodge, and Schmitz's (1983) study on voluntary quotas, that the cost of the pre-1975 sugar program had to be sizable—because part of the rents went to exporters due to the quota. However, with a tariff in place as the main protective device, the U. S. government collects the revenue—not the exporter. In addition, because of the fluctuating sugar price, the CCC could actually make money by buying low and selling high. Note from Figure 1 how prices soared in the early 1980s.

For the period 1977–78 to 1980–81, two years of low prices and CCC purchases were followed by two years of high prices and a roughly equivalent quantity of CCC sales. The relevant data are presented in Table 5. Note that, in the 1978 and 1980 periods, CCC made sizable sales of sugar roughly equal to their earlier purchases. However, because prices were higher when CCC sold the sugar (roughly 10 cents per pound higher), it actually made money from its stock operation.

**TABLE 4. Number of Farms and Per-Farm Averages of Acreage Harvested, Sugar Sold, Government Payments, and Income from Sugar, 1972[a]**

| State | Number of farms | Acreage harvested per farm | Income from sugar per farm: Sugar sold | Payments | Total |
|---|---|---|---|---|---|
| | | | dollars | | |
| Florida | 140 | 1,739 | 576,385 | 43,471 | 619,856 |
| Hawaii | 410 | 264 | 285,853 | 26,758 | 312,611 |
| Louisiana | 1,438 | 217 | 69,092 | 5,997 | 75,089 |
| Puerto Rico | 3,535 | 43 | 9,495 | 1,143 | 10,638 |
| Arizona | 54 | 178 | 67,611 | 6,759 | 74,370 |
| California | 1,423 | 235 | 96,391 | 10,731 | 107,122 |
| Idaho | 2,228 | 74 | 25,256 | 3,372 | 28,628 |
| Oregon | 266 | 76 | 29,969 | 3,590 | 33,559 |
| Washington | 866 | 90 | 42,324 | 4,561 | 46,885 |
| Colorado | 1,785 | 78 | 23,970 | 2,950 | 26,920 |
| Iowa | 19 | 94 | —[b] | 2,632 | — |
| Kansas | 220 | 177 | 55,345 | 6,150 | 61,495 |
| Minnesota | 990 | 113 | 30,077 | 4,024 | 34,101 |
| Montana | 649 | 72 | 23,305 | 3,179 | 26,484 |
| Nebraska | 1,135 | 68 | 22,350 | 2,977 | 25,327 |
| New Mexico | 9 | 67 | — | 2,778 | — |
| North Dakota | 761 | 97 | 26,833 | 3,628 | 30,461 |
| Texas | 226 | 89 | 39,823 | 3,903 | 43,726 |
| Utah | 670 | 37 | 12,294 | 1,640 | 13,934 |
| Wyoming | 658 | 94 | 30,340 | 4,122 | 34,462 |
| Michigan | 2,050 | 40 | 10,184 | 1,515 | 11,699 |
| Ohio | 936 | 43 | 13,590 | 1,577 | 15,167 |

[a] Data on farm numbers and acreage harvested per farm for beet areas are for 1971. Income from sugar per farm calculated form 1972 income and 1971 number of farms.

[b] Dashes indicate no data available.

Source: Johnson (1974, p. 56).

**TABLE 5. U. S. Sugar Market Statistics, Crop Years, 1977–78 to 1981–82**

| Crop year[a] | Demand[b] | Imports | Supply[c] | Acres harvested | Yield per acre | Price[d] | CCC purchase (sale) |
|---|---|---|---|---|---|---|---|
| | 1,000 metric tons[e] | | | 1,000 acres | metric tons[e] | cents per pound | 1,000 metric tons[e] |
| 1977 | 10,820 | 5,567 | 5,436 | 1,935 | 2.809 | 12.99 | 183 |
| 1978 | 9,389 | 4,248 | 5,557 | 1,982 | 2.804 | 14.99 | 416 |
| 1979 | 9,954 | 4,560 | 5,186 | 1,809 | 2.867 | 23.36 | (208) |
| 1980 | 9,796 | 4,078 | 5,345 | 1,873 | 2.854 | 26.63 | (373) |
| 1981 | 10,090 | 4,536 | 5,554 | 1,944 | 2.857 | 18.39 | 0 |

[a] Beginning September 1.

[b] Utilization plus change in private stocks. (Demand + CCC purchase = imports + supply + CCC sale.)

[c] Production.

[d] C.i.f. duty paid, New York.

[e] Raw value.

Source: U. S. Economic Research Service (various issues).

**TABLE 6. Changes in U. S. Welfare from Actual and Hypothetical Sugar Programs, Linear Assumption, 1977–78 to 1980–81 (Constant Elasticity Assumption)**

| Programs | Average annual change | | | |
|---|---|---|---|---|
| | Expected producer surplus | Expected consumer surplus | Expected tariff (equivalent) revenue | Total |
| | million dollars | | | |
| Actual duties and fees | 217.182 (206.509) | –451.055 (–425.780) | 264.633 (282.918) | 30.760 ( 63.647) |
| Actual duties and fees and loan purchase program | 217.182 (196.642) | –451.970 (–408.780) | 294.347 (314.893) | 59.559 (102.755) |
| Optimal variable tariff | 207.097 | $414.676 | 204.640 | 33.061 |
| Optimal fixed tariff | 207.097 (271.568) | –412.359 (–524.928) | 238.283 (323.791) | 33.021 ( 70.431) |
| Optimal import quota | 185.223 | –372.422 | 219.747 | 32.548 |
| Fixed tariff equivalent to actual duties and fees | 217.182 (206.509) | –430.221 (–410.025) | 245.728 (269.093) | 32.689 ( 65.577) |
| Import quota equivalent to actual duties and fees | 217.182 (206.509) | –450.805 (–430.791) | 262.768 (274.875) | 29.145 ( 50.593) |
| Variable tariff equivalent to actual duties and fees which maximize net welfare | 217.182 | –432.538 | 248.085 | 32.729 |

Source: Allen (1983).

Table 6, based on Allen (1983), gives further results for the 1977–78 to 1980–81 sugar programs. Note an important result:

Net U. S. welfare is improved from the sugar program. This is the case even excluding the CCC's profit on sales just discussed. Without CCC sales, the net gain was $30.76 million; when the loan purchase program was included, the gain was $59.5 million.

Recall that, in the optimal tariff case, a country gains from tariffs (see Carter and Schmitz, 1979, for a test of the optimal tariff model in the world wheat economy). The net producer gain from the tariff, plus the tariff revenues, has to exceed the consumer loss in the optimal tariff case where the world price is affected by the importer's buying power. Rents go to the importer—not the exporter. (With quotas, however, rents go to the exporters.) In fact, as Table 6 shows, the United States was pursuing close to an optimal tariff policy. The optimum policy would have yielded $32.7 million net gain to the United States while the actual program (excluding the loan purchase program) generated $30.76 million of benefits. The effect of the tariff and net producer welfare was greater than the effect of the quota in the pre-1974 era, but the consumer effect of the tariff was roughly the same as the quota effect.

## The Post-May, 1982, Period

Since one can gain by tariffs and lose with quotas, why did the government reintroduce the quota in 1982 as its main sugar policy instrument? Clearly, from Figure 1, sugar prices were dropping and, hence, tariffs had to be increased in order to prevent falling producer incomes. But tariffs are inconsistent with the General Agreement on Tariffs and Trade (GATT) and are very visible to consumer groups. As a result, the government tends toward quotas rather than tariffs to protect producers.

It is clear that in this case maximizing producers' welfare is the key to the sugar program—not maximizing society's welfare. We have made some calculations on the effects of the sugar programs which were put into existence in 1982 with quotas as the key ingredient (Table 7). Results show that the net cost of the program in 1982 was roughly $1.321 billion; in 1983, $1.347 billion. These costs are substantially greater than for the period prior to 1975 partly because of the wide gap between the world price and the U. S. sugar price and also because imports are substantially less than in the 1970s.

**TABLE 7. Estimates of the Costs and Benefits of the Sugar Program, 1982 and 1983**

| | 1982 | 1983 |
|---|---|---|
| | million dollars | |
| *Consumer cost* | | |
| Quota premium cost | | |
| Tariff cost | 1,830 | 1,740 |
| Fee cost | | |
| *Gross producer gain* | 1,160 | 1,088 |
| Government revenue collected | 161 | 67 |
| Gross cost | 509 | 584 |
| *Net producer gain* | 348 | 326 |
| Net cost | 1,321 | 1,347 |

Sources:

The quota premium price for 1982 was $4.81 per pound; for 1983, $6.93 per pound (U. S. Economic Research Service, various issues).

The tariff in 1982 was $2.81 per pound; in 1983, $2.83 per pound (*ibid.*).

The fee cost in 1982 was $2.38 per pound; in 1983, it was zero (*ibid.*).

In 1982 the United States imported 1,000,547 tons subject to duty and fee; 1,000,181 tons were imported in 1983 subject to duty (*ibid.*).

*Note*: The CCC has roughly 1 million tons under the loan program; but as of yet, the sellers have not defaulted on the loan.

## *Supply Response*

The above discussion might lead one to believe that sugar production is a profitable business and that supply should have increased due to high prices in the early 1980s accompanied by the removal of domestic production quotas. However, this has not been the case (Figure 2). Total production has fluctuated since 1972 around a trend of 6 million tons. (However, sugar beet production

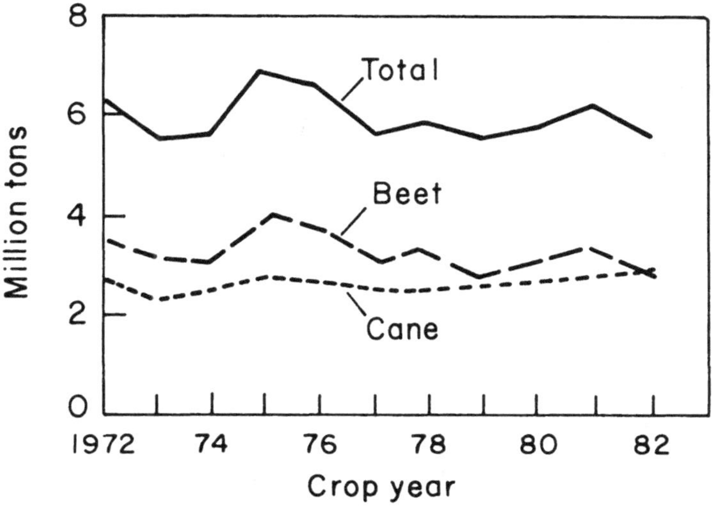


FIGURE 2. U.S. Sugar Production (Raw Value)

has declined while cane production has increased.) Thus, supply response in sugar was nothing like the positive response, for example, in wheat and feed grains. Producers respond to relative product prices where multiple cropping is possible as well as to the degree of price and production risks asssociated with each crop. In terms of relative prices, grain markets followed a price pattern similar to sugar (Figures 3 and 4). Prices soared in grains in 1974–75 and in the early 1980s just as in sugar. Hence, relative product prices did not change a great deal and sugar production did not respond to high sugar prices in absolute terms.

This raises the issue of farm programs and their effects in a macro setting. All producer groups believe in the functioning of the market when prices are high but clamor to the government for help when their industry is depressed. Unfortunately for the government and the agricultural sector, many of the specific industries within agri-

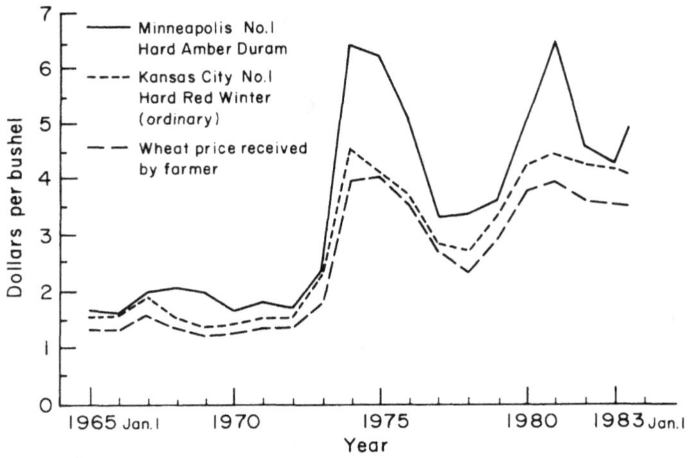


FIGURE 3. U.S. Wheat Prices, 1965–1983

culture (especially those with traded goods) move together as is certainly the case for sugar, wheat, and feed grains. In the 1960s sugar producers were protected by quota while grain producers had acreage set-asides and deficiency payments. In the early and middle 1970s, the market essentially allocated resources due to the sharp rise in prices. However, in the 1980s grain producers and sugar producers once again wanted a farm program because of low prices. Sugar producers received a quota and grain producers received payment in kind (PIK).

First, programs in each sector affects its profitability so the shift in resources among sectors is not nearly as great as might be expected because of programs simultaneously existing in almost all sectors. Second, fluctuations in government outlays from year to year are great because of the positive correlation in profitability among sectors (e.g., high sugar prices and high wheat prices). Unfortunately, when one major sector of agriculture is doing poorly, so are the others. Thus, one sector cannot be taxed to subsidize the other.

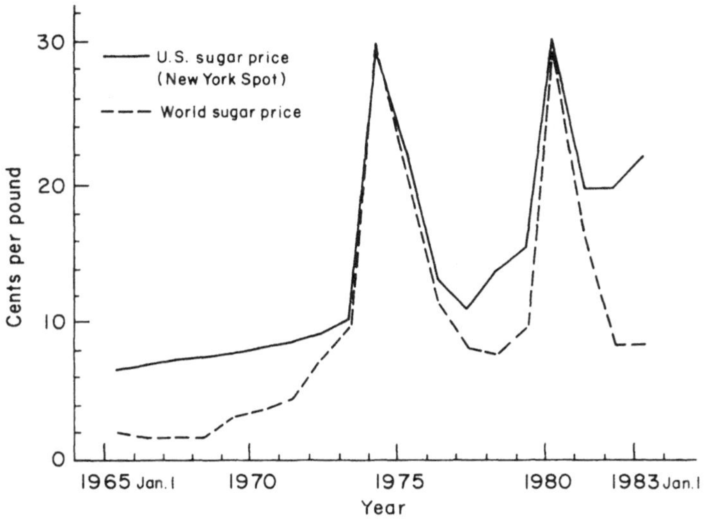


FIGURE 4. U.S. and World Sugar Prices, 1965-1983

## Conclusions

We have shown that generally the U. S. sugar program imposes a net cost to society, i.e., those who are hurt by the program lose more than those who gain. However, there have been times when society actually gained from the program. The years 1977–78 to 1980–81 was such a period. Tariffs replaced quotas and, in addition, the CCC became active in the buying and selling of sugar. But policy in place then was much different from that which existed before 1977 or after 1981 when major protective devices for U. S. sugar producers were greater but cost society money.

Quotas are not a least-cost way to protect major producers since the rents go to exporters while, with tariffs, rents go to importers. If "optimal tariffs" to protect producers are not feasible either because they are too visible or they cannot be imposed at a level sufficient to satisfy producers, then the least-cost way to protect producers is through a deficiency payment scheme. The producer receives the world market price plus the difference between the support price applied to the production produced. In view of the history of production response, it appears that acreage set-asides would not need to accompany any reasonable deficiency payment scheme. The issue of the survival of processors and refiners, however, still has to be faced.

In the choice of policy instrument, no mention was made of the effect of exchange rates. Due to the continued strengthening of the U. S. dollar in the 1980s, imports of sugar from abroad clearly became cheaper relative to domestic sources. Thus, *ceteris paribus,* as the dollar appreciates, the tariff stated in U. S. dollars has to increase to maintain a given U. S. price. In the 1980s the exchange rate movement was certainly a factor in moving away from tariff protection to a quota system. Tariffs had to increase as the dollar appreciated to keep producer welfare unchanged, but this would have been politically unacceptable. Quotas are much less visible than tariffs even though they cost society more.

## References

Allen, Roy. "Recent Protectionism in Agriculture: The U. S. Beef and Sugar Cases." Unpublished Ph.D. dissertation, Department of Agricultural and Resource Economics, University of California, Berkeley, 1983.

Allen, Roy, Claudia Dodge, and Andrew Schmitz. "Voluntary Export Restraints as Protection Policy: The U. S. Beef Case." *American Journal of Agricultural Economics* 65, No. 2 (May 1983):291–296.

Bates, Thomas H., and Andrew Schmitz. *A Spatial Equilibrium Analysis of the World Sugar Economy.* University of California, Giannini Foundation Monograph No. 23 (Berkeley, 1969).

Carter, Colin, and Andrew Schmitz. "Import Tariffs and Price Formation in the World Wheat Market." *American Journal of Agricultural Economics* 61, No. 3 (August 1979):517–522.

Conway, R. K., M. D. Hammig, H. Shapoure, and J. F. Yanagida. "A World Sugar Model." U. S. Economic Research Service, National Economics Division, Staff Report. Washington, D. C., 1978.

Flores, Arturo Sarte, Jr. "Spatial Equilibrium Analysis of the United States Sugar Industry Under Alternate Policy Measures." Unpublished Ph.D. dissertation, University of Hawaii, 1972.

Gemmill, Gordon. *The World Sugar Economy: An Econometric Analysis of Production and Policies.* Department of Agricultural Economics,

Michigan State University, Agricultural Economics Report No. 313, 1976.

International Sugar Organization. *Sugar Yearbook.* London, 1973 and 1981.

Jesse, Edward V. *Beet Sugar Supply Response in the United States.* U. S. Economic Research Service, Agricultural Economic Report No. 371. Washington, D. C.: U. S. Government Printing Office, 1977.

Jesse, Edward V., and Glenn A. Zepp. *Sugar Policy Options for the United States.* U. S. Economic Research Service, Agricultural Economic Report No. 351. Washington, D. C.: U. S. Government Printing Office, 1977.

Johnson, D. Gale. "Sugar Program: Costs and Benefits," *Foreign Trade and Agricultural Policy.* National Advisory Commission on Food and Fiber, Technical Papers, Vol. 6. Washington, D. C., 1967.

______. *The Sugar Program, Large Costs and Small Benefits.* Washington, D. C.: American Enterprise Institute for Public Policy Research, 1974.

Snape, R. H. "Some Effects of Protection in the World Sugar Industry." *Economica* XXX, No. 117 (February 1963):63–73.

______. "Sugar: Costs of Production and Taxation." *Economica* XXXVI, No. 141 (February 1969):29–41.

U. S. Economic Research Service. *Sugar and Sweetener Outlook and Situation Report.* Washington, D. C.: U. S. Government Printing Office, various issues.

# 11 DAIRY


## Emerson M. Babb*


Performance in the dairy sector has placed stress on all current dairy policies and programs, especially the price-support program.[1] Milk production has increased 13.5 percent since 1979, while commercial disappearance has increased only 2.2 percent (Table 1). As a result, the U.S. Department of Agriculture (USDA) purchases of dairy products under the price-support program have reached record levels (Table 2). In the fiscal year 1983, the USDA purchased the equivalent of 16.6 billion pounds of milk or 12.1 percent of all milk marketed. The net expenditures for dairy programs reached $2.6 billion. Some improvement in the production-consumption balance is expected during the fiscal years 1984 and 1985, partly as a result of the Dairy and Tobacco Adjustment Act of 1983. Government purchases may drop to 12 billion pounds and cost less than $2 billion during the fiscal year 1984 and may drop to about 8 billion pounds at a cost of about $1 billion in the fiscal year 1985. These levels of purchases are high by historic standards and are not

---


*The author is Professor of Agricultural Economics, Purdue University.


Valuable suggestions were made by J. L. Blum, W. D. Dobson, H. Lough, M. A. Martin, A. M. Novakovic, and D. C. Petritz, but they are absolved of any defects in the paper.


[1]The first three sections of this paper are modifications of *Dairy Policy* (Council for Agricultural Science and Technology, 1983).

**TABLE 1. U. S. Dairy Production, Consumption, and Prices, 1970–1983**

| Year | Production | Consumption[a] | Milk prices[b] All | Milk prices[b] Grade B | Retail dairy price index[c] |
|---|---|---|---|---|---|
| | billion pounds | | dollars per hundredweight | | |
| 1970 | 117.0 | 110.8 | 5.71 | 4.70 | 111.8 |
| 1971 | 118.6 | 111.0 | 5.87 | 4.86 | 115.3 |
| 1972 | 120.0 | 113.2 | 6.07 | 5.08 | 117.1 |
| 1973 | 115.5 | 113.2 | 7.14 | 6.20 | 127.9 |
| 1974 | 115.6 | 113.7 | 8.33 | 7.13 | 151.9 |
| 1975 | 115.3 | 114.2 | 8.75 | 7.63 | 156.6 |
| 1976 | 120.2 | 117.2 | 9.66 | 8.56 | 169.3 |
| 1977 | 122.7 | 116.2 | 9.72 | 8.70 | 173.9 |
| 1978 | 121.5 | 118.8 | 10.58 | 9.65 | 185.6 |
| 1979 | 123.4 | 120.2 | 12.03 | 11.09 | 207.1 |
| 1980 | 128.5 | 119.2 | 13.05 | 12.05 | 227.4 |
| 1981 | 133.0 | 120.5 | 13.76 | 12.73 | 243.6 |
| 1982 | 135.8 | 122.4 | 13.59 | 12.66 | 247.0 |
| 1983 | 140.0 | 122.8 | 13.56[d] | 12.62[d] | 249.9 |

[a] Domestic disappearance from commercial sources.

[b] Prices per hundred weight at average fat test.

[c] Price index for all dairy products, 1967 = 100.

[d] Preliminary.

Source: U. S. Economic Research Service (various issues).

likely to relieve the pressure to make changes in dairy programs at the time the 1985 farm bill is drafted.

Relief from current problems in the dairy sector will not be easy or painless. Reductions in milk production will be associated with a sharp decline in the number of dairy farms in the next five years. Time lags in adjusting production and marketing patterns in re-

**TABLE 2. U. S. Department of Agriculture Removals and Net Expenditures for Dairy Products, Fiscal Years, 1969–1982**

| Year | Removals[a] | | |
|---|---|---|---|
| | Milk equivalent | Share of marketings | Total net expenditure |
| | billion pounds | percent | billion dollars |
| 1969–70 | 4.4 | 4.4 | 393.8 |
| 1970–71 | 7.2 | 6.6 | 513.6 |
| 1971–72 | 6.6 | 5.9 | 431.8 |
| 1972–73 | 5.0 | 4.5 | 243.6 |
| 1973–74 | 0.7 | 0.7 | 121.1 |
| 1974–75 | 2.4 | 2.2 | 619.0 |
| 1975–76 | 0.9 | 0.9 | 290.5 |
| 1976–77 | 3.4 | 2.9 | 824.0 |
| 1977–78 | 3.2 | 2.9 | 589.2 |
| 1978–79 | 1.1 | 1.0 | 384.7 |
| 1979–80 | 8.2 | 6.6 | 1,436.6 |
| 1980–81 | 12.7 | 9.9 | 2,093.5 |
| 1981–82 | 13.8 | 10.5 | 2,267.3 |
| 1982–83[c] | 16.6 | 12.1 | 2,620.5 |

[a] Based on milk fat.

[b] Total expenditures for support program, school lunch and food aid, export assistance, and special milk program.

[c] Preliminary.

Source: U. S. Economic Research Service (January, 1984).

sponse to economic incentives make it difficult to attain policy objectives. It is not probable that dairy policy objectives can be achieved unless (1) there is some consensus about objectives for dairy policies, (2) the origins of the current problem are known, (3) the probable consequences of suggested options are understood, and (4) emphasis is placed on policies that could resolve the problems. There is no assurance that policy objectives will be realized even under conditions of perfect information. The current troubles in the dairy sector were projected accurately in a variety of studies (Manchester, 1978; Milk Industry Foundation, 1979 and 1982; Novakovic and Babb, 1979; Vertrees, 1981; and Vertrees and Emerson, 1979). There are several more recent analyses of dairy policy which could be used by those who make or influence policy (Carley, 1983; Council for Agricultural Science and Technology, 1983; Hutton and Helmberger, 1982; Novakovic, 1984; Thraen and Hammond, 1983; and U. S. Economic Research Service, 1984).

## *Programs and Objectives*

The dairy sector is affected by many regulations and policies. The major economic policies are the price-support program (and related restrictions on imports), milk marketing orders, and policies aimed toward cooperatives. Prices of manufacturing grade milk are supported through the USDA purchases of cheese, butter, and nonfat dry milk powder. Since 1970, dairy support prices have almost tripled (from $4.66 to $13.10 per hundredweight in 1983) while the U. S. general price level (as measured by the gross national product implicit deflator) increased 2.3 times. Since 1970, prices received by farmers for all milk have increased 2.4 times, while retail prices paid by consumers for dairy products have increased 2.2 times (Table 1).

Federal milk marketing orders use the manufacturing grade milk price supported by the USDA as the basis for all milk prices in all orders. To this base price is added a Class I differential to derive the price of milk used for bottling (fluid) purposes. Class I differentials were based primarily on transportation costs from the upper Midwest to milk orders in other regions; but since they have not been changed significantly since 1967, they now reflect less than half the cost of transportation. Class I differentials in several orders have been adjusted up and down by minor amounts to achieve a desired alignment of prices among orders, but the average Class I differential has ranged from $2.08 to $2.11 since 1970. Thus, changes in all milk order prices since 1970 were the result of changes in the price of manufacturing grade milk supported by the

USDA. If the price-support program had the effect of increasing the price of manufacturing grade milk by 50 cents per hundredweight in a given year, it also had the effect of increasing by 50 cents the price of all milk sold under marketing orders.

The absence here of much further discussion of marketing orders does not mean that they are unimportant or that changes in them do not merit attention. In the past 10 years, considerable debate has been focused on them. Changes in marketing orders to make reconstituted milk more competitive with fresh milk have been proposed, and further modifications will be advanced. The trend in marketing orders, which is expected to continue, has been toward deregulation; and market forces have been establishing prices above the administratively set minimum prices. The price-support program has clearly dominated other dairy policies and programs since 1970, and it is expected to be the focal point of dairy policy debate on the 1985 farm bill. Therefore, attention is focused on the price-support program.

The stated objectives of the dairy price-support program are to (1) assure an adequate supply of milk, (2) establish prices that reflect changes in the cost of production, and (3) assure a level of farm income that will maintain needed production capacity. The stabilization of prices and production is at least an implied objective, and other dairy programs have similar objectives. A brief assessment follows of the extent to which these objectives have been accomplished during the past 15 years. It should be noted that there is little agreement on the objectives of dairy policy among various groups that seek to influence policy even within dairy farmer or dairy processor organizations. Clarification of dairy objectives might do a lot to improve the quality of debate on policy alternatives.

## *Achievement of Objectives*

### Milk Production

**Adequate Supply.**—Since 1970, milk production has been at levels considered at times to be inadequate and, at other times, to be excessive. During 1973 and 1974, imports of dairy products were greatly expanded because of high domestic retail dairy product prices caused by low levels of U. S. milk production. Since 1980, the USDA purchases have been burdensome.

**Cost of Production.**—Based on the USDA cost of milk production studies, correspondence between milk support prices and production costs has been erratic. Net income from dairy farming varied from a loss of 62 cents per hundredweight in 1974 to a profit of $2.81 in 1979. Price supports were low relative to production costs in 1973–1975 and were high relative to production costs in 1978–1983.

**Level of Income.**—Since 1970, the support program has not prevented dramatic changes in dairy farmers' income (Babb and Boynton, 1981). Family income from dairying in the past decade probably reached its low point in 1974 when the $4,842 family income from dairying for a farm with 48 cows was below the poverty level for a family of four ($5,038). Off-farm income probably brought the average dairy farm family up to or above the poverty threshold. At the other extreme, family income from dairying in 1980 for a 54-cow herd ($28,983) exceeded the income of the average nonfarm family. This farm-nonfarm income difference would be greater if off-farm income earned were added to the farm family income.

**Price and Production Stability.**—Milk prices and production normally do not change much from one year to the next. In the decade of the 1970s, however, there were large changes in both reflected partly in the annual USDA removals which varied from 0.7 billion to 16.6 billion pounds of milk equivalent (Table 2). The annual net government expenditures for all dairy programs varied from $121 million to $2.6 billion.

Retail prices for daily products during the 1970s were slightly more stable than were either the prices for all food or the consumer price index. The increase in retail dairy prices was slightly less than for all food.

## *Low Achievement of Objectives*

### Factors Affecting Achievement

Many factors have contributed to the difficulty of achieving dairy policy objectives since 1970. Some of these same factors will affect goal accomplishment in the 1980s.

**Natural Causes.**—During the 1970s, droughts, corn blight, shortfalls of grain production in other countries, and other uncontrollable events shocked the dairy sector. These shocks may be no less frequent or severe in the 1980s.

**Other Agricultural Policy.**—Farm programs for grains and some other commodities were changed in the 1970s. The new programs were more market oriented, making feed prices more volatile. Such indirect effects on the dairy sector from policy changes in other commodities are likely to continue in the 1980s.

**Macroeconomic Policies.**—Foreign currency exchange rates, inflation, two recessions, tax policies, trade policies, energy policies, interest rates, and the like have all had major impacts on the dairy sector. In fact, these events probably had greater impacts on the dairy industry than did the dairy programs.

**Parity Formula.**—The parity formula that guided price-support decisions during the 1970s did a poor job of reflecting forces which affected milk production and consumption. For example, feed costs, which are about half of the total cost of milk production, are given a weight of only 12 percent in the parity index. Also, there was considerable inflexibility in setting price goals. Congress mandated price supports at 80 percent of parity during much of the 1970s. Some felt that performance would have been better if the Secretary of Agriculture had been given greater flexibility in setting support prices. This may be true, but the results for the last 30 years do not indicate whether Congress or the Secretary of Agriculture has used better judgment in setting support prices. Political considerations appear to have entered the decision process of both.

## *Policy Options*

The list of existing and proposed dairy policy instruments is long and varied (U. S. Economic Research Service, 1984; Manchester, 1978). Many of the proposed policy instruments are complex and contain industry jargon: trigger prices, price formulas, multiple-price plans, base plans, milk taxes, and the like. To avoid becoming bogged down in the morass of specific policy mechanisms, only a short list of selected options will be discussed.

## Current Program

The current program is the Dairy and Tobacco Adjustment Act of 1983, which became effective in December, 1983. It has aptly been described as a compromise bill and contains such features as a 50-cent reduction in the support price, a nonrefundable 50-cent assessment on all milk marketed, a diversion payment of $10 per hundredweight for voluntary reductions in production of 5 percent to 30 percent from a 1982 or 1981–82 base, and a mandatory 15-cent deduction to promote dairy consumption. Because some provisions of the Act expire on March 31, 1985, and some of its features are described in later options, the support program, as it existed during the 1970s, will be described as the current program and used as a norm for performance comparisons.[2] It should be noted that a 50-cent nonrefundable assessment that became effective in mid-April, 1983, an additional 50-cent refundable assessment that became effective on September 1, 1983, and higher production costs had set in motion production adjustments that are now taking place. The 1983 Act maintained the momentum of the adjustment process and provided incentives through paid diversion to quicken production adjustments. Promotion activity will be initiated in June, 1984; but it is not likely to have a large impact by April, 1985. Nevertheless, increases in consumption associated with economic recovery and relatively favorable prices for dairy products have contributed to a better balance of production and consumption.

In the absence of new legislation, the support program, as it existed during the 1970s (based on a percentage of parity), would be reinstated on October 1, 1985. Then what would performance of the dairy sector be for the next five years? One could be optimistic and assert that something was learned from the experiences of the past 15 years. It is certainly possible that the support program could be administered in such a way as to achieve stability goals at lower

---

[2]The paid diversion program and 50-cent nonrefundable assessment expire on March 31, 1985, and the support-price provisions expire on September 30, 1985; but the national promotional program continues until terminated by the Secretary of Agriculture or by a producer referendum. The promotion order will be terminated if not approved by a majority of dairy farmers at a referendum to be held between August 1, 1985, and September 30, 1985.

government costs. Inflexibilities, inappropriate price guidelines, ambiguous policy objectives, and political pressures all played some part in the less-than-desired performance of the dairy sector. Even without these factors, performance may not have been very good because the support program was unable to cope with the many shocks that adversely affected achievement of objectives. Problems and shocks are likely to continue, so the adopted hypothesis is that performance under the support program during the next 5 years would be about the same as in the past 15 years. Performance described above for 1970–1983 and suggested by data in Tables 1 and 2 provides a basis for comparison with other options.

## Programs More Sensitive to Economic Conditions

The main features of the current support program could be left intact with changes made in the price mechanism. The support program uses a parity price which is moved primarily by the indexes of prices paid and prices received in the total farm sector. It could be modified to recognize changes in productivity and could be made more flexible relative to the parity range used in setting support prices. A formula could be developed specifically for the dairy sector which would be more responsive to feed costs and other factors influencing milk production and consumption.

Although the mechanism for adjusting the support price can have an important impact on dairy sector performance, the level of the support price is a more critical issue. Production lags in response to price changes make it difficult to determine a support-price level that will achieve specified goals. It has been estimated that, for the rest of the 1980s, real milk prices—15 percent to 20 percent below those in 1983—would be required to clear markets (U. S. Economic Research Service, 1984). In other words, if the program goals were to stabilize production and prices around market-clearing levels, the 1984 prices would have to be more than $2.00 per hundredweight below those in 1983. For the first half of 1984, prices have been about $1.00 lower after taking into account the mandatory assessment of 50 cents. Prices required to reach market-clearing levels during the next several years may not be much lower than those in

1984 and could be reached by incremental change. However, a real price reduction of 20 percent or more may be required for market clearing by the end of the 1980s. Increases in production associated with bovine growth hormones and other technological developments are likely to cause major shifts in the supply function and significant reductions in market-clearing prices. Dairy policies will need to be flexible if they are to be useful in smoothing adjustments caused by technical change.

Suppose there were agreement on the goal of stabilizing production around market-clearing prices. What steps could be taken to give greater assurance that such a price level would be established, or how could price determination be insulated from political pressures? One approach would be to use a formula that would trigger higher or lower support prices based on USDA purchases of dairy products. A second approach would use an annual public hearing for determining the support-price level for the coming year. Government agencies and interested parties could present facts and opinions which could be subjected to examination. A third approach would involve governmental operation of a storage program with acquisition and release prices. The reserve would have an upper limit which, when reached, would end purchases; then producers would receive the market price. The government would purchase products in some years and sell them in others. Except for predetermined use of products purchased, government costs would be limited to storage costs.

Each of these approaches would make support-price decisions less sensitive to political considerations, reduce budget exposure, and improve budget control. Compared to past experience and/or prices above market-clearing levels, producer losses and consumer gains for stabilizing prices at market-clearing levels could be substantial (U. S. Economic Research Service, 1984, pp. 93–97). Market-clearing prices would probably result in fewer but larger dairy farms than exist when prices are above market-clearing levels. Given the nature of international trade in dairy products, domestic prices and production, as influenced by support prices, would have little effect on foreign sales. If policy goals involve setting support prices above market-clearing levels, the government will have to either continue operation of surplus disposal programs or institute a supply control program and be prepared to accept the public costs and economic distortions associated with both programs.[3]

## Radical Departures from Current Dairy Programs

A deficiency payment program could be substituted for the support-price program. Deficiency payments would be made to dairy farmers equal to the difference between the market price for milk and a target price whenever the market price fell below the target price. The principal advantage of this approach relates to its market orientation. The entire milk supply would move through commercial markets without government intervention. The deficiency payment program gives greater attention to demand considerations than does the support program. Compared to the current support program, consumers would benefit from lower retail prices; dairy farmers would probably experience lower returns depending on the magnitude of the deficiency payment, and there would be fewer market distortions than with direct government intervention. There are several problems with this approach including, possibly, high public costs. However, costs for such a payment program could be much lower than expected if target prices were lowered as a percentage of parity (Milk Industry Foundation, 1979) or if they were closer to market-clearing levels (U. S. Economic Research Service, 1984).

Some of the approaches suggested for making support-price decisions less sensitive to political pressure could be used for target price determination. Costs could be reduced by (1) making payment on Grade B milk only, (2) making payment on some percentage of milk marketed, (3) placing a cap on total expenditures, or (4) placing a limit on payments per farm. A deficiency-payments dairy program might cause less serious budget exposure problems than do similar programs for crops with more volatile prices. The

[3]It should be noted that the European Community may use production control to reduce its milk production. This approach can be effective in reducing production and program costs and has been used by state milk control agencies and cooperatives in the United States. Problems associated with production control, such as capitalization of base, have been well documented.

cost of administering a payments program to individual farmers might be higher than for the support program; but the cost of buying, storing, and disposing of dairy products surely is not inconsequential.

A second alternative would involve an assessment levied on dairy farmers to help pay for support program costs. Some or all of this assessment could be returned to dairy farmers who freeze or reduce their milk production. This approach was used briefly under authority of the Omnibus Reconciliation Act of 1982, and nonrefundable assessments of about $700 million in 1984 are being collected under the Dairy and Tobacco Adjustment Act of 1983. This approach is radical only in the sense that it was not used before 1983.

Each year, estimates of support-program costs could be made and part or all of these costs assessed on a hundredweight basis. Refunds of the assessment could be based on changes in production from the previous year or on a moving average of production which would minimize capitalization of a base. The decision by dairy farmers to expand or contract cow numbers or to replace cows culled would be influenced by these refunds. In effect, this approach shifts the cost of the support program from taxpayers to dairy farmers. The impacts of an assessment program would be similar to those of a modified support program except that a reduction in price support (equal to the assessment) would benefit consumers through lower dairy product prices. As with the deficiency payment plan, administrative costs might increase because of involvement with individual farmers, especially if refunds requiring maintenance of prior production records were involved.

## No Government Intervention

The only way to project accurately the consequences of no government intervention in the dairy industry would be to observe such an environment, but there is neither recent U. S. experience without intervention nor is there much experience from other countries. Some form of intervention in dairy markets occurs in nearly every country in which the dairy industry is important and in many where it is not. Given the nature of the product and the production process, it is almost certain that withdrawal of government intervention would not lead to what might be characterized as free markets with dairy price determination for transactions. The difficulty in projecting performance under no government intervention is in knowing what methods of determining terms of trade would be used as a replacement. There are many possibilities including negotiation between processors and dairy farmers or their cooperatives, increased vertical integration by cooperatives or others, and a wide variety of contractual arrangements.

Some of the consequences of stabilizing prices at market-clearing levels were discussed earlier. If the price-support program were terminated, how would performance compare with supporting prices at market-clearing levels? Given the current level of government stocks and the degree of imbalance in the sector, termination of the support program would probably require a transition period of, say, five years. Adjustments would probably be more severe in the Lake states than in other areas where transportation costs would provide some insulation and impacts would be somewhat more harsh for producers of Grade B milk.

After the adjustment period, the major difference between a support program with market-clearing prices and no government intervention would be the relative risks associated with the two environments. Price supports substantially reduce risks associated with fluctuating market prices, reducing the cost of milk production as perceived by farmers and shifting the milk supply function to the right (Dahlgren, 1980, and Thraen and Hammond, 1983). As a result, the price-support program can lead to market prices for both producers and consumers below those which would prevail in its absence or under a free market. Experiments with alternative dairy price-support policies during 1949–1978 have quantified the effects of risk reductions associated with price supports providing valuable insights about future policies (Carley, 1983).

The elimination of milk marketing orders would also increase risk; and reduced market information could have adverse effects on decision making, technical efficiency, and research. Economic power would shift away from producers; and equity problems for both producers and processors could worsen, e.g., allocating costs of providing marketwide services, sharing fluid markets, and assuring processors of equal raw-product costs. Although the institutional arrangements which would replace marketing orders are unknown, the most likely candidates would greatly reduce public involvement in and influence on the process. As indicated previously, marketing orders have moved toward deregulation since the late 1960s. Most of the problems identified with milk marketing orders would be corrected by market-clearing support prices.

## Summary

The price-support program, which has clearly dominated other dairy programs influencing industry performance, has not come close to achieving its objectives during the past 15 years. The extent to which these objectives would have been achieved without the price-support program is, of course, unknown. Although dairy policy successes or failures cannot be attributed to the price-support program alone, the program must bear some responsibility for the less-than-desired performance in the dairy sector. The support program was unable to cope with the many shocks that adversely affected achievement of its objectives. Handling exogenous shocks will be a major obstacle for any dairy policy option with an objective of stabilizing prices near market-clearing levels. This chapter did not identify a method for removing this obstacle although the public hearing process might improve the quality of economic analysis which could be used in policy decisions. Dairy programs featuring price supports above longer term, market-clearing levels in the absence of some form of supply management are not feasible without continuing purchases and out-of-Treasury costs. Higher support levels might result in less variation in price and production and reduce the number of dairy farm exits, but they would also introduce economic distortions.

## References

Babb, E. M., and R. D. Boynton. "The Accomplishments of Dairy Program Objectives." *1981 Agricultural Outlook*. U. S. Senate, Committee on Agriculture, Nutrition, and Forestry. Committee Print, 96th Cong., 2nd Sess., 1981.

Carley, D. H. *Relationships Affecting and Forecasts of Milk Production in the United States*. Department of Agricultural Economics, University of Georgia, Agricultural Experiment Station, Research Bulletin No. 306. Athens, December, 1983.

Council for Agricultural Science and Technology, Policy Task Force. *Dairy Policy.* Ames: Iowa State University Press, 1983.

———. *The Emerging Economics of Agriculture: Review and Policy Options.* Report No. 98. Ames: Iowa State University Press, September, 1983.

Dahlgren, R. "Welfare Costs and Interregional Transfers Due to Regulation of Dairy Markets," *American Journal of Agricultural Economics* 62 (May 1980):288–296.

Hutton, P., and P. Helmberger. *Aggregative Analysis of U. S. Dairy Policy.* University of Wisconsin, Department of Agricultural Economics, Agricultural Experiment Station, Bulletin No. R3191. Madison, 1982.

Manchester, A. C. *Dairy Price Policy: Setting, Problems, Alternatives.* U. S. Economics, Statistics, and Cooperative Service, Agricultural Economic Report No. 402. Washington, D. C.: U. S. Government Printing Office, April, 1978.

Milk Industry Foundation. An Analysis of Proposed Changes to the Nation's Dairy Policies. Economic Perspectives, Washington, D. C., 1982.

———. Special Dairy Policy Advisory Committee. *Examining Dairy Policy Alternatives.* Washington, D. C., 1979.

Novakovic, A. *The Dairy Industry and Dairy Policy in 1984.* Cornell University, Department of Agricultural Economics, Agricultural Economics Research No. 84-7. Ithaca, N. Y., March, 1984.

Novakovic, A., and E. Babb. *An Economic Analysis of the U. S. Dairy Price Support Program and Alternative Policies.* Cornell University, Department of Agricultural Economics, Agricultural Economics Research No. 79-27. Ithaca, New York, December, 1979.

Thraen, C. S., and J. W. Hammond. *Price Supports, Risk Aversion and U. S. Dairy Policy.* University of Minnesota, Department of Agricultural Economics, Agricultural Economics Research No. 83-9. St. Paul, June, 1983.

U. S. Economic Research Service. *Dairy Outlook and Situation.* ERS DS-395. Washington, D. C.: Government Printing Office, various issues.

———. *Review of Existing and Alternative Federal Dairy Programs.* ERS Staff Report No. AGE84-0121. Washington, D. C.: U. S. Government Printing Office, January, 1984.

Vertrees, J. G. *Food and Agricultural Policy in the 1980s: Major Crops and Milk.* Congressional Budget Office, Washington, D. C., 1981.

Vertrees, J. G., and P. M. Emerson. *Consequences of Dairy Price Support Policy.* Congressional Budget Office, Washington, D. C., 1979.

# 12 LINKAGES TO THE LIVESTOCK SECTOR

**Daryll E. Ray**
**Luther G. Tweeten**
**James N. Trapp***

Price and income instability in the U. S. livestock sector has a long history. Cattle inventory and price data show cyclical patterns as far back as the late 1800s. The traditional explanation is that producers form naive future expectations from current and lagged prices. Imperfect expectations for continued favorable (or unfavorable) economic returns, coupled with biological delays in production, cause inventory and, thus, price and income cycles.

Prior to the 1970s, prices for feed and most other inputs used in livestock production were remarkably stable. The 1970s and 1980s brought a radically different economic environment to livestock producers and all of agriculture. The sharp expansion of U. S. exports in 1973 and 1974 dramatically raised the level of grain prices. But it also started grain prices on a roller coaster ride that many grain farmers mistakenly thought went only upward. Energy prices and interest rates reached levels not considered possible just a few

*The authors are, respectively, Professor, Regents Professor, and Associate Professor, Department of Agricultural Economics, Oklahoma State University.

years earlier. As a result, livestock producers faced highly unstable input prices as well as cyclical output prices.

This chapter examines the linkages between commodity programs and the livestock economy by first considering the relationship between the feed and livestock economies. Comparisons are made between the degree of price instability for grains and for livestock. Net returns to owned resources are estimated for fed beef and hog operations. The sources of instability for these net returns are estimated with special attention to the role of feed prices, other input prices, and prices received for the livestock. A highly aggregated model of the cattle and hog sectors is estimated to test if the supply and demand structures have changed during the turbulent 1970s and 1980s. Several studies that estimate the impact of feed prices on livestock output, prices, and profits and the impact of commodity programs on crop and livestock price levels and variability are summarized. Finally, implications of feed grain-livestock-commodity program linkages are discussed.

## *Price Variability Comparisons*

Annual prices for corn, wheat, and soybeans are shown in Figure 1 and for hog and fed cattle in Figure 2. Crop prices stayed within a narrow band from 1950 to 1970. Corn prices were especially flat from the late 1950s to the early 1970s. Cattle and hog prices tended to have a slight upward trend from the mid-1950s through the 1960s. Both crop and livestock prices climbed in the early 1970s and then varied considerably to the present time.

Coefficients of variation computed for several time segments over the observation period provide one means of comparing relative variability through time and across commodities. The appropriate time interval length is debatable. Intervals should be long enough for meaningful parameters to be estimated and yet short enough to avoid marked changes in structure. Also, spanning appropriate phases of a livestock cycle is of concern. Two sets of intervals are used. The first set arbitrarily uses 10-year intervals beginning at 1950 with the period since 1970 divided into two segments of 7 years each. The decade intervals happen to coincide with the typical

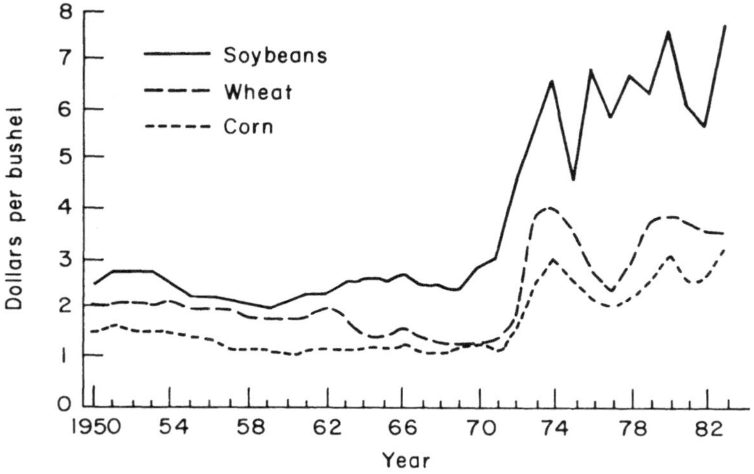


FIGURE 1. U.S. Corn, Wheat, and Soybean Prices, 1950-1983

length of the cattle cycle but do not necessarily string together years of relatively stable or unstable prices. The intervals for the second set are chosen to isolate periods of relative stability and instability. Only two intervals are used: 1957–1971 representing a period of relatively stable prices and 1972–1983 representing a period of relative unstable prices.

For most price series, the lowest coefficients of variation are in the 1960s. The highest coefficients are in the 1970–1976 interval with intermediate levels in the 1977–1983 segment (Table 1). Coefficients of variation for the 1970–1976 period are especially high because this interval bridges the steady and low prices of 1970–71 and the unstable and high prices of 1972–1976. Comparing the 1970–1976 period with the previous decade, variability of crop prices increased by as much as five times and variability of livestock prices doubled. Using the second set of intervals, variability of corn

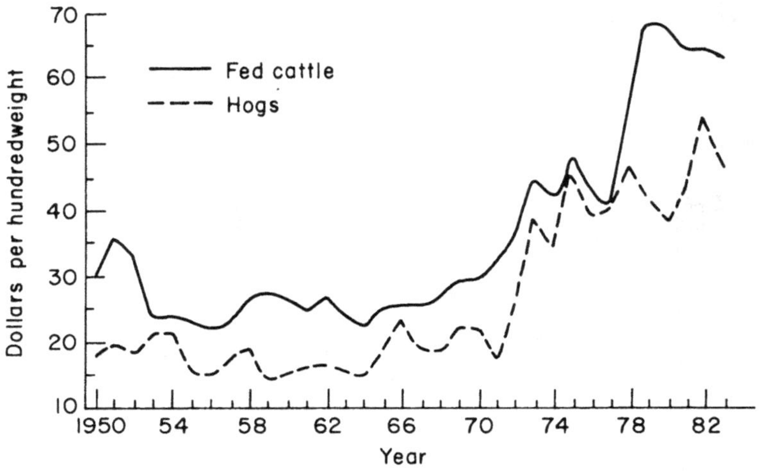


FIGURE 2. U.S. Hog and Fed Cattle Prices, 1950-1983

and cattle prices more than doubled from the more stable interval to the more unstable time interval. Variability in soybean and wheat prices increased by less than 25 percent. However, hog prices varied only slightly more in the 1972–1983 period than during the 1957–1971 period. On the whole, variabilty in livestock prices increased less than in crop prices. But this pattern may have made net returns to livestock more variable if feed and livestock prices were negatively correlated.

## *Return to Owned Resources for Cattle and Hogs*

How have the higher and more variable prices of output, feed, and other inputs affected net returns to owned resources in the cattle and hog sectors? To address this question, time series of returns to management, risk, own labor, and capital are constructed for fed cattle and hogs as the difference between market price and

**TABLE 1. Coefficients of Variation for Selected U.S. Farm Prices by Selected Periods, 1950–1983**

| Item | 1950–1959 | 1960–1969 | 1970–1976 | 1977–1983 | 1957–1971 | 1972–1983 |
|---|---|---|---|---|---|---|
| | percent coefficient | | | | percent coefficient | |
| Corn | 15.2 | 6.3 | 35.5 | 17.2 | 7.3 | 19.1 |
| Soybean | 13.0 | 7.3 | 32.9 | 12.7 | 12.9 | 16.7 |
| Wheat | 6.7 | 18.2 | 45.3 | 16.3 | 17.1 | 21.7 |
| Feeder cattle | 24.4 | 9.6 | 17.4 | 20.9 | 13.4 | 29.6 |
| Fed cattle | 17.1 | 7.6 | 15.6 | 16.6 | 10.1 | 23.6 |
| Hog | 15.3 | 16.9 | 34.3 | 11.9 | 16.4 | 17.7 |

Source: Computed from U. S. Department of Agriculture (various years).

total cost exclusive of returns to owned resources. All data are from the U. S. Department of Agriculture (1967 and 1983); U. S. Agricultural Marketing Service (1982); and U. S. Economic Research Service (1984). Cost definitions are from the U. S. Economic Research Service (1983). All prices, costs, and returns are annual dollars per hundredweight.

## Fed Cattle

In this study, fed cattle costs are grouped into categories for feeder cattle, feed, and other costs. The cost of feeder cattle is the product of the Kansas City price for 600- to 700-pound choice feeders times 55 pounds. (A 605-pound feeder is fed up to 1,100 pounds; thus, 55 pounds per hundredweight sold is purchased as a feeder, 605/11.) The 1982 feed cost ( $35.65) and other costs ( $12.83) are for farmer feedlots, all sizes (U. S. Economic Research Service, 1983, p. 117). Feed costs for other years are indexed from 1982 based on weighted averages of the season average price received by farmers in dollars per ton for feed grains and the season average soybean price per bushel received by farmers (Roberts and Heady, 1979, p. 76). The weights are based on ration requirements by type of livestock. For beef cattle, the weights are .047 for feed grain price measured in dollars per ton and .127 for soybean price measured in dollars per bushel. The 1982 "other costs" data are adjusted annually over the 1950 to 1983 period by the index of prices paid by farmers (parity index) converted to 1982 = 100. Other costs include all remaining variable costs plus fixed costs such as general farm overhead, taxes and insurance, and capital replacement costs. The sum of the feeder, feed, and other costs is subtracted from the Omaha price for 900- to 1,100-pound choice steers to obtain the return to owned resources per 100 pounds sold (Table 2).

Since 1972, only one year, 1975, provided a positive return to owned resources in cattle feeding (Figure 3). Even with zero returns to unpaid labor, land, and other capital, in 8 of the 12 years since 1972, losses exceeded $3.00 per 100 pounds. Although losses were not uncommon between 1957 and 1971, they were small in magnitude compared to recent years.

**TABLE 2. Costs, Prices, and Net Returns to Owned Resources for U. S. Fed Beef 1950–1983**

| Year | Costs | | | | Prices | Net Returns |
|---|---|---|---|---|---|---|
| | Feeder | Feed | Other | Total | | |
| | dollars per cwt. | | | | | |
| 1950 | 14.94 | 9.24 | 3.05 | 27.23 | 29.35 | 2.12 |
| 1951 | 18.51 | 10.10 | 3.36 | 31.97 | 35.72 | 3.75 |
| 1952 | 14.15 | 9.34 | 3.42 | 26.91 | 32.38 | 5.47 |
| 1953 | 9.42 | 9.12 | 3.30 | 21.84 | 23.62 | 1.78 |
| 1954 | 10.19 | 8.74 | 3.31 | 22.24 | 24.23 | 1.99 |
| 1955 | 10.45 | 8.21 | 3.29 | 21.95 | 22.59 | 0.64 |
| 1956 | 9.70 | 7.87 | 3.31 | 20.88 | 22.00 | 1.12 |
| 1957 | 12.49 | 6.85 | 3.42 | 22.76 | 23.16 | 0.40 |
| 1958 | 16.01 | 6.88 | 3.51 | 26.40 | 27.07 | 0.67 |
| 1959 | 16.37 | 6.48 | 3.55 | 26.40 | 27.67 | 1.27 |
| 1960 | 14.50 | 6.28 | 3.58 | 24.36 | 25.90 | 1.54 |
| 1961 | 14.22 | 6.89 | 3.60 | 24.71 | 24.43 | –0.28 |
| 1962 | 14.85 | 7.02 | 3.66 | 25.53 | 26.92 | 1.39 |
| 1963 | 14.18 | 7.03 | 3.72 | 24.93 | 23.58 | –1.35 |
| 1964 | 12.06 | 7.40 | 3.73 | 23.19 | 22.41 | –0.78 |
| 1965 | 13.27 | 7.32 | 3.84 | 24.43 | 24.99 | 0.56 |
| 1966 | 15.09 | 7.84 | 3.99 | 26.92 | 25.71 | –1.21 |
| 1967 | 14.67 | 6.59 | 4.07 | 25.33 | 25.29 | –0.04 |
| 1968 | 15.36 | 6.84 | 4.16 | 26.36 | 26.87 | 0.51 |
| 1969 | 17.48 | 7.24 | 4.36 | 29.08 | 29.45 | 0.37 |
| 1970 | 18.54 | 8.36 | 4.55 | 31.45 | 29.36 | –2.09 |
| 1971 | 19.18 | 7.08 | 4.77 | 31.03 | 32.39 | 1.36 |
| 1972 | 22.77 | 10.28 | 5.07 | 38.12 | 35.78 | –2.34 |
| 1973 | 29.24 | 16.12 | 5.85 | 51.21 | 44.54 | –6.67 |
| 1974 | 20.83 | 19.06 | 6.65 | 46.54 | 41.89 | –4.65 |
| 1975 | 18.65 | 15.63 | 7.31 | 41.59 | 44.61 | 3.02 |
| 1976 | 21.67 | 14.42 | 7.79 | 43.88 | 39.11 | –4.77 |
| 1977 | 22.10 | 13.34 | 8.22 | 43.66 | 40.38 | –3.28 |
| 1978 | 32.33 | 14.90 | 8.91 | 56.14 | 52.34 | –3.80 |
| 1979 | 45.69 | 16.21 | 10.14 | 72.04 | 67.67 | –4.37 |
| 1980 | 41.38 | 19.93 | 11.33 | 72.64 | 67.05 | –5.59 |
| 1981 | 36.43 | 16.00 | 12.34 | 64.77 | 63.84 | –0.93 |
| 1982 | 35.65 | 16.83 | 12.83 | 65.31 | 64.30 | –1.01 |
| 1983 | 35.04 | 21.05 | 13.18 | 69.27 | 62.52 | –6.75 |

Source: Calculated.

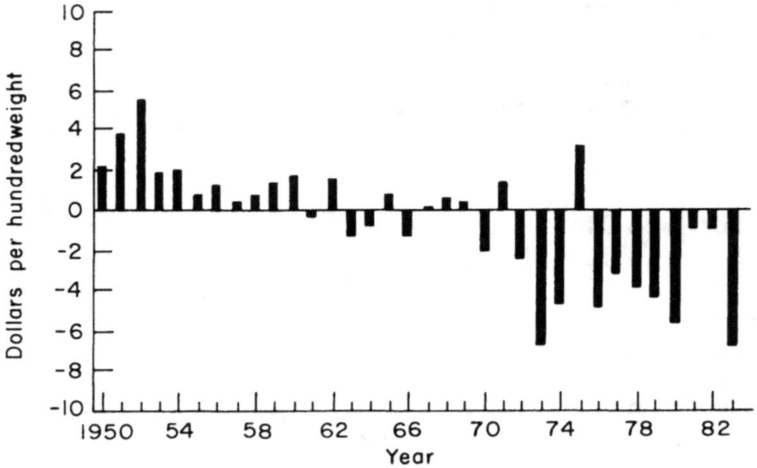


FIGURE 3. Returns to Owned Resources in Cattle Feeding, 1950-1983

## Hogs

Returns to owned resources for farrow-to-finish hog operations are computed using procedures similar to those used for beef. Only two cost categories are used: feed and other costs (defined as before). The 1982 feed cost ( $26.82) and other costs ( $20.99) are for all sizes of farrow-to-finish operations (U. S. Economic Research Service, 1983, p. 145). The weights from Roberts and Heady for estimating feed costs for hogs are .043 for feed grain price (in dollars per ton) and .308 for soybean price (in dollars per bushel). The sum of these costs subtracted from the average price received by farmers for hogs gives the return to owned resources per hundredweight (Table 3).

Returns have varied considerably since 1972 with positive returns in half of the 12 years (Figure 4). Losses were particularly severe in 1980 and 1983 as feed costs and other costs increased sharply while hog market prices declined. However, even the $6.23 positive return in 1982 fell $2.20 short of paying operator labor and a 4

**TABLE 3. Costs, Prices, and Net Returns to Owned Resources for U. S. Farrow-to-Finish Hog Operations, 1950–1983**

| Year | Costs | | | Prices | Net Returns |
|---|---|---|---|---|---|
| | Feed | Other | Total | | |
| | dollars per cwt. | | | | |
| 1950 | 13.86 | 4.99 | 18.85 | 18.00 | –0.85 |
| 1951 | 15.18 | 5.50 | 20.68 | 20.00 | –0.68 |
| 1952 | 14.21 | 5.60 | 19.81 | 17.80 | –2.01 |
| 1953 | 13.93 | 5.40 | 19.33 | 21.40 | 2.07 |
| 1954 | 13.23 | 5.42 | 18.65 | 21.60 | 2.95 |
| 1955 | 12.35 | 5.38 | 17.73 | 15.00 | –2.73 |
| 1956 | 11.88 | 5.42 | 17.03 | 14.40 | –2.90 |
| 1957 | 10.49 | 5.60 | 16.09 | 17.80 | 1.71 |
| 1958 | 10.46 | 5.74 | 16.20 | 19.60 | 3.40 |
| 1959 | 9.92 | 5.81 | 15.73 | 14.10 | –1.63 |
| 1960 | 9.82 | 5.85 | 15.67 | 15.30 | –0.37 |
| 1961 | 10.71 | 5.89 | 16.60 | 16.60 | 0.00 |
| 1962 | 10.93 | 5.99 | 16.92 | 16.30 | –0.62 |
| 1963 | 11.10 | 6.09 | 17.19 | 14.90 | –2.29 |
| 1964 | 11.66 | 6.11 | 17.77 | 14.80 | –2.97 |
| 1965 | 11.48 | 6.28 | 17.76 | 19.60 | 1.84 |
| 1966 | 12.32 | 6.53 | 18.85 | 23.50 | 4.65 |
| 1967 | 10.52 | 6.65 | 17.17 | 19.10 | 1.93 |
| 1968 | 10.78 | 6.81 | 17.59 | 18.50 | 0.91 |
| 1969 | 11.22 | 7.14 | 18.36 | 22.20 | 3.84 |
| 1970 | 13.08 | 7.45 | 10.53 | 22.70 | 2.17 |
| 1971 | 11.61 | 7.80 | 19.41 | 17.50 | –1.91 |
| 1972 | 16.83 | 8.29 | 25.12 | 25.10 | –0.02 |
| 1973 | 25.37 | 9.58 | 34.95 | 38.40 | 3.45 |
| 1974 | 29.93 | 10.89 | 40.82 | 34.30 | –6.52 |
| 1975 | 23.84 | 11.96 | 35.80 | 47.60 | 11.80 |
| 1976 | 24.18 | 12.74 | 36.92 | 43.00 | 6.08 |
| 1977 | 22.00 | 13.44 | 35.44 | 40.10 | 4.66 |
| 1978 | 24.66 | 14.57 | 39.23 | 46.60 | 7.37 |
| 1979 | 25.99 | 16.58 | 42.57 | 41.80 | –0.77 |
| 1980 | 31.83 | 18.53 | 50.36 | 38.00 | –12.36 |
| 1981 | 25.52 | 20.19 | 45.71 | 43.40 | –2.31 |
| 1982 | 26.28 | 20.99 | 47.27 | 54.00 | 6.73 |
| 1983 | 33.45 | 21.56 | 55.01 | 46.20 | –8.81 |

Source: Calculated.

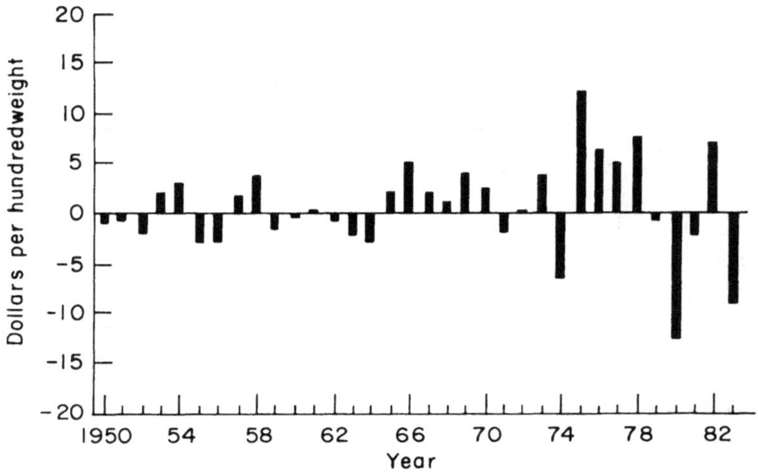


FIGURE 4. Returns to Owned Resources in Farrow-to-Finish Hog Operations, 1950-1983

percent return on capital and leaves no return to risk and management. From 1957–1971, 9 of the 15 years show positive returns; but, after 1972, positive returns were not as high—negative returns were lower.

## *Estimated Variance of Returns to Owned Resources*

This section estimates sources of instability for cattle and hog net returns over time—whether most of the variance is due to market prices for the livestock or to the price of feed and other costs. The variance of returns to cattle and hogs is decomposed using the usual formula:

$$\mathrm{Var}(T) = \sum_{i=1}^{n} \{\mathrm{Var}(Xi) + 2\mathrm{Cov}(XiXj)\} \quad i < j$$

where T is the sum of the positive or negative Xi.

Two sets of time intervals are again used with 1950–1983 first divided into four segments and then into two segments—the relatively stable years of 1957–1971 and the more unstable years of 1972–1983. The sources of variance of cattle and hog total cost and net returns to owned resources are decomposed by their cost and return components.

## Cattle

For all time intervals before 1970, market prices for fed beef were the major contributor to variance of net returns for fed beef production (Table 4). Furthermore, net return variances were comparatively low relative to intervals covering the 1970s and 1980s. Higher net return variances since the 1970s resulted from both more variable market prices and larger variances for production costs. However, in contrast to earlier years, the larger share of the variation came from the cost side. Between 1960–1969 and 1970–1976, price variance for fed cattle increased by about nine times, while the cost component increased by a factor of 20. The variance for the feed cost portion of total cost variance increased by nearly 100 times while feeder cattle price variance was up by about seven times, and other costs increased by a factor of 22.

The second set of intervals, segmented into stable and unstable time periods, again shows the price variance component to be larger than for total cost in the earlier period and the reverse in the later period. However, in contrast to the 1970–1976 interval, cost of the feeder rather than feed was the largest component of total cost variance during 1972–1983. A portion of the high variances for feeder cattle in the most recent time segments was due to the ballooning of feeder prices in 1979 and 1980. Even though feeder price variance was large during 1972–1983, feed cost variance increased more between the 1957–1971 and 1972–1983 periods: variance was 30 times larger for feed cost in the later period and 19 times larger for feeder cattle.

Feed costs directly contributed only 4.1 percent to the variance in production costs of fed cattle in the 1957–1971 period and only 5.2 percent in the 1972–1983 period. Arbitrarily attributing half of the interaction of the variance of feed costs with the other costs to

**TABLE 4. Components of Estimated Variance of Returns to Owned Resources for Fed Cattle by Selected Periods, 1950–1983**

| Variance source | 1950–1959 | 1960–1969 | 1970–1976 | 1977–1983 | 1957–1971 | 1972–1983 |
|---|---|---|---|---|---|---|
| | dollars per cwt. | | | | dollars per cwt. | |
| Var(feeder cost) | 9.3724 | 1.7735 | 12.0467 | 47.0284 | 3.8697 | 73.0856 |
| Var(feed cost) | 1.3662 | .1718 | 17.0421 | 6.3371 | .2550 | 7.8623 |
| Var(other cost) | .0172 | .0627 | 1.4035 | 3.2671 | .1549 | 7.1317 |
| 2 Cov(feeder · feed) | .1505 | – .0770 | 10.5468 | 17.7203 | .5190 | 21.1208 |
| 2 Cov(feeder · other) | .2432 | .4504 | .2769 | 10.7363 | 1.1529 | 32.6361 |
| 2 Cov(feed · other) | – .1818 | .0522 | 7.4301 | 6.6823 | .2012 | 8.8394 |
| Var(total cost) | 10.9677 | 2.4336 | 48.7461 | 91.7715 | 6.1527 | 150.6759 |
| Var(market price) | 18.8762 | 3.4353 | 30.4964 | 84.4602 | 6.5768 | 137.5878 |
| –2 Cov(cost · price) | –27.5853 | – 4.9729 | –68.8841 | –172.1901 | –11.5899 | –281.1556 |
| Var(net returns) | 2.2586 | .8960 | 10.3584 | 4.0416 | 1.1396 | 7.1081 |

Source: Computed from data in Table 2.

feed, variation in feed costs accounted for 10 percent of the variance of production costs of feed cattle in 1957–1971 and for 15 percent in 1972–1983. Inflation and nonfeed costs were the dominant sources of growing instability of fed cattle returns.

Normally, variance in net returns to livestock would be expected to decrease as feed prices become more stable. However, under certain conditions, the opposite could occur. Since feed and livestock output prices are typically positively correlated, variance of livestock net returns is less than the sum of the independent variances in market prices and production costs due to the subtraction of twice the covariance between market price and production costs. Given output prices, less variation in feed prices would reduce the correlation between feed cost and output price and, hence, between total costs and output price. Commodity programs, if used effectively to stabilize grain prices, could actually increase the variation in livestock net returns *if* twice the covariance between production costs and output prices is reduced more than the sum of variances in costs and output prices. Although the impact could be to increase variation in returns, *average* returns could be higher or lower depending on the average level of prices.

## Hogs

During the decades of the 1960s and 1970s, prices received for hogs accounted for most of the variance in net returns compared to the cost side (Table 5). Both price and cost variances increased manyfold during 1970–1976, but market prices were still the more variable. During the 1970–1976 time interval, variance of net returns increased by substantially less than the price and cost components due to the large (negative) covariance term. In each interval set, the net return variances for the last time segment show large increases; and the relative importance of market price and total cost variances reverse positions from those in earlier years. The variance of net returns increased nine times between the 1957–1971 and 1972–1983 periods; much of the increase was due to increased feed cost variance.

Feed costs are a larger percentage of production costs for hogs than for beef, and the increased variance in feed prices from 1957–

**TABLE 5. Components of Estimated Variance of Returns to Owned Resources for Farrow-to-Finish Hogs Operations by Selected Periods, 1950 –1983**

| Variance source | 1950–1959 | 1960–1969 | 1970–1976 | 1977–1983 | 1957–1971 | 1972–1983 |
|---|---|---|---|---|---|---|
| | dollars per cwt. | | | | dollars per cwt. | |
| Var(feed cost) | 2.9744 | .4215 | 40.6803 | 14.1520 | .7039 | 17.9371 |
| Var(other cost) | .0462 | .1679 | 3.7566 | 8.7446 | .4147 | 19.0882 |
| 2 Cov(feed · other) | –.4248 | .1415 | 19.3969 | 14.9662 | .6836 | 22.4710 |
| Var(total cost) | 2.5958 | .7309 | 63.8333 | 37.8628 | 1.8022 | 59.4963 |
| Var(market price) | 6.8321 | 8.3836 | 107.4196 | 23.9114 | 8.2422 | 49.3424 |
| –2 Cov(cost · price) | –4.3071 | –3.7167 | –141.7498 | –11.2989 | –4.9278 | –60.4187 |
| Var(net returns) | 5.1208 | 5.3978 | 29.5031 | 50.4753 | 5.1166 | 48.4200 |

Source: Computed from data in Table 3.

1971 to 1972–1983 contributed significantly to greater variance in production costs. "Other costs" directly contributed more to variance than did feed costs in the 1972–1983 period. And the proportion of production *total* cost variance contributed by feed cost fell whether measured directly or indirectly (including half of covariance with "other costs").

## *Supply and Demand Structural Changes*

This section focuses on whether the structural supply and demand relationships for cattle and hogs have changed as a result of the higher and and more unstable input and output prices of the 1970s and 1980s. A highly aggregated model of the cattle and hog markets is constructed. Due to its simplicity and lack of detail, conclusions from the structural analysis should be viewed with caution. Results merely provide hypotheses for testing with more detailed models.

All relationships are estimated with ordinary least squares (OLS). The iterative Cochrane-Orcutt procedure is applied when autocorrelated disturbances are indicated. Each relationship is estimated using data for 1957–1983, 1957–1971, and 1972–1983. The Chow test (Maddala, 1977, pp. 198–201) is applied to test for structural change between 1957–1971 and 1972–1983.

### Cattle

The U. S. liveweight cattle slaughter is used as the supply variable; it is related to previous years' prices for fed cattle, previous years' feed costs for cattle (as defined earlier), and livestock slaughter lagged one year. Various lags on market price and feed cost were tried with a lag of five years, which corresponds to the midpoint of the typical cattle cycle length, explaining beef supply the best. The 1975 price freeze disrupted cattle supply as producers held cattle off the market in anticipation of higher prices. A dummy variable was constructed to capture the effect or reduced marketings in 1975 followed by larger marketings in 1976 (1 in 1975, –1 in 1976, 0 otherwise).

The price-dependent demand equation is specified as a function of current year cattle slaughter and a composite demand shifter represented by time (using the last two digits of the year). Various combinations of variables representing substitutes for beef were tried without success.

An equation is also specified that relates feeder cattle prices to fed-cattle price and the feed cost in the previous year.

The estimated relationships for the three time periods are presented in Table 6. Most of the equations explained over 95 percent of the variation in the dependent variables. All coefficients have expected signs, and t values exceed 2 for all coefficients but two. The Durbin h statistic indicated nonautocorrelated disturbances for the slaughter equations. The iterative Cochrane-Orcutt procedure was applied to fed cattle price equations and the 1951-1971 equation for feeder cattle because the Durbin-Watson d statistic suggested autocorrelated disturbances in the respective OLS equations.

## Hogs

The variable specifications for the hog supply and price-dependent demand equations are similar to those used for cattle. The lag structure for the price and feed cost variables in the supply equation are, of course, different. In contrast to the time lags used for cattle, which were a year or two longer than expected, the lags in the hog supply equation were somewhat shorter than anticipated. Liveweight hog slaughter is related to the average price of hogs lagged one year, the cost of hog feed lagged two years, and hog slaughter lagged one year. Regressions using corn price in place of feed cost and replacing both price and feed cost with the hog-corn ratio or the hog-feed cost ratio were tried. These regressions gave acceptable results but did not improve the explanation of hog slaughter.

The hog price equation corresponds directly with the cattle equation with hog slaughter and time as the independent variables. Here, again, time is used as a proxy for all demand shifters.

Results are given in Table 7. Autocorrelation of the disturbances was rejected for the supply equations based on the Durbin h, but OLS estimates for the price equations resulted in significant Durbin-Watson d statistics for two of the equations which were then rees-

TABLE 6. Ordinary Least-Squares Estimates for Cattle Equations, United States, Selected Periods

| | 1957–1983 | 1957–1971 | 1972–1983 |
|---|---|---|---|
| | Cattle slaughter (liveweight, million pounds) | | |
| Intercept | 2,058.74 | 9,978.64 | 8,504.83 |
| Fed cattle price (–5)[a] | 218.46<br>(3.34)[b] | 55.21<br>(0.55) | 246.16<br>(2.40) |
| Cattle feed price (–5)[a] | –654.80<br>(–4.83) | –1,067.85<br>(–3.36) | –669.33<br>(–3.45) |
| Freeze dummy | –2,332.30<br>(–2.70) | | –2,114.24<br>(–2.09) |
| Cattle slaughter (–1)[c] | .939<br>(19.24) | .913<br>(14.98) | .752<br>(4.73) |
| $R^2$ | .965 | .970 | .830 |
| ESS | 3.235E +7 | 1.064E +7 | 1.380E +7 |
| | Fed cattle price (dollars per cwt.) | | |
| Intercept | –113.94 | –70.23 | 9.92 |
| Cattle slaughter[c] | –0.0027<br>(–8.48) | –0.0014<br>(–3.54) | –0.0024<br>(–6.68) |
| Time | 3.15<br>(14.33) | 2.14<br>(3.98) | 1.80<br>(2.38) |
| Rho | .657 | .660 | .714 |
| $R^2$ | .980 | .755 | .963 |
| Durbin-Watson d | 1.69 | 2.40 | 2.10 |
| EES | 120.47 | 24.21 | 60.09 |
| | Feeder cattle price (dollars per cwt.) | | |
| Intercept | –.77 | –7.41 | 3.78 |
| Fed cattle price[a] | 1.477<br>(23.20) | 1.449<br>(16.02) | 1.452<br>(13.74) |
| Cattle feed prices (–1)[a] | –1.437<br>(–6.77) | –0.441<br>(–0.99) | –1.635<br>(–4.44) |
| Rho | —[d] | –.319 | — |
| $R^2$ | .980 | .928 | .956 |
| Durbin-Watson d | 1.75 | 2.69 | 1.80 |
| EES | 159.85 | 13.78 | 128.18 |

[a] In dollars per cwt.

[b] Figures in parentheses are t statistics.

[c] In million pounds.

[d] Dashes indicate no value.

**TABLE 7. Ordinary Least-Squares Estimates for the Hog Equations, United States Selected Periods**

| | 1956–1983 | 1957–1971 | 1972–1983 |
|---|---|---|---|
| | Hog slaughter (liveweight, million pounds) | | |
| Intercept | 5,205.22 | 5,416.37 | –1,878.41 |
| Hog price (–1)[a] | 191.76<br>(4.43)[b] | 315.26<br>(4.28) | 240.61<br>(3.69) |
| Hog feed price (–2)[a] | –307.26<br>(–3.98) | –485.00<br>(–1.63) | –257.84<br>(–2.72) |
| Hog slaughter (–1)[c] | 0.720<br>(5.40) | .712<br>(3.89) | .912<br>(4.46) |
| $R^2$ | .638 | .764 | .736 |
| ESS | 2919E +7 | 7.871E +7 | 1.173E +7 |
| | Hog price (dollars per cwt.) | | |
| Intercept | –44.16 | –1.04 | 73.64 |
| Hog slaughter[c] | –0.0032<br>(–9.63) | –0.0028<br>(–6.86) | –0.0028<br>(–5.59) |
| Time | 1.90<br>(10.87) | .98<br>(5.65) | 2.21<br>(7.85) |
| Rho | .659 | .447 | d |
| $R^2$ | .969 | .843 | .877 |
| Durbin-Watson d | 1.98 | 2.08 | 1.92 |
| EES | 133.96 | 19.40 | 72.60 |

[a] In dollars per cwt.

[b] Figures in parentheses are t statistics.

[c] In million pounds.

[d] No value.

**TABLE 8. Chow Test for Structural Stability**

| Equation | Computed F |
|---|---|
| Cattle slaughter | 5.89* |
| Fed cattle price | 3.01 |
| Feeder cattle price | .88 |
| Hog slaughter | 7.82* |
| Hog price | 3.66** |

* Significant at the .01 level.

** Significant at the .05 level.

Source: Based on regressions presented in Tables 6 and 7.

timated using the autocorrelation algorithm. The $R^2$ values, ranging from .64 to .97, are not as high as for the cattle equations, but the coefficients are highly significant for most variables.

## Structural Stability

Responsiveness to changes in feed prices in the livestock supply equations remained relatively stable from the 1957–1971 to the 1972–1983 periods, especially for hogs. The hog supply elasticity with respect to feed price, computed at variable means, was –.28 for the 1957–1971 period and –.30 for the 1972–1983 period. For cattle, the supply elasticity with respect to feed price declined from –.27 to –.20 between the two periods. Hence, it appears hog and cattle producers tend to respond similarly to a 1 percent change in feed prices whether feed prices are relatively stable or volatile.

Even though the supply elasticities with respect to feed prices showed little change, viewing equation coefficients as a set suggests that, overall, cattle and hog supply and demand relationships have changed between the 1957–1971 and 1972–1983 periods. The results of the Chow structural stability tests for the three cattle and two hog equations are presented in Table 8. The test of no change in structure between 1957–1971 and 1972–1983 is rejected for all relationships at the 5 percent or above significance level except for the cattle price equations. However, the computed F statistic for the fed cattle price equation is significant at the 6 percent level. Although the highly aggregated nature of the relationships demands caution, results suggest that future estimated cattle and hog relationships should account for a change in structure beginning in the early 1970s.

## *Feed Price and Livestock Connection: A Summary*

Grain and livestock variability increased concurrently from the 1960s to the 1970s and 1980s. Evidence of causality is, of course, circumstantial. It is apparent, however, that increased variance of net returns to fixed resources in fed cattle and hog operations in recent years is partially attributable to larger feed price variances. The increased variability of the agricultural environment beginning in the 1970s also coincides with an apparent change in the structure of the supply and demand relationships for cattle and hogs.

The importance of feed prices on livestock supply and thus on livestock prices is evident from numerous studies—some using multiequation disaggregated models and others using highly aggregated models (Arzac and Wilkinson, 1979; Folwell and Shapouri, 1977; Freebairn and Rausser, 1975; Ospina and Shumway, 1981; Roberts and Heady, 1979; Ray and Trapp, 1977; and Salathe, Price, and Gadson, 1982). The Ospina and Shumway study, which specifically addresses the impact of corn price on the beef economy, found short-run beef supply to be more responsive to a 1 percent change in corn price than a 1 percent change in beef price.

The livestock industry tends to be the buffer that is forced to absorb year-to-year variability in grain production and exports (Plaxico, 1983). Moderate changes in grain supplies and prices can be accommodated in the livestock sector by marginal adjustments in selling weights and numbers, but sharp changes can cause large short-run losses for livestock producers and exacerbate the inherent cyclical volatility of the livestock industry.

Furthermore, effects of a one-time corn price shock tend to dampen more quickly in the feed-grain sector than do the repercussions to the livestock sector. For example, Salathe, Price, and Gadson (1982) report the impact on livestock prices of a 500 million bushel increase in 1980 corn exports with prices shown oscillating above and below baseline levels. Corn prices increase 75 cents per bushel the first year, dampen to 9 cents above the base in the second year, and then fall 3 cents below the base by the sixth year. Slaughter steer prices show no effect the first year but decrease $4.29 per hundredweight below the baseline in the third year, increase by $4.14 the sixth year, and continue to show significant but dampened effects for over 10 years. Impacts on hog and poultry prices were sizable for over six years and were moderately apparent even after 10 years.

Repeated shocks that cause sharp oscillations in feed prices confound the problem of income instability in the livestock sector. However, a new higher feed price plateau results in a more orderly adjustment in livestock supplies and prices (Ray and Trapp, 1977). As expected from theory, empirical studies that show higher sustained prices (a once-for-all-time increase) cause higher equilibrium prices in the livestock industry. In the case of beef, Ospina and Shumway (1981) show changes in the composition and relative prices of beef slaughter categories resulting from an increase in corn prices. Choice beef supplies decline, but lean beef production increases causing widened price differentials.

The livestock industry can adjust to a high sustained feed price or low sustained change in the level of feed price, but unstable feed prices cause gyrations in livestock production and prices that disrupt the industry for many production periods into the future. Unstable feed prices that first rise and then fall, or vice versa, cause an initial action which must be reversed later. Because of the biological nature of livestock production, once production decisions are started, they are not easily reversed. Sustained feed price changes are easier to cope with since there is no reversal of previous decisions.

Hence, the effect of government grain policy on the livestock sector centers on whether feed price variability is reduced by government farm programs and how grain price levels have been affected by commodity programs. Grain price instability is of more concern than the level of grain prices. Over the long run, the livestock industry will adjust to changes in price level and maintain relatively constant profit levels, but price volatility can lead to a compounded misallocation of resources, large losses, and inefficient production.

## *Commodity Program Impacts*

Numerous studies have estimated the effects on agriculture of eliminating commodity programs (Nelson and Cochrane, 1976; Ray and Heady, 1972; and Tyner and Tweeten, 1968). A recent study by Morton, Devadoss, and Heady (1984) investigates price instability and the short-run and long-run price levels of ending commodity programs compared to continuing the 1977–1980 policy. They conclude that average *long-term* grain and livestock prices would not differ significantly from those under the 1977–1980 policy. However, during the transition to no programs, prices would fall sharply forcing many farmers out of agriculture. Furthermore, price instability would increase dramatically compared to current programs.

Morton, Davadoss, and Heady also note that, while, in general, farm programs reduce price instability, inconsistent administration of farm programs can be a destabilizing influence on prices. Farm program management during the 1980s has been erratic. Although the farmer-owned reserve was intended as a price stabilization tool, high reserve loan rates during 1980 and 1981 served to increase the level of prices and incomes. Farmers responded to the price signal which, coupled with good weather and an ineffective acreage reduction program, caused record stocks and sagging prices. Program managers then set the brakes with large-scale acreage reduction incentives under the payment-in-kind program. With reduced acreage and drought, production plummeted and grain prices surged. "Thus, agriculture has been subjected to a whipsaw effect on prices that has adversely affected livestock producers and consumers. Moreover, recent policy changes have created uncertainty about the direction of future policy, perhaps adding instability rather than reducing it" (p. 88). Late program announcements, especially wheat provisions which typically are finalized months after winter wheat is in the ground, and uncertain availability of grazing and haying provisions further aggravate decision making and are a source of instability in both cattle production and prices.

## *Conclusions*

Commodity programs, on the whole, reduce variability of grain supplies and prices both in the short and long run. Reduced variability of feed prices benefits livestock producers.

Large year-to-year changes in commodity programs, late program announcements, and uncertainty about future program direction hinder systematic planning and can offset intended stabilizing aspects of commodity programs. These program changes and uncertainties disrupt both crop and livestock markets.

The impact of a grain price shock irrespective of its source tends to work through crop markets faster than through livestock markets. Sharp changes in feed prices have significant effects on livestock production and price patterns for many periods into the future.

Commodity program configurations that raise the level of grain prices above market levels reduce short-run but not long-run livestock profits after livestock producers make desired adjustments. Commodity programs have little long-run effect on feed prices and quantities and, hence, on livestock output and prices.

To minimize disruptions to the livestock industry, it is important to articulate policy direction, to use policy instruments consistently and predictably, and to announce specific program provisions well in advance. The collection of instruments and their specific levels should focus on stabilizing agricultural markets within a price corridor. The rules for defining the corridor should be spelled out clearly and be known by all. The farmer-owned grain reserve should be used as originally intended to stabilize prices within the price corridor and not as a substitute for acreage reduction. Timely acreage reduction programs should be used to *prevent* excessive stock buildups. Use of other policy instruments should be structured so as not to interfere with the price corridor concept.

Large reserve stocks of grains are associated with low and stable grain prices; small reserve stocks are associated with high and unstable grains prices (Tweeten, 1979, p. 226). Grain farmers prefer high grain prices, livestock feeders prefer low grain prices, and both groups prefer stable prices. It is apparent that difficult trade-offs are unavoidable between competing interests which the market and the political process, rather than economists, must resolve.

## References

Arzac, E. R., and M. Wilkinson. "A Quarterly Econometric Model of United States Livestock and Grain Markets and Some of Its Policy Implications." *American Journal of Agricultural Economics* 61, No. 2 (May 1979):297–308.

Folwell, R. J., and H. Shapouri. *An Econometric Analysis of the U. S. Beef Sector.* Washington State University, Technical Bulletin No. 89, 1977.

Freebairn, J. W., and G. C. Rausser. "Effects of Changes in the Level of U. S. Beef Imports." *American Journal of Agricultural Economics* 57, No. 4 (November 1975):676–688.

Maddala, G. S. *Econometrics.* New York: McGraw-Hill Book Company, Inc., 1977.

Morton, A. S., S. Devadoss, and E. O. Heady. *Potential Grain Price Instability Under Alternative Farm Policies.* Center for Agricultural and Rural Development, CARD Report No. 129, Ames, Iowa, April, 1984.

Nelson, F. J., and W. W. Cochrane. "Economic Consequences of Federal Farm Commodity Programs, 1953–1972." *Agricultural Economics Research* 28, No. 2 (April 1976):52–64.

Ospina, E., and C. R. Shumway. "Impact of Corn Prices on Slaughter Beef Composition and Prices." *American Journal of Agricultural Economics* 63, No. 4 (November 1981):700–703.

Plaxico, J. S. "International Markets and the Wheat-Feed Grain Industries." Paper presented at the National Wheat Pasture Symposium, Stillwater, Oklahoma, October, 1983.

Ray, D. E., and E. O. Heady. "Government Farm Programs and Commodity Interaction: A Simulation Analysis." *American Journal of Agricultural Economics* 54, No. 4 (December 1972):578–590.

Ray, D. E., and J. N Trapp. "Grain-Livestock Interrelations and Trade-Offs with Implications for the Structure of the Livestock Industry." In *Proceedings, Farm and Food Policy Symposium.* Great Plains Agricultural Council Publication No. 84, February, 1977, pp. 25–49.

Roberts, R. K., and E. O. Heady. *A Five-Commodity Econometric Simulation Model of the U. S. Livestock and Poultry Sector.* Center for Agricultural and Rural Development, CARD Report No. 83T, Ames, Iowa, March, 1979.

Salathe, L. E., J. M. Price, and K. E. Gadson. "The Food and Agricultural Policy Simulator." *Agricultural Economics Research* 34, No. 2 (April 1982):1–15.

Tweeten, L. G. *Foundations of Farm Policy.* Lincoln: University of Nebraska Press, 1979.

Tyner, F. H., and L. G. Tweeten. "Simulation as a Method of Appraising Farm Programs." *American Journal of Agricultural Economics* 50, No. 1 (February 1968):66–81.

U. S. Agricultural Marketing Service. *Livestock and Meat Statistics.* AMS Statistics Bulletin 333. Washington, D. C.: U. S. Government Printing Office, 1963, and supplements for 1965 and 1970 and AMS Statistics Bulletin 522 supplement for 1982.

U. S. Department of Agriculture. *Agricultural Statistics.* Washington, D. C.: U. S. Government Printing Office, 1967, 1982, and 1983.

U. S. Economic Research Service. *Agricultural Outlook.* Washington, D. C.: U. S. Government Printing Office, March, 1984.

———. *Economic Indicators of the Farm Sector: Costs of Production, 1982.* ECIFS 2–3. Washington D. C.: U. S. Government Printing Office, 1983.

# CONSISTENCIES, INCONSISTENCIES, AND OTHER VIEWS ON COMMODITY POLICIES: A REVIEW AND CRITIQUE OF PRECEDING PAPERS

**Alex F. McCalla***

## *Introduction*

I have, without a doubt, the worst task of the conference. First, I come on after you have listened to nine papers and just before the cocktail hour. Thus, it is unlikely that you are terribly interested in what I might say. Second, I have had to read nine papers and try to distill out of them what is generally applicable and relevant to this conference. I have worked harder doing this than I have in writing many previous papers. Third, I am to be a critic of not just the usual one, two, or three papers—which is dangerous enough—but of nine!

The design of this conference was apparently based on the notion of matrix organization or management. In the first day we would have two general critiques of commodity policy and six or seven

*The author is Professor of Agricultural Economics and Economist in the Agricultural Experiment Station and on the Giannini Foundation, University of California, Davis.

specific analyses of past policy and suggestions for the future. The second day we would look across commodity programs at general issues such as resources, food and nutrition, macro policy and trade, and structure. So perforce today's program was commodity oriented and that is reasonable, given that actual policy is focused on individual commodities. We have had some good papers, and we probably know more about past performance (which is, in the author's view, generally bad) than we really care to know. We have heard some suggestions for change, but I think it is fair to say we have not been overwhelmed with innovative new ideas. Most argue for modest changes. They cannot really conceive of no programs or radical alternatives. This may be realistic in that I would place a 95 percent probability, as did Firch, on the fact that 1985 will not be significantly different from 1981, 1977, 1973, 1965, 1948, 1938, and 1936. Namely, we will see marginal adjustments within the basic framework. So why should we engage in a conference like this? First, there is a 5 percent chance of significant change so we should have thought about it. Second, if we are going to go on, we should know what we are doing and how much it costs us. Third, Berkeley is nice in June.

In trying to review and critique the nine papers, I proceed as follows: First, I present a capsule of each paper in case you have forgotten. Some of these may contain an occasional critique. Second, I ask what can be learned from the papers. Third, I point out some weaknesses in the papers. Fourth, I attempt to identify from my perspective the important issues left partially or completely untouched by the papers that bought to be the focus of our review and the discussions of the 1985 bill.

## *Brief Reviews of the Papers*

Today has consisted of two general retrospective papers, two prospective papers on general policy (i.e., grain), and five commodity papers.

### William E. Lesher

In the beginning there was Lesher and his perspective on recent policy performance. The early part of his paper presents powerful evidence that the long-range policy plans of mice and men "gang aft agley." He cites no less than 19 shifts in policy direction since 1977 (his tenure in Washington). Guilt lies about equally with the Executive Branch and Congress. The remainder of the paper looks at loans, acreage reduction programs, target prices, and commodity reserves. He argues that, however well intentioned, the reality has, in general, been bad. Of particular interest is his evaluation of payment in kind (PIK) which comes off as necessary and the least-cost alternative at the time. This evaluation is at considerable variance with conventional wisdom and most other papers. The bottom line is that, judged by his particular narrow view of economics—namely, that it deals only with prices—all past policies have performed badly and that a big change is necessary. In doing so he laments that agricultural economists do not always produce analysis that supports his particular perception of what ought to be, but he provided a comprehensive and useful administration perspective.

## D. Gale Johnson

Next was Johnson who, as usual, presents a clear and perceptive critique of past performance. Based on political economic analysis, Johnson deduces seven lessons. The most important among them are:

1. The level of price supports influences how much we produce and where we sell it. It also influences how much our competition produces.
2. Supply management programs have little impact on output unless they are massive, e.g., PIK.
3. The level of price supports and supply management have "almost no long-run influence upon the prosperity of agriculture." Off-farm employment alternatives influence this. (This is the most controversial conclusion.)
4. Rather, as almost everyone agrees, benefits go basically to landowners or quota owners.

Thus, our two general papers, which reviewed past performance, both conclude that past programs get poor marks for performance:

- Land management as a supply control mechanism is very leaky.
- Price supports increase the value of farm assets but not necessarily farm income.
- They are expensive.
- They have tended to price us out of world markets at times.
- They result in the ultimate neoclassical economic sin—they cause resource misallocation.

The next five papers were to look at the past performance of a particular commodity policy and project to the future. Each was asked to analyze the impacts of four options:

1. Continuation of the 1981 Act.
2. Modifications of the 1981 Act, e.g., to make the levels of policy variables more sensitive to economic factors such as the exchange rate or the rate of inflation.
3. A "radical" alternative with no commodity programs or land control. Needless to say, most authors were puzzled by this alternative because they were asked to look at specific commodities. Therefore, in the time-honored academic tradition, most ignored this option.
4. No government intervention. Some considered this alternative briefly.
5. The final paper of the group looked at the impacts of farm programs on the livestock sector.

All of the papers are different so I will make a few comments about each before trying to look across the papers.

## Bruce L. Gardner

As is his usual approach, Gardner cuts to the heart of the issue with back-of-the-envelope analysis. He looks at five variations on the theme of the 1981 Act. He concludes that target prices and voluntary diversions are the most efficient means for income redistribution and stability. But, clearly, he does not like current programs. After a brief and nonanalytical look at the no-program alternative, Gardner pushes the notion of put options which essentially allow farmers to buy price insurance. His most telling point is that, if farmers do not buy these, perhaps all our concerns about stability are misplaced.

## Richard E. Just and Gordon C. Rausser

Just and Rausser's paper was also supposed to focus on grains but is, instead, a general case for simple conditional policies. They argue that, because our ability to predict levels of economic variables in the future is not very good, we should not choose absolute levels of support but rather tie them to moving variables. Two quick comments: First, they dismiss quickly an income objective for farm programs—this is a major assumption which I will return to later. Second, their policy proposal is potentially radical—get rid of the farmer-owned reserve, loan rates, set asides, and deficiency payments (most of the current instruments). In their place they recommend a linear stock/purchase rule based on a notion of desired stocks which, in turn, are based on budget costs. The stock adjustment would be accomplished by the government taking positions in the futures market. It sounds neat and simple, but my concern is that the first time someone's ox is gored, Congress and/or the Administration will respond by changing the policy (see Lesher's section). Both morning discussants called for political reality. The next four papers are more conventional.

## Robert S. Firch

This paper presents a detailed and useful analysis of past cotton programs. In Firch's judgment, they have worked well primarily because loan rates and target prices have been kept at "reasonable" levels. Firch does not address the no-policy or radical options but argues for fine tuning of loan rates.

## Dale N. Hoover and Daniel A. Sumner

This paper is an interesting review of complex policies. Both commodities are involved with rigid allotments, and they have been adjusted recently—tobacco to allow the sale of quotas and peanuts to a two-price scheme. In the cases of both programs, these recent adjustments look like closing the barn door after the export market has been lost. The paper does not look at the no-program option but does contain an excellent review of past policies.

## Andrew Schmitz, Roy Allen, and Gwo-Jiun M. Leu

This paper reviews past programs in a general sense and presents the beginnings of an interesting welfare comparison of three different policy regimes:

- Quotas—domestic and foreign—plus tariffs (1934–1974).
- A loan rate-stocks program (1978–1982) when there was a net societal gain.
- 1982 to now—quotas, tariffs, and no domestic supply control; clearly, a very costly program.

If this approach were pursued further, it could be useful. There is no discussion or comment on policy options.

## Emerson M. Babb

Dairy policy, the most malignant of all, has performed poorly—partly because we have no clear objectives for a policy and because support levels at times have been elevated too far beyond market-clearing prices. While Babb did address the various options and sees a need for change, in general, he predicts a gradual change toward a more market-oriented policy.

## Daryll E. Ray, Luther G. Tweeten, and James N. Trapp

The paper by Tweeten, Trapp, and Ray is different from the preceding ones. There is no livestock program except for import quotas which are not mentioned except in passing. The paper is an empirical analysis of price and return instability. There are three important conclusions:

> Programs that stabilize grain prices, in most cases, benefit livestock producers.
>
> Large year-to-year program changes destabilize the livestock sector and cause problems which take longer to work out than in the grain sector.
>
> It is instability in grain prices that is the problem—not their level.

Perhaps, the question that should have been addressed is impact of no programs on the livestock sector.

Included in this paper are numbers which suggest that, since 1972, beef feedlots have lost money in every year but one. This raises questions about feedlot operator rationality and the correctness of the analysis, or perhaps there is a third explanation. On the surface, it seems a little strange.

## *What Can We Learn From These Papers?*

1. Most agree that past policies (except for cotton) have not worked well given some notion of economic efficiency and stability. However, are these the proper goals? (I will return to this later.)

2. Almost all agree that the long-term benefits of commodity programs are capitalized into either land or the right to produce (e.g., quotas). Stated conversely, programs do not have much impact on income; but they do on wealth.

3. Rapidly changing economic conditions, particularly economywide (macro) and international, are increasingly frustrating attempts to chart longer term policy courses. When a policy crisis arises, quick action is taken which creates policy uncertainty. This may be more deleterious to agriculture than weather uncertainty. It is even worse when the two come together as they did under PIK in corn.

4. Policy rigidities are to be avoided. One solution (Just and Rausser) is for conditional policies. A more likely outcome is for a shorter term focus and continuous political adjustment—a conclusion supported by Lesher's 19 changes since 1977.

5. Most agree that macro variables are more important than farm policy variables, but few see how to deal with them.

6. The papers are a rich set of analyses with potential for comparative policy analysis. The commodities discussed have been subjected to almost every conceivable type of policy instruments—production quotas, loan programs, deficiency payments, two-price schemes, and tariffs and quotas. Surely, there is a challenge to economists to do what Bruce Gardner wants done on a much broader scale.

7. While some papers addressed options, most concentrated on adjustments to current programs. Is this because agricultural economists have no imagination? Is it because there are no new options? Is it because those who study commodities closely become so enmeshed that they become part of the policy? Or is it because we are so conditioned by the past that we consider discussion of radical options to be irrelevant!

8. Finally, most of the papers provide much material on the nature of past programs and their performance. This is clearly a first step toward better prospective policy analysis.

## *Weakness and Limitations of Papers*

Surely there must be weaknesses and inconsistencies in the papers. Looking hard, here is what I have found:

1. What is the policy framework we are dealing with? When I teach policy, I argue that policy is the selection of policy instruments—and the choice of their level—to accomplish goals. Clearly, there was not a clear discussion of goals. Most seem to see the goal as efficiency of resource allocation, but is this really the goal? Few distinguish between the choice of the instruments (e.g., target prices, loan rates and storage programs, two-price schemes, etc.) and the level of the instrument selected—the target price, the loan rate, and the rate of storage payment. Have past policies failed because we had the wrong goals, the wrong instruments, or simply the wrong levels of instruments? If it is only the latter, then fine tuning is appropriate as opposed to throwing the baby out with the bath water. What is needed is a policy analysis framework clearly spelled out.

2. Economists seem to be hung up on comprehensive long-term empirical policy analysis that leads to efficient policy. We favor com-

prehensive policy over a significant forward period. Thus, we like a Tinbergen kind of policy approach. But what is the actual policy process? It appears to be short-term, disjointed, and responsive to short-term economic and political crises as perceived by interest groups, i.e., Lindblom's "Policy of Muddling Through." This poses a dilemma: Should we economists continue to preach to the policy-makers that they are screwed up and continue to produce much irrelevant stuff? Or should we attempt to do policy analysis more in line with political system?

3. Few of the papers pay particular attention to distributional aspects (e.g., the bimodal nature of agriculture)—big *vs.* small or regional distribution issues. Yet, both are crucially important in the political economy of agriculture.

4. The papers address only price and income policies. What about all the rest, e.g., marketing, water, research, and credit?

5. Finally, there is little attention paid to interprogram effects. Suppose we took a random draw from the papers and put together a composite policy of the following sort:

- *Grains*—eliminate current program and use put options or linear storage rules.
- *Cotton*—leave as is.
- *Dairy*—adjust prices down with coresponsibility levies.
- *Tobacco and peanuts*—leave alone.
- *Sugar*—pick a program from the spinning wheel.

What would be the aggregate impact, i.e., what are the interprogram linkages? This requires knowledge of the cross elasticities of supply and demand and some understanding of interprogram impacts on uncertainty, but it is perhaps unfair to blame these papers for this weakness given the organization plan of the first day.

## *Questions for Discussion*

So what questions, in my opinion, should we be discussing for the remainder of the day?

- What are the goals: stability, equity, income transfers, or all of the above? For whom are we doing it? How are various groups weighted? These are crucial questions which, if not addressed, allow bright people like Gardner and Just and Rausser to talk by each other as they did in today's discussion.
- What are the prospects for longer term flexible (conditional) policies operated by mechanical (i.e., apolitical) decision rules in an intensely political system? If chances are low, perhaps we should do more analysis like Gardner's on options within the 'general framework of current policy.
- What difference does international market exposure, whether as exporter or importer, make to policy options? We still talk as if we can choose our own destiny irrespective of our trading partners and competitors.
- How should we deal with interprogram, interregional, and intergroup distribution effects?

What these questions all point to is linkages and interdependencies, but perhaps we should not fault the papers because the United States does approach agricultural policy on a commodity basis, and this was the way the discussion of the day was formulated. Perhaps tomorrow will rectify the situation and consider cross-program effects.

# ADDITIONAL DIMENSIONS OF FOOD AND AGRICULTURAL POLICIES

*13. Structural Dimensions of Agricultural Policies*

*14. Trade and Macroeconomic Dimensions of Agricultural Policies*

*15. Resource Dimensions of Agricultural Policy*

*16. Food and Consumer Dimensions of Agricultural Policy*

# 13 STRUCTURAL DIMENSIONS OF AGRICULTURAL POLICIES

David Zilberman
Harold O. Carter*

## *Introduction*

Structural considerations have had secondary importance in the formulation of agricultural policies in the last half century. Recent agricultural policies have been focused primarily on problems of sagging farm prices and incomes and, to a lesser extent, on income stability. This approach is in sharp contrast to the 19th Century when the preeminent policy issues were structural; for example, the distribution of public lands was a national issue dating back to the beginning of this nation. The debate hinged on whether to distribute the land in small parcels at modest prices to homesteaders or in large grants with the purpose of maximizing government revenue. Over a period of 80 years and many compromises, federal lands were transferred to private landholders. A second major structural issue dealt with the question of using black slave labor—the plan-

*The authors are, respectively, Associate Professor of Agricultural and Resource Economics and Economist in the Agricultural Experiment Station and on the Giannini Foundation, University of California, Berkeley, and Professor of Agricultural Economics and Economist in the Agricultural Experiment Station and on the Giannini Foundation, University of California, Davis.

tation system of the South—in the new western territories; and resolution of this issue came at a high price.

The structural policies of the 19th Century were successful as U. S. agriculture developed, for the most part, a structure consistent with the ideology of the family farm. Once achieved, structure lost its prominence as a policy objective; thereafter, price and income issues associated with the farm problem received the major policy attention.

It is incorrect to assume that the recent evolution of the structure of the farm sector is desirable or consistent with the ideology of the family farm. As recent studies indicate (Tweeten, 1984, and Penn, 1979), the number of farms in America is declining, the average farm size is increasing, and there is an increasing tendency toward separation of landownership and operation. Vertical coordination mechanisms for production control, as well as vertical integration, have become more and more prevalent over time resulting in some farmers giving up their decision-making powers to processors, input suppliers, etc. The broiler industry is an extreme example of the "industrialization" of agriculture; it is, as Reimund *et al.* suggest (1979), a trendsetter.

Farm-size distribution is becoming increasingly bimodal. On one hand, there is a large number of small farms, the owners of which receive a significant share of their income from nonfarm sources. On the other hand, a relatively small number of large commercial farms control a significant share of farm output. According to Tweeten (1984), traditional family farms (i.e., medium-size farms) are declining in both absolute numbers and terms of the share of all farms. Their future seems to be less promising than that of either larger or smaller farms. Over time, farm enterprises have relied more on debt financing and thus have become vulnerable to cash-flow shortages. Medium-size family farms are more likely to encounter this problem which may lead to foreclosure.

Despite the structural transition undergone by U. S. agriculture, questions of structure now emerge only peripherally, if at all, when farm policy is being formulated. Concerns are voiced occasionally about factory farms, foreign ownership of farmland, developers using prime agricultural lands, and the demise of the family farm. Legislative response to such concerns takes the form of payment limitations on commodity programs, homage given to family farms in the preamble to most farm legislation, and directives to the U. S. Department of Agriculture to prepare reports on the status of . . . Analysis takes the form of extrapolating historical trends well after the fact rather than anticipating possible structural impacts when price and income programs are being considered.

Why is this so? Apparently, it is easier to obtain political consensus on income support than on what structural attributes are desired for the farm sector. Moreover, size, ownership, organization, and control are influenced by a myriad of factors within and outside of the farm sector. Even if we could obtain consensus about the desirable properties of farm-sector structure to pursue these structural goals, we would need a framework to analyze the direct and indirect structural impacts of various farm policies. Concern for and dismay over the lack of such a framework were raised by Gardner (1978).

In an attempt to fill this void, Just, Rausser, and Zilberman (1983) developed several conceptual formulations for addressing structural implications of sectoral and general policies. This paper presents some of their qualitative findings with respect to agricultural policies. First, their assumptions regarding key factors and relationships governing the farm sector are introduced. Second, the structural impacts of two generic forms of government intervention—diversion programs and price-support policies—are presented. The paper concludes with a discussion of some policy implications for the 1985 farm bill.

## *A Framework for Structural Impacts of Agricultural Policies*

Agricultural policies affect the entire food sector—input suppliers, farmers, livestock producers, processors, distributors, wholesalers, and retailers. For simplicity and ease of exposition, the analysis presented here will focus on the impact of agricultural policies on the farm sector. The rationale for agricultural policies, after all, has been to alleviate the farm problem and safeguard farmers' welfare. Nevertheless, recent tendencies of increased integration of ownership, management, and control between various components of the food chain suggest the need for a more comprehensive structural analysis of agricultural policies beyond the farm gate addressing impacts on the food sector as a whole.

The structure of an industry is the institutional setting for its economic activities. Analysis of structural impacts of government policies consists of the evaluation of the relative and absolute effects that policies have on different economic agents and institutional arrangements within an industry. Some of the main structural dimensions affected by agricultural policies include (1) farm-size distribution and the relative and absolute welfare of farms of different sizes; (2) the relationship between landownership and operation; (3) returns to agricultural inputs and assets—in particular, to land, agricultural labor, management, and technology; (4) geographic and demographic distribution of farm assets, ownership, and production control; and (5) control of production by nonfarm stages of the food chain (input suppliers and processors).

Because of space and time limitations, impacts of agricultural policies on only the first three aspects of structure, which seem to be subject to the most controversy and interest, will be addressed. The analytical framework used here attempts to capture some of the main features characterizing agricultural production—features that have been widely documented in the agricultural economics literature (Brandow, 1977). These characteristics include rapid rate of technological change embodied primarily in innovations; low income and price elasticities of demand; competitiveness; price and yield uncertainty; asset fixity; heterogeneity of input quality (variability of land quality and management ability); risk-averse behavior of farmers; and institutional limits to credit.

The first five features were documented by Schultz (1953), Cochrane (1958), and Johnson (1958). Schultz called attention, also, to the large differences in the rates of return to resources among regions as well as among producers. Most of this variation emanates from differences in land quality, production techniques, human capital, and wealth controlled by individual producers. Just (1974) documented the importance of risk-averse strategies in farmers' behavior. Limitations of credit availability for producers of different size classes have been noted in recent empirical evidence. This evidence suggests strongly that larger farmers borrow more in general; they borrow more to invest in capital. Their ability to borrow more stems, in part, from their higher repayment capacity and greater asset holdings (Baker, 1977; Quinn, 1975; and Riboud, 1977).

The increased importance of credit financing in agricultural investment and its significant distributional implications were emphasized by Carter and Johnston (1978). The basis for their concern is the evidence that "the proportion of (farmland) transfers on which debt was incurred rose from 58% in 1950 to 88% in 1977 and the ratio of debt to purchase of price credit-financed transfers rose from 57% in 1950 to 77% in 1977" (p. 744). Farmland debt has increased from approximately $30 billion in 1971 to about $72 billion in 1979.

A specification of the technical aspects of agricultural production, which admits all of the special features noted above, is the so-called putty-clay framework (Johansen, 1972, and Salter, 1966). This framework recognizes the technically embodied capital available for adoption by farmers. Moreover, at least some of the "new" capital is indivisible which, in turn, leads to unequal degrees of returns to scale in using the new technology for large- versus small-scale producers. Particularly in the case of machinery, these capital goods are often so specialized that their input-output ratios cannot be altered. Of course, prior to such investment decisions, producers can select an alternative technology described by a conventional neoclassical production function. This latter specification was used extensively in the framework reported here. A generalized mean-variance approach, derived from the expected utility approach by Just and Zilberman (1983), was used to analyze risk-averse behavior.

This paper presents results pertaining to two generic commodity policies: price supports and diversion payments. A generic price-support program is defined as a policy that raises the farmers' average returns per unit of production (acreage) and reduces the variability of the returns. This definition emphasizes the variability and uncertainty associated with agricultural production and the role of price-support policies in reducing downside risk; it allows the results obtained by the analysis to apply (at least partially) to a wide range of existing and past price-support, income-support, and crop-insurance schemes. A generic diversion program consists of a diversion payment in exchange for fallowing a fraction of acreage historically allocated to the crop. Again, this definition is quite general; the analytical results obtained with it pertain (at least partially) to a large variety of programs from soil conservation to the payment-in-kind program (PIK).

The range of policies encompassed conceptually by these two generic policies is wide and includes most of the existing highly visible and important programs. Just and Zilberman (1984b) addressed some of the structural impacts of other policies, i.e., public research and extension and credit and input subsidy. Many policy schemes are combinations of the generic programs analyzed here. For example, entitlement to deficiency payments or loan rates (generically price-support policies) may require participation in a set-aside scheme. However, as the objective here is to demonstrate the potential of a new framework for structural analysis and its application, the policies considered remain very general and abstract. Quantitative analysis will require more specific details in defining the policies to be analyzed.

## *Structural Impacts of Generic Price-Support Policies*

Just, Rausser, and Zilberman (1983) and Just and Zilberman (1984a) evaluated the relative welfare effects of price-support and deficiency payments. Their analysis emphasized the economic uncertainties faced by farmers and the consequent risk-averse behavior, the use of land as collateral for obtaining credit, and the simultaneity of land-allocation and technology-adoption decisions. The model assumes that farmers can allocate their land between two activities: an old one and a new one. The old technology is less profitable, on the average, but it involves less economic risk. The new technology is potentially more profitable but is more capital intensive and riskier; it requires not only higher per acre costs but also an additional fixed cost associated with acquiring special equipment and/or knowledge. A farmer has to decide whether or not to adopt the new technology and, if so, how much land to allocate to the new technology and how much to the old.

This model applies, for example, to the choice between irrigated and nonirrigated corn production in Nebraska where the farmer has to decide whether or not to invest in the infrastructure for irrigation and how much land to irrigate (increasing average revenue and revenue variability relative to not irrigating). Another example is the choice between two crops such as cotton and alfalfa in California (where cotton is the more capital intensive and riskier) or between cotton and corn production in the Mississippi Delta.

The analysis considers variation in farm sizes, risk preferences, and wealth among farmers but assumes that risk aversion and wealth are functions of farm size. (Wealth increases with size, absolute risk aversion declines with size, and relative risk aversion increases with size.) The conclusion is that the agricultural industry may be represented by four types of farms:

1. *Traditional small farms* that do not adopt modern, scale-intensive technologies or grow capital-intensive crops but continue to operate with low-risk, older activities.
2. *Leveraged farms* that are sufficiently large to acquire the ability to use the new activity but not large enough to generate sufficient credit to finance the new activity on all landholdings—that is, the extent of the new technology use is bounded by credit constraints.
3. *Specialized modern farms* that are sufficiently large to allow full utilization of the new activity but which are not so big that risk considerations prevent their specialization. (Recall that relative risk aversion increases with size.)
4. *Risk diversifiers* (farms) that acquire the new activity but allocate land between the two activities in order to reduce risk.

Tweeten (1984) divides farms into three size classes (by annual sales): small ($5,000–$39,999), medium ($40,000–$199,999), and large ($200,000 and over). It seems that the *traditional small-farm* category approximates that of Tweeten (1984) and consists of land owned primarily by part-time and older farmers. Most farms that belong to the *leveraged farm* and *specialized modern farm* categories fall into Tweeten's medium-size farm class and are mainly full-time family farms; the *leveraged farms* are likely to be owned by young, financially struggling farmers. Farmers in the *risk-diversifier* category correspond to large farms in Tweeten's typology—they are either large-size family or corporate farms.

The analysis shows that the impacts of policies depend on whether the two technologies are used with the same crop (irrigated and nonirrigated corn) or with different crops (cotton and alfalfa). In both cases, results depend on the degree of correlation between revenues under the two technologies. The study also evaluates how policies will affect expected income corrected for risk (henceforth referred to as risk-corrected or risk-compensated income) and weights results by farm size to analyze the *relative* impacts of the policies.

First, consider the case when the two activities pertain to two different crops. An increase in a price-support policy of the more capital-intensive crop will increase its average price and reduce its price variability. It may reduce the minimum size required for beneficial use of the new activity thus causing its adoption by some of the small traditional farms; it may also increase the use of the new crop by the risk-diversifying farms in the case in which correlation of revenues between the crops is small.

A comparison of the relative welfare effects of an increase in price support indicates that this change will not affect the risk-corrected income of small traditional farms. The risk-corrected income of leveraged farms and specialized modern farms will increase more than proportionately with farm size with stronger impacts on the leveraged farms. The risk-compensated income of the large risk-diversifying farms will increase with farm size as price support increases in the likely case that returns between the crops are not strongly correlated. If the correlation of returns between the technologies is sufficiently high, the gains from increased price support will decline with size among risk-diversifying farmers. In any case, the risk-compensated income of the risk diversifiers is not likely to increase more than proportionately with farm size as price support increases.

When the two activities are associated with two different crops, an increase in the price support for the new crop has a stronger relative impact on leveraged farms and specialized modern farms than on risk-diversifying farms or traditional small farms. Therefore, removal of price support for the newer technology is likely to have a relatively stronger negative impact on the medium-size farms of Tweeten's typology than on small or large farms. Hence, it is likely to accelerate the solidification of a bimodal farm-size distribution. Note that, even within the risk-diversifying group, large farms will suffer relatively less from the removal of the price supports than will smaller farms in the group.

When the two technologies pertain to the same crop (irrigated and nonirrigated corn), an increase in the price support for the crop is likely to decrease the minimal size of the farm which will adopt the new technology. A comparison of the relative welfare effects of an increase in price supports indicates that they will always increase the risk-compensated income more than proportionately with farm size among small traditional farms and specialized modern farms. In the case that returns under both technologies are highly correlated, the risk-compensated income of leveraged farms will increase more than proportionately with farm size. Note, however, that, for leveraged farms, the increase in risk-corrected income associated with increased price support may decline with farm size in the unlikely case of weakly correlated returns between the two technologies. Under reasonable circumstances, an increase in price support will increase the income of small traditional farms, leveraged farms, and specialized modern farms more than proportionately with farm size with full adopters making the largest gains. The gains from increased price supports among risk diversifiers, however, will not increase more than proportionately with farm size. In the case of high correlation of revenues between the technologies, gains from increased price support will actually decline with farm size among risk diversifiers.

In cases where both technologies are used to produce the same crop, an increase in price support tends to benefit the small- and medium-size farm groups proportionately more than the large farms. Moreover, among the large diversified farms, relative benefits may even decline with size.

Combining the results of these two cases, it seems that price support (and other policies that increase revenues and reduce revenue variability such as deficiency payments) tends to have relatively stronger positive impacts on medium-size farms than on large farms. Furthermore, above a certain farm size, relative gains associated with price supports may decline with size. Thus, if one considers policies that favor medium-size farms relative to large farms to be desirable from a structural perspective, generic price-support schemes have a positive impact on the structure of farming, whereas return to a free market may have undesirable structural implications.

An obvious limitation of the analysis is that it restricts farmers to diversify and allocate their resources only in crop-production activities. In reality, farmers tend to diversify by working off the farm, investing in assets outside the farm sector using financial and futures markets, and participating in more than one stage of the food chain. Nevertheless, incorporation of these real world possibilities into the analysis does not seem to alter the primary outcome. While small farms are somewhat protected from farming risk by off-farm income and choices of crops and technologies and large farms are diversified between crop production and activities outside the farm sector, middle-size farms tend to commit the bulk of their assets to agricultural production. Thus, they are most vulnerable to farming risks and, therefore, may benefit relatively more than do the other types of farms from risk-reducing, generic price-support schemes.

Another limitation of these results is that they pertain to the effects of price supports on the average welfare of different farm types within a static, production-oriented framework without taking into account the effect of policy changes on land prices and farm-size evolution over time. Further study is needed to analyze the impact of price supports within a more realistic framework.

## *Structural Impacts of Diversion Policies*

Diversion policies are the second prevalent type of agricultural policies in use. Generically, a diversion policy consists of two parameters: (1) the diversion requirement (the fraction of acreage historically allocated to a farm crop that will be fallowed or not utilized in the production of the crop) and (2) the diversion payment (the payment per acre utilized for the production of the crop). Because participation in diversion programs is voluntary, it is reasonable to assume that not all farmers entitled to participate choose to do so. The generic formulation can be applied to a wide variety of programs including the recent PIK program.

Rausser, Zilberman, and Just (1983) investigate structural impacts of diversion policies. Their model distinguishes explicitly between farm operation and ownership and recognizes the heterogeneity of land quality and the technological variability among farmers. They analyze simultaneously production, program participation, landownership, and technology adoption choices assuming imperfect capital markets and risk neutrality under several future land price expectation schemes.

The model predicts that the introduction of diversion policies will result in slippage—a tendency to divert the lowest quality land to satisfy the diversion requirement. It suggests, also, that farmers participating in the diversion program will tend to utilize higher quality land than would nonparticipants. The initial effects of the introduction of a diversion policy, a reduction in the diversion requirement, or an increase in diversion payments are to (1) increase rental rates for land of all qualities as well as to widen the gaps between rents of different land qualities (representing differences in productivity), (2) reduce the quasi rents to technology and operations, and (3) decrease the number of farm operators. The operators squeezed out are likely to be older, smaller, and less efficient. In this respect, the acreage-reduction aspect of diversion programs contributes to the reduction in farm numbers and to the increase in size of average farm operations over time. The introduction of a diversion program, an increase in diversion payments, or a reduction in the diversion requirement also has an important impact on prices. Land prices will increase, as will the gap in land prices between lands of different qualities, but the rate of increase in land prices will be smaller than the rate of increase in rental rates. Moreover, when input demand and output supply curves are not perfectly elastic, prices of some variable inputs may decline and output prices will increase.

Higher output prices will trigger second-round effects. Rental rates and land prices will rise even further in the second round as will both the gaps between the rents and the prices of different land qualities. Some of the primary effects of participation in the program on quasi rents for operations and technology and on the number and size of farms may be reversed; nevertheless, the overall impact of diversion policies will be to reduce both the number of farms and the relative profitability of farm operation.

The differences in the direction of effects on returns from landownership and those from its operation suggest that the introduction of diversion policies or an increase in effective payment per diverted acre will tend to increase the separation of ownership from operation. Some small landowners with relatively inefficient operations will continue to hold their land by renting it to others for operation and, in the long run, even for diversion.

In evaluating the impacts of diversion policies on technology adoption, one has to distinguish between yield-increasing technologies (irrigation) and cost-reducing technology (machinery) as well as between landowners and renters as investors in the new technology. If the modern technologies require a larger scale operation than do older ones, the introduction of a diversion policy or an increase in effective payment per diverted acre will make the adoption of cost-reducing technologies less attractive but may make the adoption of yield-increasing technologies more attractive. Of course, the impact on the actual adoption rate depends not only on pure profitability but, also, on credit considerations. Hence, the difference between owners and operators is quite striking. For landowners, an initial increase in land prices followed by a secondary effect associated with higher output prices will result in a better wealth position which will increase their collateral and expand the availability of credit. The expanded availability of credit—along with, perhaps, better credit terms—provides further incentive to adopt yield-increasing technologies and may even result in the adoption of cost-reducing technologies. Thus, one may expect that the tendency of large landowners to adopt large-scale, modern technologies may increase because of an expansion in diversion policies. On the other hand, renters will have a relative disadvantage as a result of expansion in diversion policies. Their quasi rent from operations will decrease and their rental rate will increase; thus, their tendency to adopt the new technology will decline.

Expansion of diversion policies may reduce, in particular, opportunities for young persons without assets to enter agriculture. The increase in land prices will make it more difficult for them to purchase land, and the reduction in quasi rent from operations will make adoption of new technologies less desirable. On the other hand, some established farmers may increase the size of their farm operations by taking advantage of the better economic situation; by adopting modern technologies; and by renting land from smaller, less efficient farms. These effects combine to increase the concentration in farm operations over time.

The increased tendency of farm owners who are in better financial positions (because of diversion policies to adopt large, primarily yield-increasing technologies) has an additional effect that produces a dangerous disequilibrium. In the long run, this tendency will increase supply thereby putting pressure on markets to reduce prices further. This will either generate a need for further expansion of diversion programs or bring expanding farms to a cash-flow crisis of the type we are currently witnessing.

Our analyses have demonstrated that diversion policies tend to have negative structural implications. They tend to reduce the number of farms, encourage separation of ownership from operation, make entry to farming more difficult, and may even encourage expansion of large-farm operations. Gradual removal of diversion policies seems to have desirable structural implications, reducing land prices and increasing returns from operation. It will reduce the trend toward fewer and bigger farms and toward separation of ownership from operation, and it will ease entry into the farming industry.

## *Conclusions*

Since the introduction of commodity programs about 50 years ago, the structure of U. S. agriculture has gone through a significant transition. Farm size is approaching a bimodal distribution, and middle-size family farms—once the backbone of U. S. agriculture—are becoming the "soft belly" of the farming structure. Obviously, the correlation between this structural transition and agricultural policies does not necessarily imply a causal relationship. Moreover, there is a wide consensus that technological change, off-farm income earning possibilities, and economic growth are the major causes of the structural transition (Tweeten, 1984). Nevertheless, the results presented here invalidate the belief that changes in agricultural policies are neutral in impact on the structure of the farm sector. We have demonstrated analytically the causal links between various agricultural policies and structural change. While different policies may have opposing structural impacts, it is unlikely that the net effect of agricultural policies is nil. Thus, we conclude that agricultural policies have combined and interacted with monetary and fiscal policy, technological change, and off-farm earning possibilities to affect farm structure. Therefore, whenever there is concern about the structure of U. S. agriculture, the structural impacts of alternative legislative packages should be considered. The framework presented here is likely to be useful in quantitative estimation of the structural impacts of alternative policies.

Using the family farm ideology as a criterion for evaluating the structural impacts of agricultural policies, it seems that price supports and diversion programs affect structure in opposite directions. The risk-reducing component of price supports serves to benefit mostly middle-size farms producing positive structural effects. Diversion policies tend to benefit landowners relative to operators, reduce the number of farms, and make entry to farming more difficult. Therefore, diversion programs have undesirable structural impacts. Moreover, the analysis indicates that diversion programs do not solve the excess supply problem but, rather, augment it owing to slippage and credit feedbacks. Thus, from a structural perspective, it seems that a transition away from policies that rely on diversion would be desirable. On the other hand, structural considerations may justify maintaining price-support and stabilization policy schemes and/or crop-insurance programs, especially if the program parameters essentially reduce risk. Of course, the performance and viability of such risk-reducing schemes must be evaluated constantly, and they may be curtailed or dismantled if and when the private sector develops insurance schemes that can substitute for government programs.

## *References*

Baker, C. B. "Instability in the Capital Markets of U. S. Agriculture." *American Journal of Agricultural Economics* 59, No. 1 (February 1977):170–177.

Brandow, G. E. "Policy for Commercial Agriculture, 1945-1971." In *A Survey of Agricultural Economics Literature.* Ed. L. Martin. Vol. 1. Minneapolis: University of Minnesota Press, 1977, pp. 209–292.

Carter, Harold O., and Warren E. Johnston. "Some Forces Affecting the Changing Structure, Organization, and Control of American Agriculture." *American Journal of Agricultural Economics* 60, No. 5 (December 1978):738–748.

Cochrane, Willard W. *Farm Prices: Myth and Reality.* Minneapolis: University of Minnesota Press, 1958.

Gardner, Bruce L. "Public Policy and the Control of Agricultural Production." *American Journal of Agricultural Economics* 60, No. 5 (December 1978):836–843.

Johansen, Leif. *Production Functions: An Integration of Micro and Macro Short-Run and Long-Run Aspects.* Amsterdam: North-Holland Publishing Co., 1972.

Johnson, Glen L. "Supply Function—Some Facts and Notions." In *Agricultural Adjustment Problems in a Growing Economy.* Ed. E. O. Heady, H. G. Diesslin, H.R. Jensen, and G. L. Johnson. Ames: Iowa State College Press, 1958, pp. 74–93.

Just, Richard E. "An Investigation of the Importance of Risk in Farmers' Decisions." *American Journal of Agricultural Economics* 56, No. 1 (February 1974):14–25.

Just, Richard E., and David Zilberman. "Stochastic Structure, Farm Size, and Technology Adoption in Developing Countries." *Oxford Economic Papers* 35, No. 2 (July 1983):307–328.

———. "Equity Implications of Agricultural Policies." University of California, Department of Agricultural and Resource Economics, Working Paper No. 303. Berkeley, January, 1984a.

———. "The Effects of Agricultural Development Policies on Income Distribution and Technological Change in Agriculture." University of California, Department of Agricultural and Resource Economics, Working Paper No. 310. Berkeley, July, 1984b.

Just, Richard E., Gordon C. Rausser, and David Zilberman. "Modeling Equity and Efficiency in Agricultural Production Systems." In *Growth and Equity in Agricultural Development.* Proceedings: Eighteenth International Conference of Agricultural Economists, Jakarta, Indonesia. Oxford, England: Institute of Agricultural Economics, 1983, pp. 120–138.

Penn, J. B. "The Structure of Agriculture: An Overview of the Issue." In *Structure Issues of American Agriculture.* U. S. Department of Agriculture, Economics, Statistics, and Cooperative Service, Agricultural Economics Report No. 434. Washington, D. C.: U. S. Government Printing Office, November, 1979.

Quinn, Olin B. "Sources and Uses of Funds in Agriculture." *American Journal of Agricultural Economics* 56, No. 5 (December 1975):1063–1065.

Rausser, Gordon C., David Zilberman, and Richard E. Just. "An Equilibrium Model of Distribution Effects of Land Controls in Agriculture: Theoretical Implications of Programming Models." University of California, Department of Agricultural and Resource Economics, Working Paper No. 80. Berkeley, November, 1983.

Reimund, Donn, J. Rod Martin, and Charles Moore. "Structural Change in Agriculture." In *Structure Issues of American Agriculture.* U. S. Department of Agriculture, Economics, Statistics, and Cooperative Service, Agricultural Economics Report No. 434. Washington, D. C.: U. S. Government Printing Office, November, 1979.

Riboud, Chris. "Agricultural Credit Markets." Department of Economics, Massachusetts Institute of Technology, Cambridge, 1977.

Salter, W. E. G. *Productivity and Technological Change*, 2d ed. Cambridge: Cambridge University Press, 1966.

Schultz, Theodore W. *The Economic Organization of Agriculture.* New York: McGraw-Hill Book Company, Inc., 1953.

Tweeten, Luther G. "Causes and Consequences of Structural Change in the Farming Industry." Department of Agricultural Economics, Oklahoma State University, Stillwater, 1984.

# 13 DISCUSSION

## J. B. Penn*

The Zilberman-Carter chapter is very interesting. They note that their objective is to demonstrate the potential of a new framework for structural analysis, and I believe they achieve that objective quite well.

The cases considered—generic price support and diversion policies—are admittedly very general and abstract. The results suggested by the analysis are generally consistent with the conventional wisdom as to the past effects of these policies. The important contribution here is the initial development of a framework (drawing upon previous works of Rausser, Just, Zilberman, and others) which can be extended and elaborated to incorporate more interesting cases.

The authors acknowledge that the assumptions are quite limiting in some cases; therefore, the results are limiting. (Price-support effects are within a static, production-oriented framework which does not take into account the effect of such policy on land prices and farm-size evolution over time.) Analysis of the impacts within a more realistic framework would overcome this objection.

The two major conclusions of the paper are:

*The author is Principal Member, Economic Perspectives, Inc., Washington, D. C.

- Diversion policies tend to reduce farm numbers, encourage separation of ownership and operation, restrict entry, and encourage expansion of large farms.
- Price-support policies tend to have stronger positive impacts on medium-size rather than large-size farms.

Overall, the authors conclude that their results invalidate the belief that changes in agricultural policies are neutral with respect to the structure of the farm sector. They certainly illustrate that not all firms are affected in the same way.

The authors also observe that structural considerations have had only secondary importance in the formulation of agricultural policies in the last 50 years. I do not believe anyone would disagree with that. It is easier to obtain political consensus on price and income instruments than on any particular desired structural configuration. We have been and continue to be plagued by the resulting (and, perhaps, largely unintended) structural impacts of past policies. Those effects on the diverse economic agents still are not well understood.

There was considerable discussion in the conference on the goals of agricultural policy (McCalla). There was more agreement than not that the programs have not worked very well in the past. As we look to the future, there still is much uncertainty about what the programs are intended to do. Bruce L. Gardner pointed out that, if the programs are to transfer resources to the farm sector, they have done that in a fairly wholesale way. Why do we need to continue transferring resources? D. Gale Johnson also reminded us that, while we have a far different agriculture today, we still have the same old programs.

I tend to agree with McCalla's "goal crisis" notion or, alternatively stated, a "problem identification crisis." We cannot begin to agree on policy objectives until we can agree on what the problems are that warrant public intervention in the farm sector.

It has been said many times that the initial objective of the farm programs simply was to transfer income to the farm sector to improve the economic status of farmers—to raise their incomes to levels more commensurate with nonfarm incomes. Most analysts would agree that that has been achieved—by growth in off-farm income earned largely by smaller farmers and by growth in farm size and farm incomes. Even though we no longer have that problem, we continue transferring resources to the sector.

Still, problems remain in the farm sector—problems that have structural dimensions although they might not involve an explicit structures policy. Aggregate indicators for the sector (net farm income, net cash income, debt-to-asset ratio, rate of business failures, etc.) are not consistent with other reportings of anguish from the farm sector. These indicators do not suggest there is a depression in the farm sector. We continue to hear of a farm sector on the verge of bankruptcy and of proposals for massive "bailouts."

The farm sector is so varied today—so heterogeneous—that there is no one common problem that affects, to the same extent, the vast majority of farm businesses. Some farms are in dire straits (inadequate cash flow with an eroding asset base) to be sure; others are doing quite well, receiving an attractive return on their investment. Some of these successful farms are large; some are of medium size; and, undoubtedly, there remain some cases of chronic poverty among the smaller farms.

Without more solid, reliable indications of the nature and extent of the problems, we have little basis for informed prescriptions. The danger is that anecdotal reportings become the conventional wisdom. Policy prescriptions are then formed on that basis, and Congress will again legislate multibillion dollar programs that, in four years, we will again conclude have not worked very well.

Such programs are becoming less and less acceptable as we recognize that the number of beneficiaries has become quite small. Today, 29 percent of the farms produce 88 percent of the output and receive most of the farm income. The remaining 71 percent receive most of their income from off the farm.

| *Sales* | *Number* | *Percent* | *Output (percent)* |
|---|---|---|---|
| More than $40,000 | 700,000 | 29 | 88 |
| Less than $40,000 | 1,700,000 | 71 | 12 |

With the numbers so small, perhaps it is already too late to focus on an explicit structural objective such as maintaining a large number of diverse family farms. Focusing on the nature of the problems of farms across the various sizes and types will help define the objectives for the programs and enable more efficient programs to be designed.

# 14 TRADE AND MACROECONOMIC DIMENSIONS OF AGRICULTURAL POLICIES

G. Edward Schuh*

Both the U. S. economy and the international economy of which it is a part have undergone dramatic changes these last 20 years.[1] In fact, these past two decades may have witnessed the greatest change of any comparable period in modern history. We have witnessed a remarkable increase in the openness of our economy or in our dependence on trade (whichever way you prefer to view it)—with that dependence doubling during the decade of the 1970s and *tripling* since 1965. Today, approximately 25 percent of this nation's gross national product is attributed to trade. Consequently, it no longer makes sense for us to view ourselves as *Fortress America*, or to ignore the constraints the international economy puts on our economy and our economic policies.

These last 20 years have also witnessed the emergence of a remarkable international capital market—one that ties the economies of the world together in ways every bit as important as international

---

*The author is Professor and Head, Department of Agricultural and Applied Economics, University of Minnesota, St. Paul.

[1]For more detail on these changes and the implications for U. S. agriculture, see my testimonies before the Joint Economic Committee (Schuh, 1982 and 1983b).

trade. Perhaps more important, the international capital market provides a significant link among the economic policies of our respective countries.

The last 20 years also witnessed the shift from the old fixed exchange rate system to one of flexible exchange rates—a system which might best be described today as a system of block floating. Finally, the last 20 years have witnessed a very significant increase in monetary instability at the very time that other changes in the system made that instability particularly important. For our purposes today, it does not matter whether that instability is due to our own monetary policy or to autonomous shifts in capital. The important thing is the instability.

These transformations in how we relate to the international economy have changed greatly the economics of agriculture. They have also changed the way we have to think about much of our national economic policy. Unfortunately, I fear we have not fully recognized the effects on the economics of agriculture, nor have we fully recognized how we need to think differently about our national economic policies. As a consequence, I am going to spend most of my time today talking about things we tend to leave out when we think about agricultural policy. In the time allotted me, I can do little more than touch on some of the important issues. But in so doing, I hope I can at least get them on our agenda.

## *A Macroeconomic Perspective on Agricultural Policy*[2]

It should be obvious to most observers that the overwhelming share of work on agricultural policy in this country has been done from a sectoral perspective. Agricultural policy has not come out of a planning ministry or a ministry of finance, nor has much research on agricultural policy been done by macroeconomists. Instead, policy has, for the most part, come out of the U. S. Department of Agriculture and the two agricultural committees in Congress. (It is only recently that the Office of Budget Management has been put-

[2]For a more in-depth discussion of some of these issues, see Schuh (1976).

ting significant constraints on policy choices.) The policy *research* has been done, for the most part, by agricultural economists who have had very little training or interest in macroeconomic issues. There are exceptions to both of these assertions but, for the most part, they are exceptions that prove the rule.

A macroeconomic perspective on agricultural policy brings out a number of important issues. I would like to review these briefly for they are really what my paper is all about. First, in taking a macroeconomic perspective, we need to give more attention to what I call national or global resource efficiency. This means that (1) agricultural policies have to be evaluated in terms of making more efficient use of our national resources and (2) we have to view our national efficiency criteria in the context of the international economy and our international comparative advantage. This issue becomes increasingly important as our hegemony over the international economy declines.

Second, we need to understand the linkages among monetary and fiscal policy, the exchange rate, the international capital market, and international adjustment. If we have grossly underestimated one thing of importance, it is the issues surrounding these macroeconomic policies.

Third, the domestic terms of trade become a critical analytical issue. It is not the parity ratio that matters as has long been recognized by agricultural economists. It is the price of agricultural commodities relative to the price of all other goods and services that determines the relative social profitability of agriculture. That means that our real exchange rate is important. It also means that trade policy is important, and it is not just agricultural trade policy that is important. It is the *relative* protection among sectors that matters.

Fourth and finally, there is the issue of food as a wage good. Here we have somewhat of a paradox. We tend to generate a great deal of rhetoric about how cheap food is in this country and about how U. S. consumers spend a smaller share of their budget on food than consumers in any other country in the world. But at the same time, we give little attention to the significance of this as a factor affecting our overall competitive position in the international economy. In my judgment this will be an increasingly important issue in the years ahead.

With this as background, I would like to turn now to a discussion of the major issues before us as we consider the trade and macroeconomic dimensions of agricultural policies.

## *The Major Issues Before Us*

There are some six issues I want to address.

### Macroeconomic Policy Issues

The *substance* of what I want to address under this rubric has to do with monetary policy, fiscal policy, and the exchange rate. The institutional changes that have brought these topics higher on our policy agenda are the emergence of a well-integrated international capital market and the shift from a system of fixed exchange rates to a system of flexible exchange rates.

I have discussed most of these issues elsewhere and do not want to dwell on them here today (Schuh, 1979). There is much less controversy now among agricultural economists than there was 10 years ago when I first raised the issue as to whether the value of the dollar matters to U. S. agricultural trade. But at the practical level of making agricultural policy, the issues still are not fully recognized, nor is the significance for agricultural policy of the changes in this part of our economic world fully appreciated.

A number of things are important here. First, what we have witnessed in the 1980s is a perfect example of a major change in one of our macroeconomic variables—the exchange rate—completely swamping a reasonably well-designed commodity policy. The rise in the value of the dollar from 1980 through 1983 was far greater than was imagined when loan rates, target prices, and release and call prices were established with the 1981 farm bill. Consequently, we ended up with the payment-in-kind program (PIK) which has proven to be very costly.

Second, such a major realignment in the value of our currency brings about major international adjustments. It sends signals that say we should be transferring resources out of agriculture and other export sectors but that resources should be transferring into these same sectors in other countries. Nothing has been more frustrating than to see our general failure at the policy level in understanding the nature of the adjustment problem we faced. Nothing has been more embarrassing than to see the complaints this nation has made to other countries about their failure to adjust in the same way we have tried to adjust. It has been clear that policymakers in this country simply have not understood the kind of economic world in which we now live.

Then, in general, we have not fully understood the significance of the international capital market in the adjustments we have faced. It is the international capital market that is now driving things. But back of the capital market is our fiscal policy. The combination of our large budget deficits, the proper unwillingness of the Federal Reserve to monetize our federal debt, and our own low savings rate has caused us to induce savings from abroad to finance that debt. The capital inflow is now on the order of $60 billion to $70 billion a year. That helps make the dollar strong and penalizes agriculture, other export sectors, and sectors that compete with imports.

In the last two years, we have become sensitive to the international capital markets as a consequence of the debt crisis in countries such as Mexico, Brazil, and Argentina. But our understanding of these capital markets and their importance is still rather superficial. For example, we need to be more sensitive to the adjustments other countries are making in response to these crises. For example, Brazil has devalued its currency tenfold over the last two years while experiencing a quadrupling of its price level. That constitutes a major decline in the real value of the cruzeiro. If Brazilians make that realignment stick—and there is every reason to believe they will—Brazil will very likely take away a major share of our soybean markets. Will we be ready for that kind of adjustment? What will be our response?

Similarly, incredible as it may seem, if present U. S. fiscal and monetary policies continue, this nation will at some time in 1985 become a net debtor country. In other words, we will become like Brazil, Mexico, and Argentina. If the policy response is proper, that will be a good thing for agriculture for, at some point, it will lead to a weakening in the value of the dollar. But for the nation as a whole, it will mean a reduction in our real income as the external terms of trade shift against us. Are we prepared for that development?

Finally, we need to recognize that both (1) the weakness of the dollar during the 1970s and the export boom it helped induce and (2) the great rise in the value of the dollar in the 1980s and the shock it has imposed on agriculture have been as much a consequence of our petroleum and energy policy as of our monetary and fiscal policies. To implicitly subsidize the imports of petroleum, as we did during the 1970s, can only be described as foolhardy. It imposed major adjustments on the rest of the economy. The deregulation of the petroleum industry by President Reagan has brought about a reversal of these adjustments. This only shows what an interdependent world we now live in and the importance of these macroeconomic variables.

Our failure of analysis is demonstrated by the fact that discussion of the so-called energy crisis always focused on the rise in price of petroleum as a cost item in the production process. The real significance of that rise in price was its consequences for our balance of payments, the value of the dollar, and, in turn, the relative price of agricultural products.

## Trade Policy

Our approach to trade policy in this country can, in my view, only be described as naive. We willingly negotiate voluntary export agreements seemingly not recognizing that that works to the benefit of other countries (Allen, Dodge, and Schmitz, 1983). We do not seem to recognize that, in a flexible exchange rate regime, the efficacy of tariffs and export subsidies is called into serious question. Also, we seem to act as though we had never heard of the theory of second best.

Let me, for now, put most of my emphasis on the theory of second best. Our starting point has to be that, despite our general free trade stance, we do make significant use of measures to protect certain sectors of our economy—both within agriculture and in the economy as a whole. We need to recognize that such protection constitutes discrimination against our export sectors including agriculture. In fact, protection raises prices of both importables and home goods relative to the price of exportables and is equivalent to an export tax on the export sector.

To what extent do we take account of this when we think about agricultural policy? In my judgment, very little. Also, the issue is not a special-interest plea for agriculture. It is one of global efficiency. A little use of the theory of second best would tell us that, once we intervene, we should have the same degree of intervention in all sectors of the economy. Yet, we seldom ask that kind of question. The truth of the matter is that, with present protection for the automobile sector, the steel industry, and the textile industry, a strong case can be made for the use of export subsidies. Such subsidies would offset the consequences of current protection of the other sectors of the economy.

## Food as a Wage Good

We have not in this country given much attention to food as a wage good unless one wants to argue that early policymakers sensed this importance when they created the original Land-Grant System. Such an argument is hard to make, however, since the intent of the system, as I read our history, was to do something for the farmer and not the worker or consumer.

In any case, food as a wage good was not an important issue as long as the performance of our national economy did not rise or fall on our international trade performance. However, our growing dependence on trade will make this issue increasingly important in the future. Recent steps toward deregulation have shown the extent to which regulation of the domestic economy was protecting labor. Our need to compete internationally has brought forth the importance of our wage rates as a factor influencing our competitive ability. As we move into the years ahead, this issue will be increasingly important. Our eventual ability to compete internationally will be determined in an important way by what we do about food as a wage good.

## Srategic Issues

Strategic issues can be defined in a number of ways. In their discussion of strategic U. S. trade policy, Grossman and Richardson

(1984) describe a strategic *environment* as one in which the number of economic agents making interdependent decisions is relatively small. The idea behind their perspective is that participants in the environment are enough to be able to influence outcomes. This contrasts with a perfectly competitive market in which the numbers of individual participants is so small as not to be able to influence outcomes. The dimension I would like to add to their perspective is that we talk about something other than the short term where the emphasis is on strategy rather than tactics.

We have at least three strategic issues before us when we think about agricultural trade and our position in the world in this larger context. The first is raised by our relative dominance in the trade of selected agricultural commodities per se. The issue is raised by the seminal paper of Carter and Schmitz (1979) which argues that the major wheat importers may be using a scientific tariff. It is also raised by our West Coast colleagues in their discussion of cartels (Schmitz *et al*, 1981). And beyond this, there are the various issues surrounding state trading companies, state-owned enterprises, and other forms of state activities.

I, personally, am not persuaded that we have much leverage in these markets if for no other reason than that most countries are only marginal importers of the commodities we export. Our experience with the embargo on sales to the Soviet Union provided us strong evidence that we have little leverage in the international grain markets. But my skepticism aside, we do need to keep this set of issues on our agenda.

The second set of issues has to do with our Research and Development policy (R&D). In the past we have rarely looked at this as an international issue or tried to channel R&D efforts so as to assure our continued international competitiveness. We no longer have that luxury. The growing system of International Agricultural Research Centers is now starting to generate a sustained flow of new production technology for tropical agriculture. Many countries, such as Brazil, are making sustained commitments to their agricultural research systems. We no longer can afford the complacency about our agricultural R&D system that we have tolerated up to now (Schuh, 1984).

Third, there is the issue of the immigration policy which Congress attempts to do something about but continues to be stymied by it.

The issue is rapidly coming down to whether we import labor and do the value added here with our other resources or whether we import labor-intensive products. The costs to our consumers of failing to do one or the other can be quite high.

## Adjustment Policies

When agriculture was producing primarily for the domestic economy and the domestic economy was relatively stable, the major adjustment problem agriculture faced was the secular one of adjusting labor out of the sector. That problem is largely behind us now or at least made more tractable by the small share that agriculture makes up of the total economy. In its place is the shorter term problem of adjusting to changes in foreign demand induced by changes in the real exchange rate and other factors in foreign markets.

The problems agriculture now faces are an important example of such adjustment needs. Agriculture in the United States experienced an export boom in the 1970s motivated, in large part, by a decline in the real value of the dollar (see Figure 1). A net flow of resources was pulled into the sector for the first time in 50 years.[3] The dramatic rise in the value of the dollar in the early 1980s (see Figure 1) sent signals to transfer resources out of agriculture. Unfortunately, rather than facilitating that adjustment so that supply could be brought back into balance with demand, the commodity programs actually impeded the adjustment here at home while sending even stronger signals for producers abroad to increase output.

Under present arrangements of exchange rates that undergo such large swings, this nation needs positive adjustment policies to facilitate the required adjustment as well as more flexibility in loan and target prices so that the needed adjustment is not as great. It is doubtful whether voluntary set-asides will be adequate to bring about the needed adjustment. Paid diversion programs may be able to accomplish it but will require more political and administrative agility than appears to have been present in the past. Mandatory

[3]The total stock of resources in agriculture increases approximately 8 percent in the second half of the 1970s.

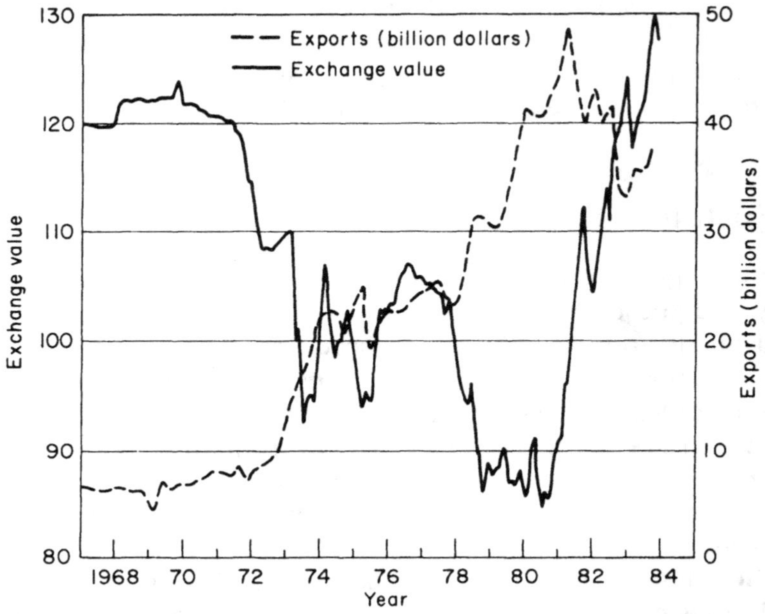


FIGURE 1. Exchange Value of the Dollar versus U.S. Agricultural Exports

production controls suffer from the same defects as paid diversion programs with the added caveat that, at least in the past, farmers did not opt for them when given the choice.

Providing more flexibility in loan and target prices will reduce the need for adjustment policies. The remaining policies needed will have to focus on facilitating labor adjustment including the usual instruments of training programs, job locations, and support of relocation costs.

## Reform of Our International Institutions

This topic deserves a paper in itself. We now find ourselves in a situation in which the economic integration of the international economy has far outpaced its political integration. In addition, most international institutions were designed at the end of World War II. Some of these institutions, such as the General Agreement on Trade and Tariffs (GATT), have grown increasingly irrelevant to our international trade. Others have broken down and disappeared such as the Bretton-Woods provisions regarding fixed exchange rates. Also, in some cases, we never did have any effective international agreement such as rules for preventing distortions in exchange rates.

It is clear to me that the problems of U. S. agriculture will not be solved by domestic policies alone because U. S. agriculture is now too much a part of the international economy. With present institutional arrangements, shocks emanating from abroad can be offset or attenuated only at very large costs. Hence, we need to be designing and helping to establish new international arrangements at the same time as we undertake the reform of our domestic policies.

In seeking to establish new international arrangements, two issues should have high priority. The first is the need for a more robust system to manage the international monetary system. In my judgment we are playing Russian roulette with a system that is precarious, fragile, and haphazard. Given that the world is basically on a dollar standard, the United States acts essentially as a central banker. We reap a modest seigniorage from providing this service. But at the same time, we impose significant costs on important sectors of our economy.

I have argued elsewhere that we need to establish an international central bank (Schuh, 1983a). This could be done in a rudimentary form by phasing out the dollar as the international reserve currency and substituting Special Drawing Rights (SDR) in its place. Then, the International Monetary Fund should be given a mandate to keep the stock of SDR growing at a constant rate.

The United States has opposed such a solution largely out of a lack of confidence that any international institution can be made to work effectively. It is difficult to imagine what an alternative arrangement would be because some means are needed to keep international monetary reserves growing to facilitate trade and to reduce the international monetary instability that has characterized the last 15 years. It is not clear to this observer that it is in the best interest of the United States or of the rest of the world to continue with the present system. The consequences of a collapse of this system would be quite great. Similarly, the consequences of our pumping a great deal of U. S. money into the system, should it start to collapse, would have a high price to our economy.

The second institution needing reform is the GATT. The GATT needs to include countries reflecting our changing patterns of trade toward the centrally planned and less-developed countries. Agriculture should be a more integral part of the GATT, and the rules of trade need to be extended to cover nontariff barriers to trade, state enterprises, distortions in exchange rates, and barriers to exports imposed by exporting countries themselves. The settlements mechanism could also stand a great deal of improvement.

## A Concluding Comment

As World War II drew to a close and the postwar period emerged, the international economy could best be described as a collection of national economies tied together with a little bit of trade. Today, this nation is an integral part of a highly interdependent international economy in which the international capital market is as important a link, tying national economies together, as is trade.

Commodity policies designed for that earlier era no longer serve us well. The costs of the PIK program should have effectively made that point.

Unfortunately, a fundamental difficulty this nation now faces is that, in Congress, agricultural policy is still made by the agricultural committees. The issues of trade, monetary, and fiscal policies are largely beyond the responsibility of those committees. Consequently, we can expect them to do little more than tinker with the commodity programs. Surely, we do not expect the members of these committees to vote themselves out of business. Until they do, however, we can expect to see little more than a fine tuning of our commodity programs. The issues that really matter to agriculture will largely go by default to others, and the costs of our commodity programs will continue to be high.

## *References*

Allen, Ray, Claudia Dodge, and Andrew Schmitz. "Voluntary Export Restraints as Protection Policy." *American Journal of Agricultural Economics* 65, No. 2 (May 1983):291–297.

Carter, Colin, and Andrew Schmitz. "Import Tariffs and Price Formation in the World Wheat Market." *American Journal of Agricultural Economics* 61, No. 3 (August 1979):517–522.

Grossman, Gene M., and J. David Richardson. *Strategic U. S. Trade Policy: A Survey of Issues and Early Analysis.* National Bureau of Economic Research, Research Progress Report, Cambridge, Massachusetts, 1984.

Schuh, G. Edward. "What Agricultural Exports Mean to Agricultural Research." Paper presented to the National Industry-State Agricultural Research, Arlington, Virginia, February, 1984.

———. "Towards Reform of our International Monetary and Trade Institutions." In *Issues in Third World Development.* Edited by Kenneth C. Nobe and Rajan K. Sampath. Boulder, Colorado: Westview Press, 1983a, pp. 419–434.

———. "U. S. Agricultural Policy in an Open World Economy." *Toward the Next Generation of Farm Policy.* Hearings before the Joint Economic Committee, U. S. Congress, Washington, D. C., May 26, 1983b.

———. "U. S. Agriculture in Transition." *Changing Economics of Agriculture: Review, Evolution, and Future Direction.* Hearings before the Joint Economic Committee, U. S. Congress, Washington, D. C., April 28, 1982.

———. "Floating Exchange Rates, International Interdependence, and Agricultural Policy." In *Rural Change: The Challenge for Agricultural Economists.* Proceedings, International Conference of Agricultural Economists, Banff, Canada, 1979, pp. 34.1–34.7.

———. "The New Macroeconomics of Agriculture." *American Journal of Agricultural Economics* 58, No. 5 (December 1976):802–811.

Schmitz, Andrew, Alex F. McCalla, Donald O. Mitchell, and Colin A. Carter. *Grain Export Cartels.* Cambridge, Massachusetts: Ballinger Publishing Co., 1981.

# 14 DISCUSSION

## Robert G. Chambers*

Professor Schuh's chapter adumbrates six major issues that he feels are particularly relevant to current and future policy analysis. The six issues are macroeconomic policy, trade policy, food as a wage good, strategic issues, adjustment policies, and the reform of existing international institutions. Actually, each of these issues are major areas of policy concern; and I wholeheartedly agree with Schuh's general evaluation that, in all of these areas, our knowledge is lacking. However, I do not necessarily agree with the topics he chooses to emphasize within each of these broad categories, and I feel that his paper overlooks some important problems. Given the confines of a short paper that Professor Schuh faced, this is not only understandable, it is necessary. In what follows, I shall attempt to outline briefly some areas of concern that I think deserve emphasis. The major policy issue areas are the same as those used by Schuh.

*The author is Associate Professor, Department of Agricultural and Resource Economics, University of Maryland.

## *Macroeconomic Policy Issues*

Recent work on the macroeconomics of agriculture, spawned by the seminal efforts of Schuh, has revolved almost totally around stylized questions of the type: "How do macroeconomic variables affect U. S. agriculture?" When one replaces macroeconomic variables with exchange rates, interest rates, and inflation rates, one has a fairly comprehensive catalogue of the recent work in this area. To my mind, this is only half the question; and it is that half of the question that a profession, accustomed to viewing agriculture as a closed sector, would ask when it first became interested in such issues. But it seems to me that our knowledge has matured to the point where we can now attempt to look beyond such gropings to the much more important question of how agriculture fits into the overall macroeconomy. In economics, there is a tendency to be circular and not unidirectional. A clear grasp of such interrelationships is important and necessary to the formulation of sound agricultural policy. It is here that I disagree with Schuh's emphasis somewhat. While, as he says, "the international capital market is now driving things . . .," this does not mean to me that agricultural economists should spend their time attempting to decipher this market. I believe it would be better to devote more time to how agriculture fits into the international economy of which the international capital market is only a part—a very important part but still a part.

## *Trade Policy and Strategic Issues*

I could not agree with Professor Schuh more than when he suggests that we do not have a coherent agricultural trade policy. He is correct, and no amount of gainsaying will change that fact. We, in the United States, constantly employ agricultural policy tools that ultimately transfer domestic resources abroad in the form of implicit and inefficient export subsidies eventuated by deficiency payment programs and the like. But I do not necessarily agree with his assertion that the theory of "second best" calls for the same degree of intervention in all sectors of the economy. In fact, just the opposite can be true depending upon the policy goal. For example, it is well known that a Ramsey optimal tax scheme does not imply equiproportional intervention. So before we advocate the use of export subsidies in agriculture, we should be clear in our minds just what are our goals.

## *Miscellaneous*

Following is a brief survey of some important points that do not fit easily within any of Schuh's six categories. One of the most pressing problems for agricultural policy analysts is our relative lack of knowledge on how agricultural prices adjust. A key to much of the literature on the effects of macrovariables on agriculture is the implicit assumption that agricultural prices adjust more rapidly than nonagricultural prices. If and why this may be true are important issues, and existing explanations of this phenomenon are completely sterile for several reasons. First, they explain only absolute and not relative price variability. Second, the more modern explanations do not recognize the peculiarities of agriculture's institutional structure. Because of its key importance, much more needs to be done along these lines.

Finally, we do not have a clear understanding of what determines capital formation in agriculture. This is in stark contrast to the nonagricultural sectors where information on investment abounds. Until we have a clear understanding of this mechanism, we cannot hope to have a clear grasp of why U.S. agriculture seems to be so slow to adjust to many of the signals it seems to be receiving from international markets.

# 15 RESOURCE DIMENSIONS OF AGRICULTURAL POLICY

Charles M. Benbrook
Pierre R. Crosson
Clayton Ogg*

## *Introduction*

Agricultural policy analysts and observers in Washington are struck by the rapid and relatively unencumbered emergence of major resource management initiatives in the last year or so. Legislative ground, once thought to be reinforced in concrete, has been broken; the Secretary of Agriculture regularly affirms his strong commitment to a sizable conservation reserve and supports adoption of such a program in the 1985 farm bill; and virtually the entire delegation of congressional leaders from both parties in the House and Senate committees on agriculture has endorsed major new initiatives to link conservation and commodity programs more effectively.

It is clear that some of the new enthusiasm for conservation stems from political discomfort about the cost and distribution of payments under recent supply control programs. However, there is now

---

*The authors are, respectively, Executive Director of the Board on Agriculture, National Academy of Sciences; a Senior Fellow at Resources for the Future; and an economist with the U. S. Department of Agriculture, Economic Research Service.

The views expressed are those of the authors, not necessarily those of the National Science Foundation, Resources for the Future, or the U. S. Department of Agriculture.

more genuine concern among agricultural policymakers about resource management issues than there was before. These issues include the impacts of commodity policies on soil erosion, pesticide contamination of surface and groundwater, and food safety. Deep-seated environmental and public health concerns, coupled with steady scientific progress in the monitoring and estimation of hazards, have generated sufficient pressure to sustain several major regulatory actions entailing sizable economic impacts. In contrast to earlier pesticide controversies, the consequences of recent regulatory actions have been spread unpredictably throughout the agricultural sector. Major input suppliers and food processors have incurred substantial losses and are devoting considerable effort to identifying and mitigating future crises. The agricultural community is paying attention to environmental and health matters because it cannot afford to do otherwise.

It remains to be seen what type of specific initiatives will be adopted in the next farm bill. We are confident in predicting that the bill will incorporate at least a few substantive resource management initiatives—some of them novel. We expect some special provisions addressing groundwater contamination and, perhaps, some innovative policies encouraging more cost-effective and safer pesticide use patterns.

The issues we address in this paper are those most likely, in our view, to play an important role in shaping the 1985 farm bill. Some contemporary resource issues, such as agricultural land retention, are not treated because they are unlikely to figure prominently in the farm bill debate or legislation. We give attention to water management issues in the arid and semiarid West but not so much as would be appropriate otherwise because we do not expect these issues to command major attention in the farm bill. Groundwater overdraft, salinization, and reclamation issues will probably remain beyond the reach of public policies for several more years.

Our main focus is on soil erosion and its consequences in lost productivity of the land and in damages to water quality, loss of reservoir capacity, and so on. As stated earlier, linkage of erosion control with commodity policies will receive much more attention than in previous farm bills—attention we view as appropriate and timely. We also address some environmental aspects of pesticide use because the farm bill debate may include some novel strategies for encouraging safer use patterns of certain problematic materials. Farm bills have not historically been vehicles for pesticide policies. However, in recent years, Congress has experienced difficulty in passing amendments to the federal Insecticide, Fungicide, and Rodenticide Act; and congressional leaders now are considering the 1985 farm bill as a way around these obstacles.

We take it as axiomatic that present practices for management of agricultural land, pesticides, and water for irrigation impose excessive social costs and that a primary objective of policy is to bring costs within socially acceptable limits. No attempt is made to justify this position vigorously beyond noting that off-farm damages of erosion and pesticides and the common property management of surface and ground water impose uncompensated external costs. In principle, policy should, as a first-order condition, aim toward measures that equate the full costs of resource conservation practices with the full value of the benefits stemming from adoption of the practices. Conservation costs, of course, must include the costs of organizing and administering the conservation effort. However, the information needed for such benefit-cost calculations is almost never available; even if it were, the calculations probably would not weigh decisively in the policy process. So we take a different tack. We assume that the federal government will spend a certain amount of money annually to deal with these problems (e.g., several hundred million dollars a year in the case of soil erosion), and we look for politically feasible policies that promise the greatest social payoff per dollar spent. We are not looking for first-best or even second-best solutions—significantly better solutions are our goal.

## *Issues Affecting Resource Policies*

### Policy-Relevant Aspects of Erosion Damage

The two kinds of erosion costs—on-farm losses of productivity and off-farm damages to water quality—have quite different policy implications. The main objective of soil conservation policy has traditionally been maintenance of soil productivity on grounds of intergenerational equity. We readily accept this rationale, but it provides little practical guidance to policy. No one knows how much erosion can be tolerated without violating the commitment to intergenerational equity. Conventional soil-loss tolerances, or T values, have little scientific standing as indicators of productivity-threatening erosion. The concept of T values, however, has a great deal of merit and could, in principle, be utilized in a rigorous analytical framework to estimate "socially optimal" erosion rates. We know that these rates will differ markedly across soils and can change as a result of technological innovation and that intergenerational equity will never require zero loss of soil productivity as long as nonland inputs can be substituted economically for soil at the margin.

Policy issues are complicated further by the fact that farmers have a direct stake in protecting the productivity of their land. Without government intervention, the cost-conscious, well-informed farmer will generally conserve soil whenever the cost of protection is less than the estimated value of productivity loss. Policies promoting conservation would be justified only where the social cost of productivity and other losses is higher than the conservation costs the farmer is willing to bear. We accept that intergenerational equity may make a case for such policies. In practice, however, the concepts are slippery, the data poor, and the analysis necessary to address this issue rigorously has just begun. We generally know where and how we can cost-effectively reduce erosion, but we are hard pressed to say just how much soil loss is tolerable.

The situation with off-farm costs from erosion is quite different as the farmer has little or no incentive to control these. There is a prima facie case that these costs are socially excessive although the case requires showing that, at the margin, the costs exceed the cost of policies to reduce them.

The erosion-imposing excessive off-farm costs are not necessarily the erosion-imposing excessive productivity losses. For example, the deep loess soils of western Iowa and other Corn Belt soils have suffered erosion losses of 10 to 20 tons per acre per year for many years with no detectable decline in soil productivity. But the off-farm damages of erosion on these soils may be significant. On shallow southeastern soils, on the other hand, much lower erosion rates may significantly impair productivity while posing no threat of off-farm damages. Consequently, policies to control one kind of erosion damage will not necessarily contribute much to control of other kinds of damage.

The difference in the two types of erosion damage complicates policy in yet another way. Where productivity loss is clearly a major concern, the damage can be reduced reliably by controlling erosion at the threatened site. The land requiring such treatment is relatively easy to identify. This is not true of off-farm erosion damages. Waterways, harbors, reservoirs, ponds, and lakes where sediment and sediment-related damage are occurring can be identified, but knowledge of overland and instream sediment transport processes is often inadequate to trace the damaging sediment and attached nutrients to their place of origin. In these cases, it is difficult to target accurately the erosion control efforts to reduce off-farm damage.

The chemical and energy dynamics of moving water present yet another problem. Reducing cropland erosion may increase stream bank and bed scouring but result in little or no reduction in sediment delivered downstream, at least in the short run. In such areas, policies to control erosion on the land are clearly inappropriate for reducing downsteam sediment damage.

Finally, recent research suggests that, for the nation as a whole, the costs of off-farm erosion damage are greater than the costs of productivity loss, perhaps substantially greater. Analysis undertaken at the Conservation Foundation (report forthcoming) estimates conservatively that annual costs of off-farm damage are between $2 billion and $6 billion, with a point estimate of $3.1 billion. By way of contrast, soil scientists at the University of Minnesota have developed a model (Pierce, *et al.*, 1983) for estimating the long-term physical effects of erosion on soil productivity and have applied it to the Corn Belt (Pierce, *et al.*, 1984). They found that corn yields would decline unavoidably on average only 4 percent from what they otherwise would be if erosion in that region continued at the 1977 per acre rates for 100 years. Soybean yields would be affected proportionately.

If we assume that over the 100-year period corn is priced at $3.00 per bushel and soybeans at $7.00, that 70 million acres will be planted in each crop annually, that the yield decline on the 140 million acres will average 7 percent rather than 4 percent (because not all land planted to corn and soybeans can withstand erosion as well as can the 98 million Corn Belt acres), that initial corn yields are 110 bushels per acre and soybean yields are 32 bushels, and that the yield loss is spread over the 100 years in equal annual arithmetic units, then the first-year yield loss is about $30 million. The loss largely reflects diminished water-holding capacity of the soil and, for practical purposes, is permanent. Consequently, the annual loss is cumulative: $60 million the second year, $90 million the third year, and so on. Of course, if one assumes higher prices and amounts of land in each crop, the estimated loss will be higher, and vice versa. If technology increases average yields, the assumed 7 percent loss caused by erosion would mean a greater absolute reduction in bushels per acre, but the effect on the economic loss would be offset if crop prices are lower because of the technology-induced increase in production. The possible combinations of assumptions are limitless. However, as long as the loss in yield averages no more than 7 percent over 100 years, the annual value of permanent productivity losses on corn and soybean land is probably in the tens of millions of dollars rather than in the hundreds of millions (or more) dollars.

Whatever the value of the economic loss based on the Minnesota model, it is an understatement because it does not include the cost of things farmers do to compensate for the effects of erosion. These costs include applying additional fertilizer to replace lost nutrients, liming to maintain favorable soil pH, subsoiling to improve infiltration, and so on. The Minnesota model assumes that farmers do these things. Inclusion of the associated costs would surely increase—probably substantially—the estimate of erosion damages derived from the model. Nevertheless, it is likely that total annual productivity costs of erosion still would be less than the $3.1 billion for off-farm damages estimated by the Conservation Foundation (report forthcoming). U. S. Department of Agriculture (USDA) analysts, led by Dr. Roy Gray of the Soil Conservation Service, also estimated the economic value of erosion-induced productivity losses. The analysts developed a macro erosion-productivity model using data from the 1,100 published soil surveys. The following paragraph and Table 1 summarize the USDA methodology:

> "The soil is subjected to an average erosion rate of 15 tons per acre per year; corn has a value of $2.00 per bushel; in its uneroded state, the soil is deep and has a potential yield of 143 bushels per acre; and a soil loss of 15 tons per acre per year amounts to 1 inch of erosion in 10 years. These assumptions reflect typical conditions throughout the Corn Belt" (U. S. Department of Agriculture, Soil Conservation Service, 1981, pp. 3–41).

**TABLE 1. Effect of Erosion on Cumulative Corn Production Over 50 Years**

| | Number of years erosion continues | | | | | |
|---|---|---|---|---|---|---|
| | 0 | 10 | 20 | 30 | 40 | 50 |
| | bushels per acre | | | | | |
| Per acre yield at start of period | 143 | 140 | 137 | 134 | 131 | 128 |
| Production loss from erosion during 10-year period | | | | | | |
| First | | 15 | 30 | 30 | 30 | 30 |
| Second | | | 15 | 30 | 30 | 30 |
| Third | | | | 15 | 30 | 30 |
| Fourth | | | | | 15 | 30 |
| Fifth | | | | | | 15 |
| Cumulative production loss for each 10-year period | | 15 | 45 | 75 | 105 | 135 |

Source: U. S. Department of Agriculture, Soil Conservation Service, 1981, pp. 3–41.

Based on these assumptions, the USDA estimated an undiscounted annual value of $1.00 per ton of soil loss. The USDA also analyzed the key impacts of the rate of discount on the economic losses to the farmer from erosion. Using the example noted above, the USDA concluded that the value of soil is about 50 cents per ton based on a 7 percent interest rate. A table from a draft of the Resources Conservation Act Appraisal Document, Part II, reproduced below (Table 2), illustrates the relationship between the rate of discount and the economic losses from erosion. The 50-cent per ton value of topsoil indicates that the 1.8 billion tons of sheet rill and wind erosion on cultivated cropland in 1977 cost about $900 million annually in lost productivity. This is considerably more than estimates derived from the Minnesota model but still less than the Con-

**TABLE 2. Discounted Value of Production Loss for a 50-Year Planning Horizon[a]**

| Interest rate | 50-year amortization factor | Present value of increment annuity | Annual increment | Average annual value of loss | |
|---|---|---|---|---|---|
| | | | | dollars per acre | dollars per ton |
| percent | | dollars | | | |
| 0 | 0.020 | 1,250.0 | 0.60 | 15.00 | 1.00 |
| 4 | 0.097 | 382.6 | 0.60 | 10.69 | 0.71 |
| 7 | 0.072 | 186.7 | 0.60 | 8.07 | 0.54 |
| 10 | 0.100 | 104.8 | 0.60 | 6.29 | 0.42 |
| 12 | 0.120 | 76.1 | 0.60 | 5.48 | 0.37 |
| 15 | 0.150 | 50.8 | 0.60 | 4.57 | 0.30 |

[a] Assumptions noted in quoted passage in text.

Source: U. S. Department of Agriculture, Soil Conservation Service, 1981.

servation Foundation (report forthcoming) estimates of the annual costs of off-farm erosion damage.

The difference between on-farm and off-farm erosion costs clearly has important implications for erosion control policy. One is that off-farm costs should be given relatively more attention than heretofore has been the case. Another is that the cropland targeted to reduce off-farm damage will not necessarily be that targeted to reduce productivity loss. A third implication is that, where the policy objective is to reduce off-farm costs, instances might arise where control efforts should focus on preventing sediment from reaching places where it causes damage rather than on reducing on-farm erosion.

These policy implications of the distinction between on-farm and off-farm erosion costs may have little impact on the erosion control provisions of the 1985 farm bill. It is not at all clear that policymakers in the USDA or Congress are aware of the implications. Even if they are, however, present knowledge of the magnitudes and spatial locations of the two kinds of costs is too limited to permit the policy fine tuning needed to target them effectively.

Until more knowledge is available, optimal strategy remains to focus erosion control efforts primarily on the relatively few highly erosive cropland acres [the 1977 National Resources Inventory found that 70 percent of erosion over 5 tons per acre came from 8.6 percent of the cultivated land (U.S. Department of Agriculture, 1978)]. Our discussion of erosion control policies is based on this practical realization. We consider both the technical and political feasibility of the policies; and we are satisfied that, if adopted, they will significantly reduce erosion. Indeed, we argue that a 50 percent reduction in U. S. cropland erosion is within reach at little or no extra cost. This would carry us a sizable first step toward the modest goal of doing significantly better in reducing costs of erosion. We think the policies for achieving this deserve attention in connection with the 1985 farm bill; but we emphasize that reducing erosion is only a proxy (and not everywhere a good one) for reducing erosion damage—the real objective. At the same time that we mount efforts to reduce erosion, we should also devote more resources to developing the technical information needed to locate and measure the two kinds of erosion damage and to devising policies and procedures for bringing them within socially acceptable limits.

## The Policy Environment

Policies for managing agricultural resources in the 1980s and 1990s will be affected profoundly by the general economic environment at home and abroad. Both short-term perturbations and long-term trends will prove to be of great importance. A concensus has developed that, because of their high and increasing dependence on foreign markets, American farmers should anticipate more short-term instability in commodity prices than they experienced in the relatively calm 1950s and 1960s. In this respect, so the thinking goes, the experience of the 1970s is likely to be a better guide than were the preceding two decades. The genesis of increased price instability is not our concern here, although major factors must include floating international exchange rates and the practice of many countries to insulate their agriculture against price swings and competition.

American farmers and commodity policy have traditionally responded to price fluctuations by varying the amount of land in crops. More land is cultivated when prices are rising; less, when prices fall. We assert that one basic objective of erosion control policies should be to work toward a set of responses whereby the land coming into production is the least erosive available and land taken out is the most erosive. Because policies aimed toward stabilizing commodity prices also rely mainly on adjustments in planted acreage, there is a strong case for formal integration of conservation and commodity price-support program functions. The most appealing proposal now noted as a key possible 1985 farm bill initiative is a soil reserve or targeted multiyear diversion program. We discuss such a program later and point out that the optimal land to idle for promotion of price stability is not necessarily the optimal land to remove for erosion control purposes. Long-term trends in crop demand and supply also play a critical role in shaping resource policy needs. On the demand side, the key is the growth of export demand. Of U. S. agricultural output, 25 to 30 percent is exported; and 35 to 40 percent of land in crops is devoted to production for export. On the supply side, trends in technology and in prices of farm inputs are dominant. If the combination of demand- and supply-side factors is such that crop prices rise, it is likely that farmers will bring more land into crop production (probably resulting in more erosion), use more water for irrigation, and utilize more pesticides. Pressure on the resource base will rise, and the policy burden will increase. Erosion control policy probably will be aimed toward putting more emphasis than now seems appropriate on protection of soil productivity relative to reduction of off-farm damage.

If, however, demand and supply trends are such that crop prices decline, land in crops probably will diminish; land, water, and related environmental pressures will be weaker; and the policy burden will be reduced. Control of off-farm erosion damages will probably receive more attention relative to protection of productivity of the soil than if crop prices are rising.

We cannot see any more clearly than anyone else where demand and supply trends will take American agriculture over the next decade or so. However, it is important to note that, just as it was a mistake to conclude that high crop prices in the mid-1970s portended a future of chronic scarcity, it is premature to conclude that recent price experience portends chronic surplus. It is reasonably clear that the weakness in grain and soybean prices since 1981 was not because of exceptional supply-side performance. World production of wheat and rice in the five years, 1978–1982, totaled a modest 1.0 to 1.5 percent above the trend levels for those crops established for 1960–1982. Coarse grain production for the five years was 0.2 percent above the trend (U. S. Department of Agriculture, Foreign Agricultural Service, 1983). In the United States, per acre yields of wheat in every year from 1973 to 1983 were below the trend values of yields established for 1950–1972. Corn yields were below the trend in every year (except 1979) from 1973 to 1983, and soybean yields were below the trend in three of the five years during 1979–1983 (U. S. Department of Agriculture, 1984).

The recent weak prices for these crops reflected the worldwide recession and, probably to some extent, the associated balance-of-payments problems of developing countries. On both counts, world trade in grains declined after 1980 and slowed markedly in soybeans. The U. S. exports were further disadvantaged by the rise in the foreign exchange price of the dollar in this period and by the competitive opening given to other grain exporters by the partial embargo on grain exports to the Soviet Union following that country's invasion of Afghanistan. As a consequence of these developments, the U. S. share of world trade in grains and soybeans was reduced after 1981; and the volume of U. S. exports of these commodities declined.

We believe the recent poor performance of the world economy is temporary and that a return to growth similar to that of the 1970s is not unlikely. This would strengthen world trade in grains and soybeans. We have no special expertise in exchange-rate forecasting; among those who do have expertise, the majority appears to expect the dollar to fall during the next few years. This, of course, would improve the competitive position of the United States in world grain and soybean markets.

In short, it is at least arguable that the recent weak prices for grains and soybeans in world and U. S. markets reflected temporary demand-side factors, not a permanent shift toward slower demand growth or accelerated growth of supply. In thinking about resource dimensions of U. S. agricultural policies, therefore, we believe it would be a mistake to rule out a scenario in which exports return to a growth path similar to that of the 1970s. Recent experience demonstrates that, in judging the likelihood of such a scenario, one should consider geopolitical and macroeconomic policies not only in the United States but elsewhere as they bear on rates of economic growth and foreign exchange rates.

A return to more rapid export growth would not necessarily put increasing pressure on the agricultural resource base. Much depends on the pace and nature of technological change. In the 1970s, input prices, particularly of fertilizer and energy, favored more land-using technologies than in the 1950s and 1960s (Crosson and Brubaker, 1982, Chapter 3). After 1972, farmers brought some 60 million additional acres under crops and, in 1982, total harvested cropland acreage set a record. Whether the increased acreage also reflected reduced yield potential of the technologies then in use is uncertain. As noted above, the growth of corn and wheat yields slowed after 1972, but this could have been a result—not a cause—of the shift toward more land-using technologies (Crosson and Brubaker, 1982, Chapter 4).

If future technological change is similar to that in the 1970s, growing export demand could stimulate increased demand for cropland, and erosion problems would probably become more severe. However, if technological change is more like that of the 1950s and 1960s, any likely growth in crop demand could be accommodated without any increase and, perhaps, a decrease in harvested acreage.

Apart from how yield levels will impact on total acreage needs, the reliance of production practices on pesticides will have increasingly important implications for resource policy. Pesticide regulatory actions up to and through the early 1970s were driven primarily by adverse environmental impacts. More recently, public health concerns, first for the occupationally exposed and now for the general public, arose in connection with some pesticide use patterns. A major new problem area—pesticide contamination of groundwater—will probably have profound consequences on innovation in chemical pest control technology. While some new problems are commanding attention, some older problems are abating. Most of the problematic, persistent organochlorine compounds are no longer used because of regulations and a lack of efficiency caused by genetic resistance. Integrated pest management techniques have spread rapidly in cotton production, and production has been shifting from the southeast and Mississippi Delta (where they are relatively severe) to Texas (where insect problems are relatively slight). New compounds are far more active and site specific. The total poundage of insecticide used is down sharply from the level of only five or six years ago while the percentage of acreage treated has remained stable.

Herbicide use, however, has increased steadily and rapidly and now far surpasses insecticides in tons of active ingredients applied. In large measure, this trend reflects the spread of conservation tillage practices and seems likely to continue. Most widely used herbicides are generally of low mammalian toxicity and are not regarded as serious environmental or health threats. However, substantial toxicological data gaps persist; and not all the avenues by which herbicides may impact the environment, particularly at lower trophic levels, have been investigated. Consequently, the steady trend toward increased reliance on herbicides deserves close attention, especially in light of the importance of chemical weed-control technology in publicly supported soil conservation programs.

For example, no-till systems, advanced by some as a best management practice for nonpoint pollution control, are good examples of the subtle, yet potentially significant, trade-offs that can exist between on-farm and off-farm resource management needs and objectives. No-till systems dramatically reduce erosion and, hence, instream sediment loads. They also significantly increase, at least in some watersheds, the load of toxic materials and, perhaps, nutrients washing off planted fields and into waterways. Those pollutants pose different problems that may prove as serious as sediment, especially since a great deal of no-till practice occurs where there is little or no erosion problem. Of course, there are generally alternative ways to control sediment without increasing other pollutants. Some attention during the farm bill debate might be directed toward the research and program innovations needed to focus public soil and water conservation expenditures on more robust solutions to on-farm and off-farm externalities.

A major state-federal program underway in the watersheds surrounding the Chesapeake Bay is confronting these issues at this time. In light of the lack of knowledge regarding the no-till/sediment-toxic load trade-off, questions are bound to arise over the current plans to expend some $20 million or more in state and federal funds in the next three years in and around the Chesapeake Bay with the majority of the money likely to go toward incentives for adoption of no-till systems. Some skeptics are speculating whether a clear, biologically compromised Bay will be much of an improvement over a sediment-bound Bay.

Enough has been said to make the point that agricultural resource policies cannot be addressed sensibly without awareness of the macroeconomic environment within which American agriculture evolves and awareness of trends in technology bearing on agricultural performance. Our discussion of policy issues and alternatives is shaped by that awareness.

## Resource Management Policy Options

We anticipate and argue herein that resource concerns will play an important role in shaping certain key features of the 1985 farm bill. The critical resource management issues—soil erosion, groundwater overdraft and contamination, nonpoint pollution, salinization, and pesticide externalities—will persist regardless of the policies adopted in the next several farm bills. It is interesting, however, to speculate on how alternative directions for the farm bill will impact resource policy and, conversely, how resource issues will impact policy directions.

Although resource concerns are important, we do not expect them to drive the farm bill process. Indeed, it is very clear to us that the high costs and limited effectiveness of commodity programs will be the dominant forces shaping change. Politically, the dairy and wheat programs are likely to be the main events. For both, price-support reduction is destined to be the primary avenue to cost savings.

In the near term, land retirement and supply control measures in the farm bill will probably have the most profound impact on resource management. Of course, the budget process early in 1985 may make the farm bill moot on this point (Congressional Budget Office, 1984). Consider the following scenario. The 99th Congress opens shop in a period of stagnant (if not declining) economic prospects. Deficits emerge after the election as Public Enemy No. 1, and pressure mounts for a three-pronged program: substantial restructuring of the tax system, reduction of the rate of increase in defense spending, and further cuts in outlays for domestic programs. Agricultural program costs will, to put it mildly, prove to be a particularly inviting target for savings.

Now, suppose further that the economic problems of agriculture remain acute. How will Congress reconcile pressures to keep net farm income from totally collapsing while substantially reducing spending? Perhaps Congress will resolve this conundrum in the same way it dealt with the costs of the food stamp program—costs that seemed to be escalating out of control a few years ago. Congress could, as part of the budget process, establish an annual cap for spending on commodity programs providing the Secretary of Agriculture with the mandate to administer existing programs to achieve its stated purposes as effectively as possible consistent with available funds.

The House and Senate agriculture committees will begin serious consideration of the farm bill immediately after completing the initial action on the fiscal year 1986 budget. It is inevitable that the budget process will have its share of painful decisions. Facing severe constraints on public outlays for traditional supply control programs, enthusiasm could very well grow appreciably for a conservation reserve program. Such a program can be justified on the grounds of conservation needs and environmental quality goals. Even more appealing, the funds for such a program could be channeled through the conservation component of the budget thereby avoiding the cap or other constraints on spending for traditional programs.

It is impossible to predict how and when such a conservation reserve or any other substantive change in policy will be adopted. Economic prospects in the spring of 1985 will undoubtedly color the rhetoric of the farm bill debate. Still, we feel reasonably safe in predicting that some kind of a multiyear, contractual land reserve program targeted to erosive soils will be established in 1985 for implementation in the 1986 crop year. The effect of such a program on resource management will be determined by its scope and administrative provisions. In addition, the balance struck between supply control and conservation objectives remains to be seen.

**The Concept of a Targeted Diversion Program.**—Until recently, conservation program efforts and expenditures have been spread relatively evenly across the agricultural counties of the nation. The concept of targeting or redirecting acreage reduction and/or conservation expenditures according to the severity of existing resource management problems is a relatively recent policy innovation. Although the concept is hard to oppose on economic and other grounds, some political and bureaucratic opposition has surfaced. Indeed, we can expect some thorny issues related to targeting to arise in the 1985 farm bill debate. First, what specific criteria should govern the identification of "highly erosive" or "nonerosive" lands? Who should have the responsibility of classifying a given field, and what analytic tools and data sources should be relied upon? What steps should be taken in those relatively few, but by no means insignificant, regions of the country where a modern soil survey has not been completed and/or existing analytic tools poorly capture erosion and other resource management hazards? Although we firmly believe these issues can be overcome in a realistic way, it is the responsibility of proponents of targeting in concert with the USDA to complete in the next few years the analytic work necessary to develop reasonable administrative strategies.

The simplest targeting strategy would be to shift a share to, or provide new funds for, the Agricultural Conservation Program allowing more money to be allocated to conservation districts with severe erosion problems. This and several other targeting policies are being implemented on an experimental or pilot basis. Two experimental programs of special interest idled highly erosive land. The programs required establishment of permanent vegetative cover on these lands both to protect the land most threatened by erosion and to reduce crop surpluses.

The larger of the two new integrated programs was introduced as part of the 1984 acreage reduction program; the other was a small Soil Conservation Service soil reserve pilot program begun in 1983. Agricultural Conservation Program dollars were spent in the 1984 program to place commodity program acres in conserving uses over contract periods of 5 to 10 years. Cost shares for individual farmers covered 90 percent of the cost of planting cover crops. For the first time, millions of dollars from the program were targeted for land eroding at least twice the soil-loss tolerance level thereby serving the dual purposes of supply control and soil conservation.

The other new integrated program was the 1983 pilot conservation reserve project. This project was conducted by the Soil Conservation Service as an administrative initiative of Secretary of Agriculture John Block. Results from the project suggest that renting erosive land—even on an admittedly modest scale—provides a powerful targeting tool. The pilot project more than paid for itself. It retired commodity program base acres at a lower per acre cost than did the standard program. The project was limited, however, to three counties with a $500,000 funding limit in each.

The pilot project idled very marginal, highly erosive land for about $10 per acre or about one-tenth of the cost associated with the standard commodity program provisions operative in these regions. Production adjustment is clearly achieved at bargain prices even when the lower yield on the erosive land is taken into account. Moreover, there is no indication that the relative cost advantage of the pilot project was caused primarily by any peculiarity of the pilot counties as land rents in those counties were typical of, or higher than, those in neighboring counties. Available data lead us to expect that, as a greater percentage of cropland in a given county is enrolled in such a program, the cost savings would diminish but remain significant.

The Soil Conservation Service pilot project is just the latest indication that marginal land can be retired from crop production relatively cheaply—at least when commodity prices are depressed. The magnitude of the problem is sobering. There are more than 42 million acres now in crop production eroding at more than 15 tons per acre. In fact, as long as this land remains in row crop and small grain production, it will erode excessively even if traditional conservation practices are applied religiously and maintained. When used with crops leaving heavy residues, no-till systems may effectively control erosion on some of these lands; but, for most of the 42 million acres, periodic diversion into forage-based cover crops is the only viable option for reducing erosion close to current tolerance levels.

For a host of practical reasons, current acreage-reduction programs are not appropriate policy tools for encouraging forage-based rotations on erosive land. The programs are also proving to be expensive. Deficiency payments and crop diversion payments averaged $104 in direct costs per idled acre between 1977 and 1982.

**TABLE 3. Land Groups and Their Yields**

| Land group | Land capability class (LCC)[a] or erosion designation | Average corn yield |
|---|---|---|
| | | bushels |
| 1 | LCC I | 109 |
| 2 | Wet (w) and stony (s) soils | 67 |
| 3 | LCC II(e), III(e), and IV(e); RKLS under 50[b] | 97 |
| 4 | LCC II(e) and III(e); RKLS over 50 | 85 |
| 5 | LCC IV(e); RKLS over 50 | 79 |
| 6 | LCC VI(e), VII(e), and VIII(e) | 37 |
| | Weighted average | 102 |

[a] Under the LCC designations, Roman numerals I-VIII designate severity of the problem for crop uses; (w), (s), and (e) indicate whether the problem is due to wetness, stoniness, or erosiveness, respectively.

[b] An RKLS is the product of the rainfall, erodability, slope length, and slope factors in the Universal Soil Loss Equation (Wishchmeier and Smith, 1965). An RKLS of 50 implies about 15 tons of erosion per acre under average management.

Source: Ogg, Webb, and Huang, 1984.

In 1983, costs were actually much higher. Paradoxically, $104 per acre is about twice what it would cost to rent comparable land for a typical wheat and/or feed grains farm based on state data compiled by the Economic Research Service (U. S. Economic Research Service, 1983). Several analysts have concluded that payments averaging $30 to $40 per acre would draw 20 million to 30 million erosive acres into a multiyear reserve (Ogg, Miller, and Clayton, 1984). Highly erosive land (land groups 4 and 5 in Table 3) generally is only about 15 to 20 percent less productive than average cropland, yielding 20 to 25 percent more than land with wetness and other problems.

In the current economic climate, it is understandable that attention is focused on supply control and income support policies. In structuring a major, publicly funded land retirement program, it is vital to consider, also, how provisions of the program can faciliate timely and appropriate responses when economic conditions change. Specifically, what happens with erosive yet productive cropland enrolled in a 10-year reserve if crop prices recover strongly in response, say, to rising export demand and relatively slow technological change?

This issue raises a number of complex questions beyond the scope of this paper. A conservation reserve, or any other acreage-reduction program, will not be affordable if it entails permanent retirement of the land when land values are rising. By the same token, the public is not likely to accept partial payments to keep erosive land out of production. Let us suggest one strategy for dealing with these difficult issues. When a farmer enters eligible erosive land into a multiyear reserve, the contract could specify a price level at which the farmer would be free to plant the land. The contract might establish such a price level at, for example, 175 percent of the support price and require the farmer to utilize acceptable conservation practices when farming the land. When prices drop, the farmer could be required to restore permanent vegetative cover on the land. After the term of the land retirement program expires, a range of provisions could apply to future uses of the cropland. The so-called sodbuster provision, passed this session by both the House and Senate, is an example of such a provision.

The ability to adapt efficiently and quickly to changing economic conditions will be an essential attribute for future commodity programs. For this reason, programs need to accommodate logical responses to changing future conditions. If, on the other hand, no consensus emerges on how to update current programs, it is at least possible that the existing program will be eliminated effectively through establishment of low support and target price levels adjusted automatically in response to changing market conditions. The impact of such a market-oriented approach would depend heavily on the comparative profit potential for major commodities. Production of wheat and other basic commodities would probably decline in certain regions dependent on high-cost water supplies. Farmers might also diversify to hedge against risk—a management strategy with generally positive implications for erosion and pest control.

**Economic Implications of Linking Commodity and Erosion Control Programs.**—Analysts in the Economic Research Service have been investigating several critical questions regarding the mechanics and probable economic impacts of a targeted soil reserve program (Ogg, Webb, and Huang, 1984). To address these questions, they modified the yield, cost, and land group data in the Iowa State linear programming model used extensively in the Resources Conservation Act process (English, Alt, and Heady, 1982).

Wheat has been and probably will continue to be the crop accounting for the largest acreage reductions. In an average year, wheat is the least erosive of the major crops. However, in exceptionally dry years, wheat regions can suffer extremely high rates of wind erosion. Current acreage reduction programs could easily be modified to idle the 7 million critically erosive wheat acres currently in continuous use (Table 4). The modified programs also could idle an additional 5 million wheat acres eroding between 10 and 15 tons per acre. On the other hand, feed grains, soybeans, and cotton account for nearly three-fourths of the critically erosive cropland. The 1978 feed grains acreage reduction program and recent experience with surpluses suggest it would be practicable to idle on a long-term basis about 10 million acres of excessively erosive land currently producing feed grains, soybeans, and cotton. Across all program crops, an integrated program approach might treat about half of the 42 million acres of highly erosive land which, if idled, would sharply reduce erosion.

Using the data described above, the U. S. Economic Research Service analysts (Ogg, Webb, and Huang, 1984) concluded that it would cost 18 percent less to support prices by idling critically erosive land than if the pattern of the 1978 acreage-reduction program were followed. This conclusion rests, of course, on many assumptions. An important one is that farmers will submit competitive bids equal to the net return from production foregone and participate in a targeted soil reserve program when it pays to do so.

Idling the least profitable land in the United States, regardless of erosion hazard, saves some additional dollars (for a total of about 35 percent savings over the 1978 pattern). The land idled in this case follows closely the pattern of the conservation reserve of the 1950s. However, if such a program can be targeted successfully to land that is excessively erosive when in continuous crop use, the program would eliminate about 40 percent of U. S. cropland erosion exceeding 5 tons per acre while also providing most of the supply control budget savings. Thus, the soil conservation benefits from

**TABLE 4. Critically Erosive Land by Crop, 1977**

| Crop | Total acreage | Critically erosive land[a] | Critical land in crops |
|---|---|---|---|
| | million acres | | percent |
| Soybean | 59.3 | 6.8 | 11.5 |
| Corn | 93.5 | 11.1 | 11.9 |
| Cotton | 16.6 | 6.0 | 36.1 |
| Wheat | 71.6 | 6.6 | 9.2 |
| Other | 66.7 | 11.5 | 17.2 |
| Total[b] | 307.7 | 42.0 | 13.7 |

[a] Critical lands in this table erode in excess of 15 tons per acre. Erosion of more than 15 tons is difficult to treat without periodically idling the land.

[b] Total cultivated cropland in the 1977 National Resources Inventory includes land planted to row crops, close-grown crops, and summer fallow.

Source: Webb, Ogg, and Huang (1984).

retiring erosive land are substantial even though idling erosive land is not the least costly approach to production control (Ogg, Webb, and Huang, 1984).

As noted above, the largest program cost savings from a long-term acreage reduction program comes from renting land directly. Additional cost advantages from direct rental of land would result from a bid system. Under such a system, each farmer submits rental bids specifying the payment rate at which the farmer would idle eligible cropland. The U. S. Economic Research Service analysts compared the costs of a bid system with those of an offer system such as that used in the Conservation Reserve of the 1950s (Ogg, Webb, and Huang, 1984). Although many assumptions were necessary, they concluded that the offer system cost about 26 percent more than a bid system used for idling the least profitable land. Their studies are continuing—exploring the implications of changes in assumptions and in size and other program features.

## Other Resource Issues

Erosion is not the only resource problem that can be profitably addressed through a targeting strategy. For example, effective solutions to eutrophication in lakes may, in some cases, require idling land with high erosion rates (Ogg, Pionke, and Heimlich, 1983). Establishing cover crops on critically erosive land was found to be particularly effective in reducing phosphorus losses because cover crops reduce the phosphorus content of soil leaving the field as well as greatly reducing the tons of sediment entering streams and lakes.

The general concept of targeting can be applied to any resource management problem. Targeting requires the identification of lands prone to resource problems and characterization of the severity of the problems associated with these lands. Hence, the first step in targeting is nothing more than defining the problem. The second step, tailoring program activities to management needs, is more complex. In the 1985 farm bill, only the area of soil erosion control is ripe for a major targeting initiative. We believe it possible, however, that Congress will endorse the general concept and direct the U. S. Department of Agriculture to explore new ways to better target the limited conservation funds. Indeed, we see some interesting potential applications of the targeting concept in managing a wide range of off-farm externalities. Areas contributing heavily to sediment and/or toxic loads in waterways could be targeted for special treatment. States have offered additional incentives for farmers to enroll such lands in a federal soil-reserve program; alternatively, they could require farmers to join such programs or suffer certain penalties or sanctions.

The western migration of the boll weevil into California's Imperial Valley is another intriguing case in point. The profitability of cotton production in the Valley is shaky because of high pest-control expenditures. A sophisticated, integrated pest management control system was developed to keep escalating costs under control. The system has been generally successful although the arrival in 1982 of the first boll weevils west of the Colorado River posed a serious concern. The pesticides needed to control the weevil could also unavoidably kill many beneficial insect species essential to the system adopted for control of the heliothis complex.

Experts in the Valley recognized that the cotton industry would not survive if the boll weevil became established. Accordingly, a multifaceted program was initiated to eliminate the boll weevils already in the Valley and protect against future infestations.

Growers were strongly encouraged in 1983 to produce short-season cotton varieties. This action was considered the critical change in management practices essential to reduce the likelihood that boll weevils might overwinter. Indeed, short-season varieties, which yield about 30 percent less than does full-season cotton, were recommended because of economic and other pest control advantages even before the boll weevil was discovered in the Valley. As the 1983 season approached, growers focused attention on another major, unanticipated change—the payment-in-kind (PIK) program.

Some 96 percent of the Valley's cotton base acres were enrolled in the PIK program. To minimize loss of production while complying with the program requirement to idle 50 percent of enrolled acreage, long-season varieties were chosen by the majority of cotton farmers. A relatively simple change in program provisions could have virtually eliminated long-season cotton in the Valley at no extra cost to the farmer or taxpayer. Other examples involving groundwater overdraft and surface and ground water contamination could be cited.

## *Summary and Conclusion*

Congressional awareness and concerns are growing over resource management needs and opportunities. Secretary Block is personally committed to conservation, although serious economic constraints on nondefense discretionary spending will limit new Executive Branch initiatives. During the farm bill process, a multiyear soil reserve program targeted to highly erosive lands is likely to be the most significant resource management initiative given careful consideration. Although it is not the final answer for reducing on-farm and off-farm erosion damages, such a program would move us significantly toward that goal. If such a reserve is also perceived as a cost-effective supply control tool, it is entirely possible that the 1985 farm bill will mark an historic shift in resource management.

## References

Congressional Budget Office. *An Analysis of the President's Budgetary Prosals for Fiscal Year 1985*. Washington, D. C.: U. S. Government Printing Office, February, 1984.

Conservation Foundation. "Impacts of Soil Erosion on Water Quality. Washington, D. C., forthcoming.

Crosson, P., and S. Brubaker. *Resource and Environmental Effects of U. S. Agriculture*. Resources for the Future, Washington, D. C., 1982.

English, Burton C., Klaus E. Alt, and Earl Heady. *A Documentation of the Resources Conservation Act's Assessment Model of Regional Agricultural Production*. Iowa State University, Center for Agricultural and Rural Development, CARD Report 107T. Ames, 1982.

Ogg, Clayton W., Harry B. Pionke, and Ralph E. Heimlich. "A Linear Programming Economic Analysis of Lake Quality Improvements Using Phosphorus Buffer Curves." *Water Resources Research* 19, No. 1 (February 1983):21–31.

Ogg, Clayton W., Arnold B. Miller, and Kenneth C. Clayton. "Agricultural Program Integration to Achieve Soil Conservation," 1984. Manuscript in review.

Ogg, Clayton W., Shwu-Eng Webb, and Wen Yuan Huang. "Economic Analysis of Acreage Reduction Alternatives Including a Soil Conservation Reserve and Competitive Bids." U. S. Economic Research Service, 1984. Manuscript in review.

Pierce, F. J., R. H. Dowdy, W. E. Larson, and W. A. P. Graham. "Soil Productivity in the Corn Belt: An Assessment of Erosion's Long-Term Effects." *Journal of Soil and Water Conservation* 39, No. 2 (March-April 1984):131–136.

Pierce, F. J., W. E. Larson, R. H. Dowdy, and W. Graham. "Productivity of Soils—Assessing Long-Term Changes due to Erosion." *Journal of Soil and Water Conservation* 38, No. 1 (January-February 1983):39–44.

U. S. Department of Agriculture. *Agricultural Statistics, 1983*. Washington, D. C.: U. S. Government Printing Office, 1984.

U. S. Department of Agriculture. *National Resources Inventory* (Computer Tapes). Washington, D. C.: U. S. Government Printing Office, 1978.

U. S. Department of Agriculture, Foreign Agricultural Service. *Foreign Agricultural Circular: Grains*. FG-37–83. Washington, D. C.: U. S. Government Printing Office, December, 1983.

U. S. Department of Agriculture, Soil Conservation Service. *Draft Resources Conservation Act Appraisal, Part II*. Washington, D. C.: U. S. Government Printing Office, 1981.

U. S. Economic Research Service. *Farm Real Estate, Outlook and Situation Summary*. Washington, D. C.: U. S. Government Printing Office, 1983.

Webb, Shwu-Eng, Clayton W. Ogg, and Wen-Yuan Huang. "Economic Analysis of Long-Term Acreage Reduction Strategies." Paper presented at annual meeting of the American Agricultural Economics Association, Cornell University, Ithaca, New York, August, 1984.

Wischmeier, W. H., and D. D. Smith. *Predicting Rainfall-Erosion Losses from Cropland East of the Rocky Moutains*. U. S. Department of Agriculture, Agricultural Research Service, Agricultural Handbook No. 282. Washington, D. C.: U. S. Government Printing Office, 1965.

# 15 DISCUSSION

## Richard E. Howitt*

Given the problem of covering "The Resource Dimensions of Agricultural Policy" in a single paper, it is not surprising that Benbrook, Crosson, and Ogg concentrate on the single dimension of soil erosion. However, the policy analysis of soil erosion is examined in such a specific manner that concepts applicable to other resource dimensions are hard to draw. The authors anticipate some novel policies in resource management this year but only discuss two examples—targeted acreage reductions and rental-based acreage set-aside.

The introduction opens with the statement: "We take it as axiomatic that present practices for management of agricultural land, pesticides, and water for irrigation impose excessive social costs, . . ." and cites the existence of uncompensated external costs as sufficient justification. However, from a Coasian viewpoint, it is clear that, for nonzero marginal transaction costs, social optimality will always involve the presence of some external costs; whether they are compensated or not depends on the current liability institutions. It is disappointing that this chapter in its pragmatic approach, does

*The author is Associate Professor of Agricultural Economics and Economist in the Agricultural Experiment Station and on the Giannini Foundation, University of California, Davis.

not emphasize the role of transaction costs as the basic cause of externalities.

The distinction between on-farm and off-farm erosion costs is well drawn, but the wealth transfers and property rights implications are not stressed. On-farm erosion externalities are intertemporal or intergenerational and only can be explained by a divergence between private and social discount rates, a malfunctioning capital market in land, or irreversibilities and threshold effects in the soil production function. The difficulties of measuring these effects are recognized in their chapter, and a cost-effective approach for cases of reverse erosion is advocated. Given the current political emphasis for erosion control, this is a reasonable political position but one that begs for economic analysis.

The different social costs and the nonpoint source characteristics of off-farm erosion externalities are completely discussed, and estimates of the aggregate costs of these two problems are compared. The relative importance of off-farm erosion costs is emphasized. The authors conclude the second section of their chapter with the ambitious-sounding but vague statement: "Indeed, a strong case can be made that a 50 percent reduction in the total tonnage of erosion from U. S. cropland is within reach at little or no extra cost." Presumably, the cost referred to is the cost of the existing soil conservation programs.

The central concept—that of crop acreage reductions targeted to the most erosive soils—is introduced on the assumption that the long-established practice of commodity price stabilization by crop acreage reduction methods will continue. Given the current budgetary pressures and the costs of the recent payment-in-kind (PIK) program, the concensus of several speakers that future acreage programs will be reduced seems correct. In this case the use of the same policy instrument (acreage reduction) for both erosion control and price stabilization will probably be unsatisfactory.

While it is clear that agricultural resource and erosion policies and commodity price stabilization actions should be synergistic, this does not require that they use the same policy variables. The controllability principle shows analytically that a system with two goals and a single policy control takes longer to achieve given values than a system with coupled but independently manipulated controls. The authors recognize the continuation of price instability and rapidly changing commodity price stabilization policies but downplay the inherent conflict between these policies and longer term erosion control actions. There are alternative sets of policy actions that reduce the antagonism between commodity stabilization and resource policies but allow a more flexible conditional reaction.

A second problem with the targeted diversion policy is that it implicitly confers the property rights of off-farm erosion pollution on the farmer. The program bribes the farmer to reduce off-farm pollution with incentives rather than penalizing the externality with pollution charges. The bribe solution may, indeed, be the socially optimum solution if transaction costs are significantly lower than the externality charge option. However, the assignment of the pollution property rights and consequent wealth transfer should be justified on social efficiency grounds.

A tentative suggestion that reduces some of the control inflexibility and wealth transfer problems of the targeted set-aside program is a program that deducts the cost of off-farm erosion from commodity price-support payments that the farmer might receive. This program could be named the "soil in kind" (SIK) program, a suitably cryptic acronym. The authors, Benbrook, Crosson, and Ogg, state that cropland can be categorized into broad erosion classifications and cite some off-farm erosion damage estimates. These estimates could be used to define off-farm erosion costs in terms of crop yield. A possible rate for wheatland could be a SIK rate of 1 bushel per ton of soil over 5 tons per acre.

A program of this type has several advantages. It allows the commodity price stabilization policy to work through the more flexible and conditional price mechanism, yet provides a clear incentive for highly erodible land to be the least cropped. The implicit pollution charge provides an incentive for the farmer to internalize the off-farm erosion externalities but vests the decision of erosion control with the farmer. The cost of the program would be less than the targeted set-aside alternative; and the capitalized value of price supports on highly erodible land would be reduced, bringing the capitalized value of the land closer to its social capitalized value. In addition, the transaction costs of identifying highly erodible land and setting the local SIK rate would be similar to defining targeted set-aside areas. However, at high market prices and low support levels, more land would be brought into production smoothly and voluntarily. Conversely, when market prices are low and support payments substantial, the cost of opting out of a linked price-support and SIK program would be substantial.

Clearly, any practical program would require considerably more thought, but the SIK alternative does illustrate that the important point of program linkage and synergy raised by the authors does not require a reduction in policy instruments and controllability. Until the current policies that allow the capitalization of commodity support prices on highly erodible land are modified, the social cost of mitigating this erosion will be high in low commodity price periods when it is socially optimal to internalize the off-farm erosion externalities.

The authors review two innovative programs where erosive land was removed from production by renting it. Substantial savings of this approach over the conventional deficiency and crop diversion payments are claimed. However, it is difficult to see how the savings of 50–90 percent of the cost of conventional programs could be realized over larger areas in the absence of rental market distortion or excessive administration costs. Even without these dramatic savings, the use of a rental bid system for retiring highly erosive land is an attractive and logical approach.

The authors correctly identify the contamination of groundwater by pesticides and herbicides as a major concern in the future. The encouraging trend toward integrated pest management and the trade-off between erosion and herbicide residues in some areas are discussed. Although the mammalian toxicity of herbicides is lower than that of insecticides, the problems of nitrogen and heavy metal contamination of groundwater are not addressed. It is quite possible that concern by the public over agricultural toxic waste will result in constraints on the use of pesticides, herbicides, and some fertilizers. Given the nonpoint-source nature of these pollutants, socially efficient externality control can be effected by limits on agricultural inputs. A limited regional assimilative capacity can be efficiently allocated by a bid system for pollution rights, but the ability of agriculture to modify perceptions of public health standards will be very limited if these current concerns are foremost in the public perception. An illustrative case is the planned and partially constructed San Joaquin Valley master drain in California where urban concerns about agricultural chemical residues and heavy metals being discharged through the San Francisco Bay may well prevent completion of the project.

The authors mention the possibility of the 1985 farm bill modifying the federal Insecticide, Fungicide, and Rodenticide Act but do not elaborate on specific changes. One of the alternatives to more stringent input regulation is to subsidize the generation and dissemination of management information. Some ongoing work in California's Imperial Valley shows that pest management information is a substitute for insecticides on cotton. Information in the form of irrigation scheduling can also substitute for greater water application.

The seemingly obtuse reaction (cited by the authors) of Imperial Valley, California, farmers resorting to long-season cotton in 1983, despite the associated boll weevil problem, has an alternative explanation. Short-season cotton has a higher variance of yield as well as a lower expected yield; thus, the boll weevil policy required both a lower expected payoff and higher risk. Given the risk-reducing motivation of many pesticide applications, policy planners could consider an insurance scheme to bear the risk where the externalities of risk reduction by pesticide application are high.

Benbrook, Crosson, and Ogg cite groundwater overdraft and contamination as resource management problem that is unlikely to be addressed in the 1985 farm bill. With the current concern about declining groundwater levels in states overlying the Ogalalla formation and other western states, a simple proposal to encourage local management of basins would seem attractive. The problem of optimal intertemporal management of groundwater has been widely analyzed, and the potential benefits in some overdraft areas are substantial. Why then have management proposals met with widespread farmer resistance? Perhaps one reason is that every management proposal, without augmented supplies, requires that farmers accept short-run reductions in cash flow for long-run gains which have a larger present value. If a perfect capital market existed, the farmer could amortize the capitalized value of these future gains and, thus, have an increased current cash flow. In practice, the common property groundwater basin provides no asset backing, and farmers are required to take real short-run cuts in income under most groundwater management plans. This is a clear case in which the government can overcome the externality in the asset market by underwriting loans to maintain income or purchasing additional water for local groundwater management plans. Given market rates of return that appear to be obtainable according to some groundwater management studies, such a program, once initiated, could be self-sustaining when initial loans started repayment.

The consideration of "targeting" is dominant throughout the paper but is only described as a general concept toward the end. The authors specify two steps in the process: "The first step in targeting is nothing more than defining the problem." This is clearly required for most rational approaches. "The second step, tailoring program activities to management needs, is more complex." This statement is so general that it is uninformative. However, I do detect the inference in the term, "management needs," that decisions are based on fixed physical criteria specified by outside agencies rather than on marginal economic trade-offs made by the farmer after the externalities have been internalized into his cost function.

While the authors make a convincing case for more flexible and reactive policy instruments, I suspect that the fixed physical criteria that seem to form the basis of their targeting concept will not add to the flexibility of agricultural policy. They perceptively and precisely define the dominant resource problems for the 1985 farm bill, but the central policy ploy of targeting implicitly grant full property rights to the polluter and appears to be based on relatively inflexible physical criteria. The authors state: "The ability to adapt efficiently and quickly to changing economic conditions will be an essential attribute for future commodity programs." But I am not convinced that the targeted set-aside program, as described, meets this criterion; however, use of the price system and incentives for farmer decision making can reconcile commodity and agricultural resource policies.

# 16 FOOD AND CONSUMER DIMENSIONS OF AGRICULTURAL POLICY

**Dale Heien***

## *Introduction*

In discussing food and consumer dimensions of agricultural policies, it is useful to segment our national policies into those which are consumer and those which are food or nutrition policies. In recent years, there has been an increasing tendency to identify consumer policies with food policies. Our food and nutrition policies date back roughly, in concept, to the origination of the school lunch program and the initial food stamp programs. These programs began at about the same time and under the same impetus as the farm price supports.

On the other hand, consumer policy with regard to food goes back much farther. Land-grant universities were established primarily to provide educational opportunity to the broad mass of society. The funding of agricultural research, its extension into the farming

*The author is Associate Professor of Agricultural Economics and Economist in the Agricultural Experiment Station and on the Giannini Foundation, University of California, Davis.

community, and the national emphasis on adopting modern farming techniques all resulted in what some have referred to as a "cheap food" policy. The obvious beneficiary was the consumer, i.e., the broad base of the population. As Houthakker (1976) points out, the economists' perspective is also, by and large, that of the consumer. The main focus of this policy is the price of food. Subsidiary issues are quality, security, availability, safety, and variety.

In recent years, food and nutrition policy has become increasingly important—with food stamps and, to a lesser extent, nutrition and school lunch programs dominating consideration. It is worth noting that food policy caters to specific interest groups, while consumer policy is based on a wider spectrum.[1] However, benefits received per recipient are much higher for food programs than for consumer programs. Back-of-the-napkin calculations put the welfare loss of higher food prices at around $7 billion. Direct costs for these programs now run around $13 billion–$15 billion, although an exact figure is difficult to ascertain due to year-to-year fluctuations. Hence, total direct and indirect consumer costs of these farm programs are somewhere in the neighborhood of $20 billion, or about $400 for every household in the United States. Since the *de facto* tax system is somewhat progressive, this implies that the direct costs are borne more heavily by the upper income groups. Since high food prices have a regressive effect on the poor, the indirect costs are borne disproportionately by lower income individuals.

To some extent this regressivity is redressed by various programs, especially food stamps. However, within lower income groups, widely divergent effects are experienced. Many low-income individuals are not eligible for and do not receive stamps. Furthermore, present budget constraints prevent inclusion of substantially more individuals. Individuals not included are mainly lower income individuals who are employed.

Most food stamp recipients have incomes below the poverty line. The largest class of food stamp recipients are unmarried (or previously married) women. Roughly, a third of the group is over 65 or disabled; the other third is responsible for the care of children

---

[1]There is some question about the constituency for food policy. This may, to some extent, be charitable activity on the part of society (Chaves, 1984, pp. 226 and 227).

(welfare mothers); and the remaining third is split between the working poor and unemployed. It should be noted, however, that roughly two-thirds are not in the job market. Over two-thirds of the recipients are female; 70 percent are urban residents. Less than 1 percent are strikers and less than 1 percent are students. Of the population of Puerto Rico, 70 percent (around 2 million people), are on food stamps.

## Background

Since the food and farm policy scenario is to some extent a zero sum game, it is useful to document the welfare status of the major players. These are the farmer, the consumer, the taxpayer, and the food policy beneficiary. The present status of farmers has been well documented elsewhere (see, especially, Gardner, 1983). The basic findings are that net real income per farm has risen since the 1940s. This increase has far outstripped the increase in real spendable earnings per worker in the private sector over the 1960–1980 period (Table 1). During these two decades, total private real spendable earnings per week rose $4.37 per worker, an increase of 4.8 percent over the entire period. Net real income per farm rose 7.8 percent over the same period. Although not strictly comparable, because of the inclusion of off-farm income and the mixture of return on financial and human capital, the comparison is revealing. Perhaps of more interest is the increase in real earnings of government employees over this period, an increase of 22.6 percent. The increase for Social Security recipients is 64.4 percent[2] and for unemployed persons, 7.8 percent. This is not the increase in total Social Security payments; it is the increase in payments per beneficiary.

Hence, if one examines the evidence over the last 20 years, it is clear that recipients of government transfer payments—farmers, the unemployed, government employees, Social Security recipients—have all experienced much greater per capita real income increases than those who earn their income in the private sector. Income *per household* has continued to grow, mainly as a result of increased labor force participation by women. However, the main inference

[2]This increase is even greater if Medicare payments are included.

**TABLE 1. Real Income Growth for Selected Classes During the 1960–1980 Period**

| Year | Real spendable earnings: all private employees | Net income per farm | Average compensation: full-time equivalent, government employee | Social Security: retired worker and wife | Average unemployment benefits |
|---|---|---|---|---|---|
| | 1967 dollars per month | | | | |
| 1960 | 393.90 | 273.2 | 439.3 | 139.8 | 161.2 |
| 1970 | 446.48 | 350.4 | 502.3 | 171.1 | 186.3 |
| 1975 | 439.51 | 478.8 | 594.5 | 213.4 | 188.2 |
| 1980 | 412.79 | 329.6 | 538.2 | 229.8 | 173.8 |
| Percent change 1960–1980 | 4.8 | 20.6 | 22.6 | 64.4 | 7.8 |

Source: U. S. Department of Commerce (various issues).

is clear. The period from the 1960s to the present has witnessed a substantial shift in the distribution of the nation's resources between those who work in the private sector vis-à-vis those who either work in the public sector or receive part or all of their income as a result of government transfer payments. This shift in the income distribution is not simply a reflection of the growth of government. For example, government could grow as result of increased numbers of Social Security recipients. However, the rate of growth of Social Security recipient income per person could be less than that of private income. Such has not been the case.

The 1960–1980 period has also witnessed an increasingly unequal distribution of household income. (The Gini coefficients for this period are given in Table 2.) The causes behind this increased inequality are complex and beyond the scope of this chapter. However, given that the many social programs undertaken in the 1960s and 1970s were designed and promoted on the basis of reducing the inequality of income, the effectiveness of these programs must be questioned. Also, it is not beyond the pale of imagination to

**TABLE 2. Gini Coefficients Based on 25 Income Intervals for Adjusted Gross Income by Household During the 1960–1980 Period**

| Year | Coefficient |
|---|---|
| 1960 | .432 |
| 1970 | .448 |
| 1975 | .451 |
| 1980 | .455 |
| Percent change, 1960–1980 | 5.32 |

Source: Computations by author from data in the U. S. Internal Revenue Service (various issues).

attribute much of the increase in inequality to the growth of government. Smith (1977), for example, estimated that pay scales for postal workers are 50 percent higher than market conditions warrant. The vast increase in higher paying government jobs has tended to increase the inequality of income distribution. Last, the increase in government during this period has not been limited to increased employment at above-market wages.

Equally important is the vast increase in government regulation at both the state-local and federal levels. This increase in regulation has roughly coincided with the secular decline in the rate of growth of output per unit of input analyzed by students of productivity (Bailey, 1981, and Fraumeni and Jorgensen, 1981). While much of this decline is due to obsolescence of capital stock as a result of energy price increases, a major portion is due no doubt to government regulation and disincentives occasioned by widespread welfare programs. This is reinforced by the finding that the productivity decline started in the mid-1960s long before energy crises but coincidental with the growth of great society programs.

This decrease in the natural rate of growth of output is also mirored in the increase in the natural rate of unemployment. Okun's full employment GNP was computed at a 3-½ percent unemployment rate. Today, most estimates would put this rate at around 6 percent with the natural rate still higher. This increase is due, in part, to the changing demographics of the labor force. It is also due, in part, to the many welfare programs designed to assist those who are "unemployed."

These points are made in order to illuminate the extent to which government programs have (1) not achieved their goal of a more equitable distribution of income; (2) reduced the rate of growth of productivity and, hence, income; and (3) rewarded those who have been active politically at the expense of those who sought their rewards in the private economic sector.

What about the status of food policy beneficiaries? Because the main focus of food policy is the nutritional level of the recipient population, it is useful to look first at the facts regarding nutrition and income. Table 3 gives the percentage of households meeting the Recommended Daily Allowance (RDA) for food energy and 11 nutrients. If the $20,000-and-above income group is excluded, then the rate of increase of nutrient intake as a function of income is quite small; and, for three classes, the percentage falls as income rises. Perhaps more interesting is Table 4 which shows virtually no relationship between the nutrient value of household food as a percentage of RDA and income classes. For all classes of nutrients and for all income levels, these figures exceed 100 percent and decline by income for 6 of the 12 items. This basically says that malnutrition, if it is a problem, is probably not a problem of income level. It is a problem of information—information that is quite heterogeneously distributed.

It is of significant interest to note that the area which clearly has the lowest percentage of RDA per household is calcium. The postwar decline in milk consumption per capita is often attributed to the changing demographics which, no doubt, is true. However, higher milk prices, a direct result of dairy price supports and milk marketing orders, have played a much larger role in reducing consumption than is commonly recognized. Dairy product demand is subject to considerable habit effects (Heien, 1977). Hence, policies based on inelastic demand are effective but only in the short run. Over the longer period, demand becomes more elastic and consumption falls more than proportionately. Also, higher milk prices induce substitutes, such as oil-based dairy products and soft drinks with artificial sweeteners, neither of which are highly recommended by nutritionists.

**TABLE 3. Percentage of U. S. Households Using Food Providing Recommended Daily Allowance by Income Group, 1976**

| | 1976 income before taxes | | | | |
|---|---|---|---|---|---|
| | Under $5,000 | $5,000–$9,999 | $10,000–$14,999 | $15,000–$19,999 | $20,000 or more |
| | percent | | | | |
| Food energy | 74 | 74 | 76 | 76 | 81 |
| Protein | 93 | 96 | 98 | 98 | 99 |
| Calcium | 62 | 62 | 67 | 67 | 73 |
| Iron | 84 | 81 | 83 | 83 | 87 |
| Magnesium | 69 | 72 | 75 | 76 | 81 |
| Phosphorus | 91 | 94 | 97 | 96 | 98 |
| Vitamin A | 78 | 78 | 78 | 78 | 86 |
| Thiamin | 89 | 87 | 86 | 85 | 90 |
| Riboflavin | 91 | 93 | 95 | 94 | 96 |
| Vitamin $B_6$ | 59 | 63 | 67 | 67 | 73 |
| Vitamin $B_{12}$ | 79 | 83 | 89 | 91 | 94 |
| Ascorbic acid | 90 | 91 | 92 | 94 | 96 |

Source: U. S. Department of Agriculture, Science and Education Administration (January, 1981).

This is not to say that food stamps have been ineffective in providing assistance to some individuals and families in meeting their nutritional needs. They clearly have been effective, particularly for the elderly and the young. The school lunch program, also a part of the food policy scenario, provides considerable nutritional support for the young. It also provides a subsidy to middle and upper income children. However, the evidence does indicate that nutritional education and information have a return which is relatively higher than food stamps on a cost benefit basis. Also, more research needs to be done concerning the effect of food stamps on work incentives. I do not mean to single out food policy as *crowding out* consumer policy any more than transportation policy has *crowded*

**TABLE 4. Nutritive Value of Household Food as a Percentage of the Recommended Daily Allowance by Income Group, 1976**

| | Average value per nutrition unit per day as percentage of RDA | | | | |
|---|---|---|---|---|---|
| | 1976 income before taxes | | | | |
| | Under $5,000 | $5,000–$9,999 | $10,000–$14,999 | $15,000–$19,999 | $20,000 or more |
| | percent | | | | |
| Food energy | 134 | 130 | 129 | 125 | 132 |
| Protein | 217 | 212 | 217 | 213 | 232 |
| Calcium | 122 | 116 | 123 | 119 | 133 |
| Iron | 156 | 151 | 150 | 137 | 147 |
| Magnesium | 128 | 129 | 130 | 125 | 137 |
| Phosphorus | 204 | 196 | 200 | 194 | 211 |
| Vitamin A | 202 | 191 | 172 | 161 | 185 |
| Thiamin | 181 | 166 | 166 | 151 | 164 |
| Riboflavin | 201 | 189 | 191 | 183 | 201 |
| Vitamin $B_6$ | 119 | 119 | 122 | 117 | 128 |
| Vitamin $B_{12}$ | 239 | 219 | 220 | 214 | 228 |
| Ascorbic acid | 302 | 305 | 295 | 290 | 333 |

Source: U. S. Department of Agriculture, Science and Education Administration (January, 1981).

*out* consumer policy. It simply appears that no one advocates for consumer policy.

## *Policy Impacts*

Before evaluating the impacts of the various agricultural policies, it will be useful to examine the channels through which agricultural policies affect consumers and beneficiaries of food and nutrition policy. As indicated earlier, the primary way consumers benefit is through lower farm prices. Hence, farm policy generally works against consumer interests by raising food prices. Also, expenditures for price supports come out of general tax revenues paid for by the broad base of the population. Consumers benefit from agricultural research if it results in lower prices. However, although productivity in agriculture has been among the highest of the various U. S. sectors, the ratio of food prices to all consumer prices has not declined over the postwar period. This indicates that the productivity gains have not been passed on to consumers. Since the amounts spent on agricultural research are now quite small relative to the amounts spent on price supports, I will ignore these effects.

While food policy beneficiaries are direct recipients of aid in the form of food stamps, direct food assistance, etc., it is worth noting that most food stamp recipients also receive other government welfare assistance. Food aid does not appear to be related to the level of food prices but more to general welfare concerns and, perhaps, the overall rate of inflation. Food stamp expenditures are presently at about the same level as the farm subsidy, apparently as a result of a compromise struck between rural and urban interest groups in Congress.

The main purpose of this conference is to examine the impacts of alternative farm policies. Three of the proposed policies will have, by and large, the same impact on consumers—higher prices and higher tax bills. These two burdens will be distributed differently under various programs, but their total impact is roughly the same. The impact of these programs on food policy beneficiaries is somewhat different. Most food stamp recipients pay little or no tax although, as deficits increase, their tax burden increases more than proportionately if one accepts Friedman's convention of regarding deficits as a tax (Friedman, 1984). Hence, programs, which rely on heavy direct costs, are clearly of benefit to this class. These programs ease the indirect costs and are supported by direct costs borne by the tax-paying population. A program of farm income support would fall heavily on taxpayers but might have the beneficial effect of making the program costs fully visible. It is doubtful if such a program would be acceptable to farmers since it would have too many welfare aspects. Also, although present farm subsidies are running at about the level of net farm income, subsidies to enable farmers to remain in farming might be much higher.

A policy of no government intervention or subsidies would provide substantial benefits to both consumers and taxpayers. The status of food stamp recipients might be at jeopardy, especially if one accepts the notion of the rural-urban coalition. We should bear in mind that the majority of the farm policy costs are borne by taxpayers. Since food stamp recipients pay little income tax, their benefits will not be nearly as great as those going to taxpayers.

I do not consider the free-market scenario to be at all likely. First, it is not attractive to farm interest groups although it does appear attractive to some farmers. Moreover, it is not equitable for farmers to lose support while a myriad of other interest groups, no less meritorious, continue to receive support. The present farm price-support policy has welded together additional interest groups such as equipment manufacturers, bankers and lending institutions, food processors, etc. Even consumer and food interest groups are strangely quiet regarding the farm supports, lending further credence to the coalition explanation.

## *Conclusions*

In summary, the following burdens are noted with respect to the various policy options. Continuation of the present policies results in total costs per household of about $400. Of these, roughly a third are indirect costs which bear heavily on the poor. Relief for this class is provided to some extent by food stamps and other food policy actions. These recipients are mainly welfare mothers and the elderly. Lower income working families bear a substantial burden of the indirect costs of farm policy. Direct costs (two-thirds of total costs) are borne by taxpayers; with a generally progressive tax, this burden is shifted somewhat to higher income households. However, if we reapportion the deficit as a per capita indirect tax, then the distribution between direct and indirect costs is nearly fifty-fifty. Adoption of the policy option to make instruments smooth and orderly sounds attractive, particularly with respect to timeliness. Being largely untried, however, such a policy will entail considerable learning costs since it, undoubtedly, will involve errors.

There is a fundamental problem in dealing with economic policy today. It arises out of the legislative process where the policies are formulated. This problem has been long recognized by Congress, and an attempt was made to deal with it in the Budget Reform Act which, unfortunately, lacked any real discipline. The problem to which I refer is that, when viewed from any particular congressional committee, most economic policy legislation—be it food stamps, the farm bill, transportation policy, etc.—is not large as a percentage of total government spending. As a result of this process and other forces of a political nature, which are not well understood, there is at present no effective restraint on federal government spending, i.e., government spending is out of control. No tax increase will bring it under control.[3]

Curiously enough, these programs seem to make sense when viewed individually. However, each succeeding set of policies creates a new class of "walking wounded," each of whom is successively drawn into the political arena. Some of this damage is done by the policies themselves, e.g., equipment producers feel they were victimized by the payment-in-kind program. However, by and large, the far greater damage has been done by high interest rates and high inflationary expectations brought on by the huge federal deficit and by recent outsized increases in the money supply. This deficit is a direct result of the legislative process described above, while erratic money growth is to some extent associated with the notion that the Federal Reserve Board must bear the sole task of monetary and fiscal policy.

A solution to this dilemma has recently been put forth in the form of a balanced budget amendment to the Constitution. Thus far, 32 of the required 34 states have called for a constitutional convention to pass this amendment. Such an amendment recently passed the Senate but failed in the House. Other needed measures include requirements to restrict the rate of growth of the money supply, provide the President with a line-item veto, and pass a flat-rate income tax. Only when government deficits are controlled will there be an economic environment where rational policymaking, as well as economic growth and progress in general, can occur. Reductions in government spending and in the level of government regulation and involvement will restore the U.S. economy to its former levels of

[3]For a further elaboration of this problem, see Friedman and Friedman (1983).

productivity and competitive ability. Continuation of present trends means continued stagflation and heightened uncertainty regarding our economic future.

Agriculture is in trouble today not because of poor farm policy or because of unique inherent characteristics of agriculture but because of runaway spending by Congress and the high interest rates, high exchange rates, and inflation caused by this spending. The policy, which will help farmers and consumers, is a balanced budget and a reduction in government spending on all programs.

## References

Bailey, M. N. "Productivity and the Services of Capital and Labor." *Brookings Papers on Economic Activity*, No. 1 (1981), pp. 1–65.

Chaves, J. P. "Future Directions for Domestic Food Policy." *American Journal of Agricultural Economics* 66, No. 2 (May 1984):225–231.

Fraumeni, B. M., and D. W. Jorgensen. "Capital Formation and U. S. Productivity Growth, 1948–1976." In *Productivity Analysis: A Range of Perspectives*, edited by A. Dogramaci. The Hague: Martinus Nijhoff, 1981, pp. 49–70.

Friedman, M. "The Taxes Called Deficits." *The Wall Street Journal*, April 26, 1984, p. 16.

Friedman, M., and R. Friedman. *The Tyranny of the Status Quo*. New York and London: Harcourt Brace Jovanovich, 1983.

Gardner, B. "Rural Reaganomics: An Analysis and Critique." Paper prepared for Conference on Natural Resources and the Environment: The Reagan Approach. Sponsored by the Urban Institute and Resources for the Future, June 6 and 7, 1983.

Heien, D. "The Cost of the U. S. Dairy Price Support Program." *The Review of Economics and Statistics* LIX, No. 1 (February 1977):1–8.

Houthakker, H. "Do We Need a National Food Policy?" *American Journal of Agricultural Economics* 58, No. 2 (May 1976):259–269.

Lampo, D. "Hunger Task Force Misses the Point." *Policy Report* VI, No. 3 (March 1984):2.

Paarlberg, D. "Tarnished Gold: U. S. Farm Commodity Programs After 50 Years," *Food Policy* (February 1984):6–10.

Smith, S. *Equal Pay in the Public Sector: Fact or Fantasy*. Princeton: Princeton University Press, 1977.

U. S. Bureau of the Census. *Estimates of Poverty Including the Value of Noncash Benefits: 1979 to 1982*. Technical Paper No. 51. Washington, D.C.: U. S. Government Printing Office, February, 1984.

U. S. Department of Agriculture, Science and Education Administration. "Nutrient Levels in Food Used by Households in the United States,

Spring, 1977." Nationwide Food Consumption Survey 1977–78, Preliminary Report No. 3, Washington, D. C., January, 1981.

U. S. Department of Commerce. *Statistical Abstract of the United States and Survey of Current Business*. Washington, D. C.: U. S. Government Printing Office, various issues.

U. S. Internal Revenue Service. *Statistics of Income*. Washington, D. C.: U. S. Government Printing Office, various issues.

# 16 DISCUSSION

**Sylvia Lane***

Galbraith (1983) pointed out, "Economics is extremely useful as a form of employment for economists", to which must be added in this context—as is agricultural policy for agricultural economists. Parenthetically, the consumer is almost last on this program which, it seems to me, corresponds generally to her/his position on the agenda when agricultural legislation is considered in the U. S. Department of Agriculture—yet, food programs comprise about half of the budget of the Department.

I differ with Dale Heien on food and nutrition policy not being part of consumer policy, i.e., policy for—not about—consumers. It can be a significant, important, and large subset of that broad but not yet stated policy. I do think consumer policy has its advocates, but their voices are often drowned out. The cheap food policy certainly was of benefit to the consumer when food was cheaper than it otherwise would have been; but because it also expanded the quantity of agricultural products demanded and kept wages lower than they otherwise would have been, there may have been considerations involved other than the interest of the consumer.

---

*The author is Professor Emeritus of Agricultural Economics and Economist in the Agricultural Experiment Station and on the Giannini Foundation, University of California, Davis.

Food policy, as I have heard it enunciated, involves the provision of a safe, wholesome, and nutritious food supply at the lowest possible price (Foreman, 1978). Nutrition policy is a subset of food policy. However, Timmer, Falcon, and Pearson (1983, p. 9) define food policy as "encompassing the collective efforts of governments to influence the decision-making environment of food producers, food consumers, and food marketing agents in order to further social objectives. The objectives nearly always include improved nutrition for inadequately nourished citizens . . . ." Agricultural policy affecting farm products used in or as food would, therefore, be subsumed under food policy. This definition is very useful when considering trade-offs, especially with regard to subsidies to producers, marketing agents, or consumers.

Heien notes that food stamp program expenditures are presently at about the same level as is the farm subsidy. This is the price agricultural interest groups pay for urban support. Consumer food subsidies [e.g., the food stamp program, the Women's, Infants, and Children's Program (WIC), and reduced-price or free school lunches and school breakfasts] serve, in part, to counteract the regressivity of indirect costs (higher food prices) of agricultural programs (as Heien points out); but the incidence of costs and benefits of food programs and agricultural programs seems to be far from perfectly offsetting.

The food stamp program is, above all, an income-maintenance program. It and the Social Security program are two of the most important income-maintenance programs in the United States and, also, two of our best automatic stabilizers.

Food programs also serve as important public health measures for large parts of the low-income population. According to a number of authors (Sexauer, 1978; Akin, Guilkey, and Popkin, 1983; Davis, 1982; and Lane, 1978), they—especially the WIC, the school lunch and breakfast programs, food stamps, and the nutrition programs for the elderly—provide important nutritional benefits for many participants.

Nutrition is a problem associated with information; but for the lowest income categories, it is also a problem of food availability (Madden and Yoder, 1972, and Lane, 1978). In the case of all of the nutrients analyzed in the National Food Consumption Survey of 1977 (Heien, Table 3), increased income was associated with a higher percentage of households using food that provided the Recommended Daily Allowances (RDA). This leads me to question the difference between Table 3 and Table 4. Since the percentage of households using foods that provide the RDA is, in all cases, below 100 percent and since the average nutritional value of household food, as a percentage of the RDA, is always above 100 percent, some households in each income category have far more than 100 percent of the RDA and, therefore, pull up the average. The main point is that, generally, smaller percentages of households in lower income categories meet the RDA for the 12 nutrients examined.

Perhaps more interesting is what happens to the diet as income decreases or as food prices are increased. Because lower income people spend a higher percentage of their income on food, they tend, in being efficient, to use higher nutrient density, lower priced foods so that they will obtain more nutrients per dollar. The result is a diet higher in fats and starches, i.e., less healthful and, incidentally, more monotonous diet (Abdel-Ghany, 1978, and Foytik, 1981).

I am not commenting on the macroeconomic aspects of the paper because there are those attending this conference better qualified than I to question them. I do want to add my own three postulates for consumer food policy.

1. The commodities that are essential to maintain health (the list includes clean water, medical care, and nutritious foods) are merit goods in American society; therefore, theoretically, they should be available to everyone in the United States.
2. As those without sufficient funds cannot obtain the components of an adequate, palatable diet, income supplements to increase income to the required level are appropriate. Political considerations have resulted in the WIC program and the food stamp program serving this purpose. If the latter were cashed out, the program would be cheaper to administer and consumers would have more choice, but nutrition might be affected somewhat adversely for a very few. Food stamps are highly cash equivalent (Ranney, 1983).
3. Nutrition education should be a major public health program. It deserves greater emphasis.

In conclusion, I raise two more issues. First, agricultural programs serve a relatively small proportion of our population, and the proportion grows smaller over the years. Food affects us all. It is time for the U. S. Department of Agriculture to acknowledge that it will need to become the Department of Food and Agriculture if it is to visibly broaden its constituency. Before long, this may become a political necessity if the Department is to retain—or, at least, not lose—influence vis-à-vis the other Cabinet departments.

Second, the conservation-environmental issue will not, as yet, replace the food stamp program as a means of putting together a rural-urban coalition to pass legislation for the agricultural sector. I am not optimistic about the conservation-environmental issue serving as well as did the hunger issue in the past.

## *References*

Abdel-Ghany, M. "Evaluation of Household Diets by the Index of Nutritional Quality." *Journal of Nutrition Education* 10, No. 2 (April–June 1978):79–80.

Akin, J.; D. Guilkey; and B. Popkin. "The School Lunch Program and Nutrient Intake: A Switching Regression Analysis." *American Journal of Agricultural Economics* 65, No. 3 (August 1983):477–485.

Davis, Carlton G. "Linkages between Socioeconomic Characterizations, Food Expenditure Patterns, and Nutritional Status of Low-Income Households: A Critical Review." *American Journal of Agricultural Economics* 64, No. 5 (December 1982):1017–1025.

Foreman, Carol Tucker. "Human Nutrition Policy in the U. S. Department of Agriculture." *Agricultural-Food Policy Review: Proceedings of Five Food Policy Seminars.* U. S. Department of Agriculture, Economics, Statistics, and Cooperatives Service, ESCS-AFPR-2, Washington, D. C., 1978, pp.17–20.

Foytik, J. "Very Low Cost Nutritious Diet Plans Designed by Linear Programming." *Journal of Nutrition Education* 13, No. 2 (April–June 1981):63–66.

Galbraith, J. K. Quotation from *Desk Master Diary.* Omicron Delta Epsilon (International Economics Honor Society), 1983. p. 57.

Heien, Dale. "Food and Consumer Dimensions of Agricultural Policies." Paper presented at the Conference on Alternative Agricultural and Food Policies and the 1985 Farm Bill. University of California, Giannini Foundation of Agricultural Economics, and Resources for the Future. Berkeley, June 11 and 12, 1984.

Lane, S. "Food Distribution and Food Stamp Program Effects on Food Consumption and Nutritional 'Achievement' of Low-Income Persons in Kern County, California." *American Journal of Agricultural Economics* 60, No. 1 (February 1978):108–116.

Madden, J., and M. Yoder. *Program Evaluation: Food Stamp and Commodity Distribution in the Rural Areas of Central Pennsylvania.* Pennsylvania State University, Agricultural Experiment Station Bulletin No. 780. University Park, 1972.

Ranney, Christine K. "Food Stamp Program Participation and Cash Equivalent Benefits." Cornell University, Department of Agricultural Economics, Staff Paper. Ithaca, New York, December, 1983.

Sexauer, Benjamin. "Food Programs and Nutritional Intake: What Evidence?" *Agricultural-Food Policy Review: Proceedings of Five Food Policy Seminars.* U. S. Department of Agriculture, Economics, Statistics, and Cooperatives Service, ESCS-AFPR-2, Washington, D. C., 1978, pp. 39–43.

Timmer, Peter, Walter P. Falcon, and Scott R. Pearson. *Food Policy Analysis.* A World Bank Publication. Baltimore, Maryland: Johns Hopkins Press, 1983.

# A GENERAL REACTION TO THE EVALUATION OF ALTERNATIVE PROPOSALS

*General Discussion*

# GENERAL DISCUSSION

## S. R. Johnson*

### *Introduction*

The conference has provided a summary of past experiences with agricultural policy, perspectives from these experiences, and suggestions for modifications of current policies and for the design of future policy for U.S. agriculture. The suggestions for agricultural policy change have been somewhat more limited than might have been anticipated from such a conference. However, perhaps there is something to be learned from this hesitancy to depart substantially from the current U. S. agricultural policy design. Rather than radical change, the discussion has focused more on recommendations for improving the existing policy design.

Programs, such as the 1981 farm bill, have enormous inertia because of vested interests established by implicit subsidies that such legislation invariably contains. Moving to a totally new policy design is, therefore, something that must be viewed realistically as a staged process. Thus, the absence of recommendations for radical change in policy design is likely a function of a recognition of the resistance to change and, importantly, a lack of agreement on what is desired for the functioning of the agricultural sector. The latter point should be emphasized. That is, if economic performance objectives cannot be articulated for the agricultural sector, then policies designed to

*The author is Middlebush Chair Professor of Economics and Agricultural Economics and Codirector, Center for Food and Agricultural Policy Analysis, University of Missouri.

achieve them more efficiently and transition plans for moving from existing programs to an optimal policy design are impossible to formulate. We are left with the pedestrian task of fine tuning. With these general observations on the conference, the summary comments will be in three general categories. First, in the presentations there were a number of "take home" messages; the subsequent section will highlight some of these. Choices of points to highlight relate to their broad implications for agricultural policy and, as well, implications for timing of policy change. This section on Highlights is followed by one on options for change within the current agricultural program design. As already mentioned, most of the substantive suggestions for policy change were within the design of the 1981 farm bill. These options are important because it is unlikely, given the time for debate, that a wholesale change in policy design will occur in the 1985 farm bill. The final section contains a summary of suggestions for new program designs. In addition, it deals with a major shortcoming of the conference: our failure as economists to identify major categories of policy objectives. We do know these economic policy objectives. Moreover, if they are clearly articulated, we can begin to get a sense, through past policy actions, of the implicit weights that have been placed on these objectives. This information can then be integrated into the design of new policy alternatives.

## *Highlights*

The points to be highlighted from the presentations at the conference are identified more or less in the order of the presentations. The extent is not to be comprehensive but to highlight points that have broad implications for the design of agricultural policies and/or the current policy environment and opportunities for change. These highlighted observations are presented in brief form since they were already well elaborated in the presentations.

### The Current Administration Seems Receptive to Policy Change

In addition to the now tired defense of the payment-in-kind (PIK) program, the presentation by Lesher made it clear that there is considerable willingness within the current Administration to consider substantive changes in the agricultural legislation. The experience of the Administration with agricultural legislation obviously has been anything but happy. While the tone of the presentations suggested that this is related to the design of the legislation, it should be emphasized that the information base required for good decisions within the current policy framework has not been provided by the Administration or by other parties to the policy debate.

## *Congress Is Not Fickle*

In the presentation by Lesher as well as in subsequent presentations, there was an underlying theme suggesting that Congress is somehow not willing to make "good" decisions for agriculture. It is unfortunate that someone from one of the Agriculture Committees was not available to make a defense. A more compelling argument is that Congress has not had a good information base for developing policy and program decisions. Making politically tough decisions requires a solid information base. That is, if Congress is going to make nonopportunistic decisions, it is going to have to be able to tell the public why these tough decisions were made. If a reliable information base on consequences of policy decisions is not available, Congress will make what will appear in retrospect to be opportunistic decisions. However, putting all of the blame for opportunistic decisions fully on Congress is a great miscarriage of justice.

## Agricultural Legislation by Subcommittee

The one qualification on the capacity of Congress to make good policy decisions involves the organization of the agricultural committees. Increasingly, these committees have taken on a commodity subcommittee structure, with the legislator representing the voters with the vested commodity interest as chair. This organization contributes to the political forcefulness of the commodity groups. If there is a lesson for Congress from our discussions, it is to look more carefully at the evolution of the subcommittee structure and recognize that agricultural policy is for the agricultural sector and should be more than a collection of particular commodity market policies. Appropriate integrative functions need to be provided in the agricultural legislation. Agricultural legislation developed on a commodity-by-commodity basis is not likely to achieve desired sectorwide objectives.

## Income Maintenance Through Price Supports; Flawed Policy Design

This point was made with great clarity by Johnson. Specifically, in commodity markets with features like those in U. S. agriculture, price supports or benefits of price supports are quickly capitalized into the value of the scarce resource, e.g., land, quotas, and allotments. Thus, price supports affect the level of factor use and the value of the fixed residual factors used but not the returns to factors. This is particularly true for agriculture where the supplies of labor and capital are elastic. Thus, the "benefits" of agricultural price-support programs go almost totally to the holders of the scarce resources in terms of quasi rents.

## Price Supports, Barriers to Entry, and Structure

Related to the point above is an important corollary also developed by Johnson. Specifically, if price supports are used, scarce resources or factors become high priced. These high-priced fixed factors represent important barriers to entry. Thus, price supports preserve at a minimum the status quo but likely contribute to increased concentration. Some of these structural implications were developed by Zilberman and Carter. That is, the price supports and high prices for the fixed resources have resulted in an agriculture quite different from what would have existed with other policy designs. These high prices for fixed resources, of course, make it difficult to change existing agricultural programs, for program changes can bring about important changes in the wealth position of the comparatively few who control these expensive resources.

## Impracticality of Controlling Supply Through Voluntary Programs

The experience with the current agricultural legislation and with the agricultural legislation beginning in the 1950s, as indicated by Cochrane, Johnson, Zilberman and Carter, and others, suggests that it is impractical to control agricultural supply through voluntary programs. Base acreage invariably increases due to political pressures and insured higher than market-clearing commodity prices. Nonland inputs are substituted for land to result in higher yields per acre. The result is higher than anticipated government costs because production levels are large relative to predictions based on past experience.

## Four Years as a Time Frame for Policy Regimes

The paper by Rausser and Just makes a number of interesting points regarding the experience with recent agricultural programs—programs which have come to include very inflexible rules for pricing, base acreage, and other factors. But these rules require modification almost continuously. Thus, although we have a four-year agricultural legislation, Congress is involved in changing the legislation on a year-by-year basis. The point of interest is: Can policies be designed to be more adaptive? That is, since no one can predict accurately four years into the future, rules that require such predictions are impractical for agricultural legislation. In fact, in designing the 1985 farm bill, it might be interesting to turn the question around and ask: How accurately can we predict the variables conditioning agriculture into the future and then design the rules for implementing the agricultural programs consistent with this forecast horizon? The conclusion will be, as Rausser and Just suggest, to develop more adaptive types of rules. One of the reasons for the fixed rules in the 1981 farm bill was to limit discretion, but adaptive rules are not necessarily discretionary rules nor do they necessarily imply great discretion on the part of the Secretary.

## Food and Nutrition Policy

It is generally well recognized now that the food stamp program and many of the other food assistance programs are, in fact, programs for achieving broad social objectives for the low-income population. That is, these programs have come to be relied on as income transfers as well as a way to improve the dietary status of the targeted population. In addition, it is clear that, given the general size of the population targeted by these programs, impacts on prices of basic agricultural commodities made by changing the parameters of these programs are minor. This suggests two fictions. One is that food assistance programs should be tied to agricultural policy and the second is that food assistance programs should be viewed apart from general assistance programs for the targeted low-income population. Both of these points were made albeit somewhat indirectly by Heien. In food and nutrition policy, we are limited by available data bases in evaluating the nutrition status of the low-income population. A recent Presidential Task Force called for better information bases for use in understanding food consumption patterns of the population at large and implications of these patterns for agricultural and other domestic commodity policies.

## Resources and Agricultural Commodity Policy

While the resources paper by Benbrook, Crosson, and Ogg discussed the idea of a conservation reserve, the broader aspects of the presentation and associated discussion related to the social costs of current commodity programs. Because social costs are frequently not viewed as specific consequences of commodity programs, program design may even aggravate problems relating to water, soil erosion, and other conditions. Implications for agricultural policies designed in a sector are worldwide in context (Schuh).

## Specialty Crops

The observations for peanuts, tobacco, cotton, and other specialty crops emphasize different policy designs. In many cases these policies have some of the same limitations as those for feed grains, wheat, and oilseeds. Domestic prices have moved to high levels relative to world prices, fixed resources used to limit production have increased in value, world market shares have diminished, etc. This leads to important questions about the objectives of the policies. More research is required in these specialty areas on the benefits from the policies, to whom they are being distributed, and their incidence. With all these limitations, however, the specialty commodity policies are diverse and experiences with them can be helpful in guiding policies for major agricultural commodities.

## *Options for Change Within the Current Program*

Comments on changes within the existing program will be limited largely to feed grains, wheat, and soybeans. The situation for these crops is not unlike that of many other commodities. Supply is highly elastic in the long run, with a production base in the United States that far outstrips domestic and worldwide consumption potential at prices dictated by existing programs. For example, the current base acreage for wheat is about 94 million. This is, according to estimates from our agricultural sector model at Missouri, approximately 15 million to 18 million acres larger than that required to run a buffer stocks program to maintain the currently legislated prices for wheat. The base acreage for the corn situation is not quite as bad; it is 3 to 5 percent larger than required for effective operation of a buffer stocks program for feed grains at existing loan prices. For cotton, the figure is about 18 percent; for rice, it is about 20 percent. Finally, the soybean program has been viewed more as a "free market" program. Much has been made of the soybean program and the "rolling" loan rate. What is not recognized is how the prices for soybeans are determined by the feed grain program. For example, balancing the program for corn at something like the current loan rate could release about 10 billion acres for production of soybeans. Government costs of sustaining the feed grain program had a great deal to do with soybean prices.

Thus, the problem for all three of these major U. S. agricultural commodities includes base acreages too high relative to existing loan rates, a deteriorating position in world markets because of the level of loan rate relative to production costs in the rest of the world, high government costs, and longer term implications for high land prices and concentration of agricultural production. Suggestions for change of the current program are largely aimed at loan rates, base acreage, world market shares, and the financial situation of agriculture.

## Base Acreages

The base acreage for the three crops is high relative to established loan and target rates. The loan rate for soybeans is rolling and, thus, not directly a problem. However, if the base acreages for wheat and corn were to be reduced to levels consistent with existing loan rates, then the impact on soybean acreage and price could be substantial. The implication is that, if loan rates are to remain at current levels (supporting existing high land prices), then base acreages are going to have to be reduced; if not, government costs of sustaining commodity programs will be prohibitive. Options for reducing base acreage include some type of buy-out provision. Buy-outs will have to be justified by criteria other than simply the economic welfare of agriculture. For this reason, conservation-based buy-outs have been argued (Crosson, Benbrook, and Ogg). The basis of these and other schemes is to reduce the base acreage while somehow compensating landholders for a component of the price due to government commodity programs. Whatever decision is made will be highly contested politically and, of course, contingent upon loan rate decisions.

## Loan Rates

The loan rates and target prices enacted in the 1981 farm bill were based on anticipations regarding inflation, energy prices, and exchange rates that have not materialized. These loan rates, together with U. S. Department of Agriculture program operations that permitted broad additions to base acreages for the major crops, have resulted in an untenable situation. Specifically, loan rates are above world costs of production resulting in a reduced share of export markets. However, these same loan rates, with the additions to base acreage, have resulted in large capacities for production relative to domestic and foreign demand. Suggestions for reducing government cost and production levels have included rolling loan rates similar to those for the soybean program, some type of indexing for loan rates based on interest rates and other domestic prices, and, perhaps, an index for loan rates based on world market shares in export markets. The rolling loan rate has been held up as a panacea by many, but one problem is the fact that its success for soybeans is not unrelated to the large support and high loan rates for corn, feed grains, and wheat. If feed grain and wheat loan rates were reduced, the rolling loan rate for soybeans would go down as well. Reductions of loan rates will have direct impacts on land prices so resistance to reducing the loan rate can be expected.

## Two-Tier Price Schemes

Two-tier price schemes have been offered as an alternative to reducing the loan rate. Specifically, it has been suggested that holding the target price high while reducing the loan rate may be the answer to the overproduction capacity created by past programs. These two-tier price schemes represent a way of letting the world and the domestic equilibrium prices fall while still maintaining effective farm prices consistent with levels that have prevailed over the past few years. The problem with this scheme relates to the likely difference between the target price and the loan rate which could entail high deficiency payments if the programs are not accompanied by production control. However, if the programs are accompanied by high acreage reduction restrictions, participation may not be high. This could cause a sharp reduction in domestic prices and financial crises in an agriculture with land prices rooted in the commodity programs.

## Cross-Compliance

Recently, cross-compliance restrictions for major agricultural commodities have received relatively minor attention. Base acreages for commodities have been allowed to increase without regard to whether or not farmers participate in other commodity programs. In general, the failure to recognize the importance of cross-compliance may be a result of the fact that commodity programs tend to be made by subcommittees. Specifically, the program parameters for one commodity are often set without much regard to those for other commodity programs. Much of the land base can move relatively easily among soybeans, feed grains, and wheat. Thus, effective attempts to reduce the base acreage will have to include much more stringent cross-compliance provisions. Cross-compliance will achieve great importance if programs are oriented toward reduction in the base acreage and maintenance of existing loan rates.

## Interest Rate Buy-Outs

The cash-flow situation for many U. S. farmers is difficult largely because of land prices inflated by loan rate and target prices that cannot prevail in the future and relatively high interest rates. Interest rate buy-out programs have the objective of preventing default on loans held, to a large extent, by quasi-government agencies. Loan rates cannot continue at current levels, base acreages must decrease, and target prices probably will not be as high in the future as in the past. This means that land prices will decrease as they have in the last few years. Many farmers who purchased land in anticipation of prices and price increases similar to those in the early 1980s will have no equity in the land that they "own." Many financial institutions were as optimistic as the farmers and, like the farmer, are overextended. So there will be considerable pressure for such a program because farm groups will be joined by the financial community. It is important if this program is considered to evaluate carefully the implications of the nonprice rationing that will ultimately occur. Incentives in such systems are frequently difficult to tease out but quickly recognized by those to whom the programs are directed.

## Shocks to the Cattle-Livestock Industry

It has become increasingly apparent that erratic decisions on base acreage, loan rates, and target rates have had important implications for the U. S. livestock industry. For example, in many cases price changes, largely the result of government programs, have sent signals to the livestock industry not consistent with long-term stability of the feed grains, wheat, and oilseed sectors. These signals have caused increases and decreases in herd size not consistent with demand trends and longer term feed grain prices. The result has been that the insurance for uncertainty achieved through feed grain, wheat, and oils programs has transmitted uncertainty to the livestock sector. Livestock producers are becoming increasingly aware of the implications of commodity programs for their sector. Also, if the goals for the agricultural sector are articulated clearly, it is obvious that they would include incomes of livestock producers as well as soybean, feed grain, and wheat producers. More integrated agricultural policies, recognizing general goals for the sector, must be more inclusive of the livestock sector.

## Trade Promotion

Trade promotion is an option that has been held up for U. S. agriculture for some time. Unfortunately, headway in international markets is made largely by reducing prices and/or exchange rates (Schuh). But reduced prices cannot occur without large government subsidies or a reduction in loan rates. Thus, although improvement of trade is a prominent topic in discussions of changes in the agricultural program, its ultimate impact is not likely to be substantial. Exports can be increased by increasing P. L. 480 or, alternatively, by subsidizing commercial imports. Perhaps, P. L. 480 sales may be increased based on criteria other than supporting farm prices. However, it is not clear that costs of subsidizing commercial exports or the direct impact of such activities on international markets can be sustained. Thus, there will be great discussion of trade promotion and funds may be so spent, but the ultimate impact on the agricultural sector will not be large unless such rhetoric is accompanied by substantial government subsidies.

## Information Base

Many of the problems with the current agricultural program are related not so much to Congress or the interest groups but to the fact that the parties to the legislation have not had a good information base. The increased costs of the agricultural program in the past year and the generally difficult situation for agriculture are likely to draw into the arena a number of providers of information on economic consequences of alternative agricultural policies. This increase in participation in the generation of information on economic implications of agricultural policies should help policymakers develop more viable agricultural programs.

## *Options for the New Program*

At several points during the conference and to an extent in the presentation by McCalla, it has been indicated that, as economists, we are not sure of the objectives for agricultural policy. This fiction should be put to rest. As economists, we may not know how the economic consequences of policy actions are weighted by those who make them, but we do know the consequences that are attractive to utility-maximizing producers in the agricultural sector and consumers. These are stabilization, income maintenance, efficiency, distribution, structure, and something that will be called intersector balance. The last will be helpful in explaining the nutrition-food supply interests in agricultural policy—interests that have been present for some time. In the future, it is likely that direct linkages to other sectors of the economy may also be important in "balancing" agricultural policy designs.

The economic foundations are well developed for stabilization policy, income maintenance policy, efficiency in the sense of obtaining maximum price stabilization and/or maximum producer income given a particular government cost, distribution, and structure. These theories provide useful guidelines for the design of policies to achieve various stabilization, income maintenance, and other sectorwide objectives. The problem is less with the theory than with the appropriate sets of parameters for implementing policies designed to achieve a weighted combination of these objectives.

Unfortunately, the supply and demand structures in agriculture are far more complicated than those used for illustrative purposes in the presentation by Gardner. Agricultural program operations require important information on cross-elasticities of supply, substitutability, and other parameters. Thus, a more convincing reason why economists have not been more helpful to policymakers in designing appropriate agricultural policies than a lack of theory or a lack of knowledge about policy objectives is a lack of a solid information base for specializing these theories to current policy problems. More solid econometric work will be required before the theories as they currently exist can be implemented to provide quantitative information to policymakers on trade-offs relative to these goals.

A final point in relation to objectives of agricultural policy involves intersector balance. The example indicated was food and nutrition policy. Increasingly, industry programs are going to have to be justified on the basis of contributions to national economic goals. Unfortunately, the framework for contributing to these goals through agricultural policy is not well developed. Linkages between commodity market equilibria and government intervention in these markets and, say, interindustry relationships, macro linkages, trade, and regional incidence are unfortunately rather poorly developed. Before policies for agricultural or other sectors, oriented primarily at commodity markets, can be positioned to contribute to national economic objectives, these linkages will have to be developed.

From the viewpoint of our vested interests in agriculture, it is extremely important that we begin work in better developing these linkages. Clearly, the success of agriculture in attracting tax dollars will be increasingly dependent on whether or not the performance in the sector can be shown to influence national or macroeconomic targets. Thus, in the future it is likely that macroeconomic, tax policy, interindustry, and regional and international economic linkages to agricultural commodity markets will become more visible in design and implementation of agricultural commodity policies.

With these preliminary general observations, several new program alternatives or, more properly, alternatives for new program designs were mentioned and deserve attention. As indicated, these were more limited than was anticipated. Perhaps, the reason is that the objectives of agricultural policy and agricultural programs were not clearly articulated at the outset. Without a clear statement as to the criteria for evaluating policies, it is difficult to make much progress in offering alternative policy designs toward improving existing policies.

## Sharing of Exposure

The papers by Just and Rausser and Gardner all point up the fact that, with existing types of agricultural commodity programs, there is potential for substantial government exposure. This is not as true with some of the specialty commodities. In the specialty commodity areas, sharing arrangements have been developed whereby program participants are taxed if government exposure is higher than anticipated. The theme that comes from observations by Just and Rausser, Gardner, and from the "experiments" conducted with specialty commodities suggests that some type of upper limit on exposure may be incorporated in new policy designs. Thus, upper level limiting can be accomplished by a restriction on total government costs, maximum as well as minimum prices for commodity markets, taxation of land and/or other fixed factors, auctions of quotas, and other means. It would not be surprising to see a cap on government costs in the 1985 farm bill. Alternatively, it would not be surprising to find, along with loan rates, maximum prices above which participating producers are taxed as well as taxes directed at the capitalized values of the fixed resources flowing from commodity programs.

## Intervention in Forward Markets

An option suggested by Gardner and echoed in the discussions surrounding the presentation concerns the development of contingency markets by the government, making it possible for agricultural producers to buy price insurance. The price insurance could be subsidized at whatever level the government desired. This price insurance mechanism is an alternative to those now utilized for stabilizing prices, incomes, and maintaining income levels. An interesting feature of price intervention in forward markets, which was not emphasized, involves the fact that these contingency markets can be used to control government costs. Much the same as the imposition of a maximum price or tax within current program designs, the contingency markets, if created, could hold government exposure by limiting the subsidy in the insurance contracts purchased by farmers.

## Conservation

The utilization of the resources in agriculture is receiving increased attention—particularly, water and soil but not, unimportantly, agricultural labor utilization patterns that have developed as a result of the commodity programs over the past 20 years. More broadly integrated programs with respect to factor markets and resource utilization are likely to be forthcoming. These are emerging in terms such as the conservation buy-out. However, it is clear that, as agricultural policies are directed more at broad national economic objectives, the implications of, say, commodity programs for factor markets will become more of a focus for attention. Thus, conservation reserve, the conservation buy-out, programs aimed at pockets of underutilized agricultural labor, and, in general, rural development programs are likely to become more integrated into the agricultural legislation.

## Food Supply and Nutrition

As a final new program design suggestion, it would be an omission not to mention food policy and the nutritional status of the population. Clearly, the nutritional status is important to general economic objectives including reduction of health costs, improvements in productivity, etc. Agricultural policies and, in many cases, specific consequences of those policies within the livestock sector have important implications for the food supply and the nutrition status of the population. As agricultural policies are viewed more generally, nutrition, the food supply, and food assistance programs will be more tightly tied to agriculture. In the future, however, this tie will not be consistent with the connection of these programs to agricultural policies in the past. In the past and presently, they are rationalized as existing in the U. S. Department of Agriculture largely because of their importance for commodity surplus disposition; in the future, they will be tied to agricultural programs through their relationship to national economic objectives.

# RONALD D. KNUTSON*

Farm bills are written by politicians—not by economists. The economist's role is to identify policy options and evaluate their consequences. This conference has made an important contribution to that role. The Giannini Foundation of the University of California and Resources for the Future are to be complimented on taking this initiative.

While the results of this conference should be useful in a longer run context, much of the discussion centered on options or alternatives that are not likely to be seriously considered in the 1985 farm bill. For example, much of the discussion focused on the contention that past policies have not worked well and may, in fact, have been counterproductive. The implication was the need for a major policy change. Such a major change—e.g., as the abandonment of current commodity programs—is not likely to take place. However, such longer range discussion can be useful in influencing the direction of change in policy whether toward more market-oriented or government-oriented policies and programs.

In evaluating past policies, I detected a substantial bias toward the free market policies generally espoused by the "Chicago School" economists. I doubt this is a consensus position among agricultural economists. It also reflects several value judgments concerning the goals of policy, the nature of competition in agricultural markets, and the mobility of resources between the farm and nonfarm sectors. At the risk of oversimplification, specific assumptions include:

- That the goal of policy is economic efficiency.
- That agricultural markets are competitive on both buyer and seller sides.

*The author is Professor, Department of Agricultural Economics, Texas A&M University.

- That resources move freely among sectors within agriculture and between the farm and nonfarm sectors.

Each of these assumptions is either of questionable validity or, more likely, wrong.

In the face of these assumptions, there was considerable debate over the goals of farm policy. From my perspective, this debate was overdrawn. Regardless of the perspective of economists, income enhancement is likely still the major policy goal. While instability may be asserted by economists to be the major problem in agriculture, reducing instability appears to be a secondary objective. Perhaps the most important controversy in the goals involves the relative importance of the export expansion goal. While in the 1970s export expansion appeared to be a major policy concern, considerable doubt exists in the 1980s regarding its relative importance. This doubt arises because of the large number of policy initiatives, taken by both Congress and the administration, that have run counter to this goal. Examples include the Russian grain embargo, the establishment of loan rates that exceed world market prices, increasing acceptance of protectionist policies in commodities such as cotton, and the perpetuation of macroeconomic policies that result in an overvalued dollar.

The impact of macroeconomic policies on agriculture received its due recognition. Yet, we, as agricultural economists, have much homework to do before we fully understand—and certainly before farmers understand—either the nature of or the magnitude of the interactions between the macroeconomy and agriculture. It has been nearly a decade since Schuh (1976) called for a major professional effort to improve the overall level of literacy on macroeconomic and trade issues. While substantial progress has been made within the profession, much remains to be done, particularly in terms of drawing specific policy implications and in educating farmers to the realities of today's agricultural economy.

While there was much discussion of conditional policy instruments, there was not sufficient focus on specific instruments that have potential for use in the 1985 farm bill. For example, it is almost a foregone conclusion that moving-average market prices (similar to those existing in soybeans and cotton) will be seriously considered and in all probability adopted for corn and wheat. How would such a formula perform and at what percentage of the moving-average price should the loan rate be established? How should target prices be conditionally related to the loan rate? What specific conditional policies might be utilized in triggering production controls? What type of conditions might be established for setting the milk price-support level? Each of these is being actively considered at this early stage in the policy debate with little or no evaluation at this conference.

Extensive discussion of free-market options tended to overshadow concerns about the potential for substantial overproduction and even a return to the chronic surplus conditions that characterized the 1950s and 1960s. Yet, there is substantial evidence to suggest that agriculture is on the verge of a period of technological change that could rival the conversion from horsepower to engine power or the development of hybrid and chemical agriculture. Large public and private sector investments are being made in biotechnology with their implications for potential overproduction and public policy going largely unnoticed. While the biotechnology prospects and implications are currently being investigated by the Office of Technology Assessment, the development is sufficiently important to warrant a broader base of investigation among agricultural economists and policy analysts.

The amount of attention and support given to policy options involving the use of the futures market is disconcerting. Considering the relative lack of use of the futures market by farmers as a risk management tool, it seems highly unlikely that use of the futures market options holds much hope for being a substitute for farm programs. In addition, the propensity for the government to err in program administration suggests that meddling in the futures markets could cripple its usefulness and credibility as a hedging tool. Farmers are already highly suspicious of this important risk management tool.

The discussion of structure was interesting. The conclusion that price and income-support policies favor middle-size farms is consistent with empirical research completed at Texas A&M (Smith, Richardson, and Knutson, 1984). Also recognized was the reality that for most of agriculture the long-run average cost curves in production and marketing functions may continue to slope downward in the relevant range for even large-scale farms. This recognition brings into sharper focus the difficulty associated with policies designed to preserve traditional farm structures and institutions. The wide range of values and considerations in the development of structure policy makes it particularly critical that economists precisely and realistically evaluate the economic dimensions of the structure issue.

Considerable concern is warranted over the need for improved communication between economists and policymakers. Economists must share a portion of the responsibility for policy errors because either they are not doing relevant research or they are not effectively communicating the results of that research to policymakers. Both causes are difficult to defend and overcoming them will require considerable effort.

## *References*

Schuh, G. Edward. "The New Macroeconomics of Agriculture." *American Journal of Agricultural Economics* 8, No. 5 (December 1976):802–811.

Smith, Edward, James Richardson, and Ronald D. Knutson. "Impact of Farm Policy on the Structure of Agriculture in the Texas Southern High Plains." Texas A&M University, College Station, 1984.

# ALAN RANDALL*

Perhaps the majority of our speakers have had something to say on the matter of the objectives of policy. Several have spoken about the prevalence of confusion about objectives and the need for goal clarification. Some others have suggested that the objectives of agricultural policy are quite clear; they go on to enumerate a list that usually includes efficiency, equity, and something about the structure of production agriculture. It occurs to me that anyone who could be satisfied with such a list of objectives is untroubled by platitudes or by the conflicts that arise when legitimate goals are mutually inconsistent.

My point is deeper than this. The very notion of "the objectives of policy" strikes me as dubious. While a few of our speakers were explicit about their model of the policy process, I had the distinct impression that most adhered to the rational planning model. In this model a decision maker gathers needed information from experts and subject matter specialists (like us) and decides—that is, chooses policy—in the public interest. In the rational planning model, the objectives of policy are established by the decision-maker; experts limit their usefulness if they fail to interpret these objectives correctly and apply them in their analyses.

Alex McCalla was more explicit about his model of the policy process. Nevertheless, the "muddling through" model is still a planning model. Its differences from the rational planning model lie, for the most part, in its recognition that noise, confusion, and cross-purposes are prevalent in real-world planning processes. Policy goals are still characterized as things to be established by an elite corps of decision-makers.

I would propose, instead, to use an endogenous model of government. Policy is not imposed upon the system by some *deus ex machina* but emerges from within the system. The "objectives of policy"

---

*The author is Professor, Department of Agricultural Economics, Texas A&M University.

are replaced by a whole mosaic of objectives pursued by individual and group participants in the process. Policy is the process through which the conflicts engendered by diverse objectives are resolved.

It is useful to view the participants in the policy arena as rent seekers; that is, they seek to improve their wealth positions. One's wealth position can be improved by receiving economic surplus from productive activities, or transfers, or both. The rent-seeking individual is indifferent among these sources of personal benefit. Thus, in the rent-seeking model, global efficiency is a pure public good; but all is not lost. Observation suggests that societies, modern and traditional, develop institutions to provide public goods. Recent developments in the theory of repeated games have established the rationality of cooperative behavior in certain circumstances. The upshot is that public goods do get produced and distributed although not always in optimal quantities.

So it is with efficiency. We are aware that deadweight losses, due to the success of others in using policy to gain transfer for themselves, limit our own economic aspirations. Rational rent seekers also attempt to limit the transfers that others obtain, and they do this for both efficiency and distributional reasons. It is impossible for each rent seeker to police all activities of other rent seekers that might diminish the size of the economic pie. However, it is rational to seek establishment of a layer of standing institutions (watchdogs, if you will) to ensure that the rent-seeking activities of others do not get out of hand. It is rational to prefer that such institutions exist, in general, even if there are particular instances when they make it difficult for one's own rent-seeking activities.

Most of us at this conference are associated with institutions that have some watchdog functions. Thus, it is our obligation to draw attention to the efficiency characteristics of existing and alternative policies. The observation that many others active in the policy process are not speaking the language of economic efficiency—or, for that matter, of equity broadly conceived—should not lead us to doubt that our activities in these regards have relevance. It is our duty to draw attention to them, the more so because most other participants do not. The watchdog institutions in which we work emerged endogenously in response to a demand for someone to address these concerns.

Since our institutional existence is, in part, a response to a generally recognized need to have someone worrying about economic public goods, such as efficiency and fairness, it is important that we do these jobs well. In particular, I would counsel against excessive and inappropriate use of market failure concepts. Price stability, for example, could be provided efficiently by a complete set of contingent claims markets. One must wonder whether the apparent absence of such markets is a market failure or merely a rational response to the relationship between the costs of establishing and operating such markets and the benefits of the stability they would provide. Perhaps, price stability is a better deal for the pricetaker when its costs are paid by others; it is an even better deal when provided jointly with price enhancement.

The rent-seeking model of the public decision process is rich in implications. However, in my limited time, I choose to pursue just two more of them.

First, the policy decision process tends to work by successively eliminating alternatives so that the agenda of feasible options is narrowed. Further, beneficiaries of existing policies always seek to narrow the agenda of alternatives and confine public discourse to a mere fine tuning of existing arrangements. Major changes in the direction of policy typically have a rather long gestation period. Simply getting them onto the agenda of seriously considered alternatives is, itself, a considerable battle.

This is why much of the policy analysts' contribution to the 1985 farm bill will be necessarily confined to tinkering with the details of programs that the policy middlemen consider discredited. I do not mean to say that policy analyses cannot make a useful contribution to the 1985 farm bill. Rather, my message is that significant redirections have a lengthy gestation; and at least some among us should be looking toward the farm bills of 1989, 1993, 1997, and 2001. We must never be stampeded into thinking and acting as though the only significant contributions to policy are those with immediate application. While short-fuse analyses are important, some of us must be looking to the longer term when the agenda of possibilities is much broader.

Second, the endogenous policy process works through a mixture of voluntary exchange and majoritarian brute force. The latter serves to revalue or devalue the endowments (political capital, if you will) of the various participants, while the voluntary exchange aspect emerges as participants make the trades that are essential to the formation of coalitions to get their programs enacted. Those of us in the watchdog institutions should be sensitive to the revaluation and devaluation of political endowments. The political capital of any interest group consists not of what they got last time but of what they are likely to get next time. Once we have identified the next-period outcome, which is most likely in the absence of our influence, we can treat it as a status quo and seek to identify potential gains of trading therefrom. If the status quo includes deadweight losses of the magnitude that we fear, it stands to reason that there must be substantial trading possibilities that would benefit (or at least not hurt) the interest groups while increasing the overall size of the pie.

If we are serious about fulfilling our roles in the watchdog institutions of a rent-seeking society, we must grasp the opportunities offered by both the majoritarian and exchange aspects of the political process. First, exposing inefficiency and inequity, documenting its extent, and suggesting alternatives are activities that may encourage and facilitate changes in what was imposed through majoritarian brute force. Second, identifying potential gains from trade is an activity at which economists are especially adept; it encourages global efficiency in a more subtle way by harnessing the power of self-interest. To be effective, we will do both and do them tirelessly and with the self-effacing good humor that comes from the realization that, even if we win only some of the battles and we win those later rather than sooner, we will have earned our keep.

For Product Safety Concerns and Information please contact our EU representative GPSR@taylorandfrancis.com
Taylor & Francis Verlag GmbH, Kaufingerstraße 24, 80331 München, Germany

www.ingramcontent.com/pod-product-compliance
Lightning Source LLC
Chambersburg PA
CBHW071827270326
41929CB00013B/1918
* 9 7 8 1 1 3 8 9 6 2 9 0 3 *